Springer Theses

Recognizing Outstanding Ph.D. Research

Aims and Scope

The series "Springer Theses" brings together a selection of the very best Ph.D. theses from around the world and across the physical sciences. Each volume has been selected for its scientific excellence by an internationally renowned institute or department. For greater accessibility to non-specialists, the published versions include an extended introduction, as well as a foreword by the student's supervisor explaining the special relevance of the work for the field. The series as a whole provides a valuable resource both for newcomers to the research fields, and for other scientists seeking detailed background information on special questions. Finally, it serves as an accredited documentation of the valuable contributions made by today's younger generation of scientists.

How to Apply

Theses prepared at internationally leading universities and institutes may be nominated for publication by supervisors with the endorsement of the head of department. Each such department is entitled to nominate one thesis per year. Enquiries and requests for a nomination form should be sent, together with the thesis, to Ms. Nathalie Berg (see link on right).

Eligibility Criteria

To qualify for consideration, theses should fulfil the following criteria:

- They must be written in good English.
- The topic should fall within the confines of Chemistry, Physics, Earth Sciences, Engineering and related interdisciplinary fields such as Materials, Nanoscience, Chemical Engineering, Complex Systems and Biophysics.
- The work reported in the thesis must represent a significant scientific advance.
- A maximum of 30% of the text may be a verbatim reproduction from the author's previous publications.
- They must have been examined and passed during the 12 months prior to nomination.
- The theses should have a clearly defined structure including an introduction accessible to new PhD students and scientists not expert in the relevant field.
- After acceptance, theses should be supplemented by a "Supervisor's Foreword" outlining the wider scientific relevance of the work

Indexed by zbMATH.

Meng-Zheng Wang

Design Method of Concrete-Infilled Double Steel Corrugated-Plate Walls with T-section

Springer

Meng-Zheng Wang
Tsinghua University
Beijing, China

ISSN 2190-5053 ISSN 2190-5061 (electronic)
Springer Theses
ISBN 978-981-95-7788-0 ISBN 978-981-95-7789-7 (eBook)
https://doi.org/10.1007/978-981-95-7789-7

This Springer imprint is published by the registered company Springer Nature Singapore Pte Ltd.
The registered company address is: 152 Beach Road, #21-01/04 Gateway East, Singapore 189721, Singapore

Preface

By the time I finished the manuscript in English, I had obtained the PhD degree and been working in Shanghai for a year. The publication of this thesis book brought me back to the happy and fulfilling time at Tsinghua University, where I completed series of experiments and exchanged creative ideas with Dr. Jing-Shen Zhu, Dr. Xiao Yang, and other fellow students under the supervision of Prof. Yan-Lin Guo. We felt joyful for every inch of academic progress and the solution of every problem. Hereby, I would like to use this thesis to commemorate the 9 years I spent at Tsinghua University from undergraduate admission to doctoral graduation.

This thesis addresses the failure mechanism and design method of concrete-infilled double steel corrugated-plate composite walls with T-section (T-CDSCW). Due to the composite effect of steel and concrete and the constraining effect of intermediate bolts, CDSCWs exhibit excellent load-carrying performance and applicability to high-rise buildings. To satisfy the plan layout and structural requirements, shear walls are arranged in T-section or other sections. T-CDSCWs are studied in the thesis, and the design methods are proposed to evaluate the cross-sectional capacity, global stability, and seismic performance by experimental investigations, finite element simulations, and theoretical analyses.

In Chap. 2, three sets of experiments are introduced to investigate the cross-sectional capacity, global stability, and seismic performance of T-CDSCWs. The experimental results show that T-CDSCWs exhibit excellent load-carrying capacity and seismic performance.

In Chap. 3, based on the experimental results, solid–shell and shell–beam finite element models are developed and used to analyse the load-carrying capacity and seismic performance of T-CDSCWs.

In Chap. 4, the design method is proposed to evaluate the cross-sectional capacity of T-CDSCWs. To consider the local buckling of steel corrugated plates in the cross-sectional capacity, the buckling behaviour is studied concerning steel corrugated plates with the constraint of infilled concrete and intermediate bolts, and the formulas of elastic and contact buckling loads are proposed. Taking into consideration the local buckling of steel corrugated plates, the interaction curves are proposed concerning T-CDSCWs under axial compression and bidirectional bending moment based on the full-section plastic stress distribution.

In Chap. 5, the design method is proposed to evaluate the global stability of T-CDSCWs. Concerning buckling under axial compression, the formulas of elastic flexural and flexural-torsional buckling load are proposed, the critical parameters of buckling modes are determined, and the curves of the strength reduction factor are developed. Combined with the interaction curves of the cross-sectional capacity, the interaction curves are proposed concerning in-plane global stability under axial compression and bending in the web plane.

In Chap. 6, the seismic requirements for steel plate concrete composite walls are proposed. Based on the existing experimental results, the formulas are proposed concerning the ultimate drift ratio of steel plate–concrete composite walls.

In Chap. 7, future research directions are provided for reference.

Beijing, China Meng-Zheng Wang

Acknowledgements

To my wife, Jingjing.

Competing Interests The author has no competing interests to declare that are relevant to the content of this manuscript.

Contents

Abbreviations

Load

M	Bending moment (unit: kN m)
M_u	Bending moment capacity (unit: kN m)
N	Axial compression (unit: kN)
N_{cr}	Elastic buckling load under axial compression (unit: kN)
N_u	Axial compression capacity (unit: kN)
V	Shear load (unit: kN)

Geometrical Parameters

A_c, A_{cc}, A_{cf}, A_{cw}	Cross-sectional area of concrete at the whole cross-section, the boundary elements, the flange walls, and the web walls (unit: mm^2)
A_s, A_{sc}, A_{sf}, A_{sw}	Cross-sectional area of steel at the whole cross-section, the boundary elements, the flange walls, and the web walls (unit: mm^2)
b_c, d_c	Breadth and thickness of the boundary elements (unit: mm)
B_f, B_w	Overall breadth of the flange and web walls (unit: mm)
b_f, b_w	Breadth of the flange and web walls (unit: mm)
d_f, d_w	Thickness of the flange and web walls (unit: mm)
d_v	Vertical bolt spacing (unit: mm)
H_w	Wall height (unit: mm)
I_x, I_y	Second moment of area about x- and y-axis (unit: mm^4)
I_t	Torsion constant (unit: mm^4)
I_w	Warping constant (unit: mm^6)
t_c, t_s	Thickness of the rectangular steel tubes in boundary elements and the steel corrugated plates in wall elements (unit: mm)
q	Wave length of steel corrugated plates (unit: mm)

Material Properties

E_c, E_s Elastic modulus of concrete and steel (unit: N/mm^2)
f_{cu}, f_c, f_t Compressive strength of concrete cube, axial compressive strength, and axial tensile strength of concrete (unit: N/mm^2)
f_y, f_u Yield strength and ultimate strength of steel (unit: N/mm^2)

Design Index

u Displacement (unit: mm)
n Axial compression ratio
β_f Parameter to determine the first-order elastic buckling mode under axial compression
β_t Parameter to determine the elastic flexural-torsional buckling mode
δ Displacement (unit: mm)
ε Strain
θ Drift ratio (unit: rad)
λ Normalised slenderness ratio
ξ Confinement factor
σ Stress (unit: N/mm^2)
φ Strength reduction factor
ϕ Curvature (unit: mm^{-1})

Chapter 1
Introduction

The rapid economic growth of China has driven the development of infrastructure construction. At the same time, urgent issues, such as resource consumption, environmental pollution, structural safety, and service life, require structural engineers to pursue the application of high-performance structures (Tao et al. 2017). In high-rise buildings, shear walls are widely used. In long-term practical applications, various types of shear walls have emerged to meet the performance requirements of structural design. This thesis introduces a type of high-performance steel plate concrete composite wall, namely the concrete-infilled double steel corrugated-plate wall (CDSCW) (Guo et al. 2017). CDSCWs can meet the requirements of industrial production and high load-carrying performance, which conform to the direction of high-quality development in the construction industry. In this thesis, the load-carrying mechanism of CDSCWs with T-section (T-CDSCWs) will be studied, and the design method will be proposed to provide the basis for the engineering application.

1.1 Background

Due to the high load-carrying capacity and lateral stiffness, shear walls can be used as the main lateral force-resisting components in shear wall structures, frame-shear wall structures, and frame-core tube structures. According to the materials, the basic forms of shear walls can be divided into reinforced concrete shear walls, steel plate shear walls and steel plate concrete composite walls (Guo and Zhu 2020).

Reinforced concrete shear walls are mainly subjected to vertical and lateral loads. The ductility of reinforced concrete shear walls is degraded significantly under large axial compression. To ensure the seismic performance of reinforced concrete shear walls, it is often necessary to restrict the axial compression ratio

M.-Z. Wang, *Design Method of Concrete-Infilled Double Steel Corrugated-Plate Walls with T-section*, Springer Theses,
https://doi.org/10.1007/978-981-95-7789-7_1

below the upper limit value by enlarging the cross-section of reinforced concrete shear walls when they are applied in high-rise buildings. This increases the overall self-weight of structures and decrease the usable floor area. To improve the performance, enhancement measures are introduced to reinforced concrete shear walls, such as internal trusses (Cao et al. 2007; Lan et al. 2018), internal braces (Ding et al. 2008; Ji et al. 2016; Wu et al. 2016), and steel profiles (Dan et al. 2011a, b; Qian et al. 2012), which improve the load-carrying capacity and control the concrete cracking.

Steel plate shear walls are embedded in the frames and mainly subjected to in-plane shear force. Steel plate shear walls have good deformation capacity. Under the action of wind or small earthquakes, steel plate shear walls generally maintain in the elastic range and act as load-carrying components. Under the action of large earthquakes, steel plate shear walls reach the plastic range and dissipate the input energy, thus acting as both load-carrying components and energy dissipating dampers (Guo and Dong 2005). According to the types of steel plates, steel plate shear walls can be divided into steel flat-plate shear walls and steel corrugated-plate shear walls.

According to the width-thickness ratio and the corresponding load-carrying mechanism, steel flat-plate shear walls can be divided into thick steel flat-plate shear walls and thin steel flat-plate shear walls (AISC 2016; Guo and Zhou 2011). Concerning thick steel flat-plate shear walls, the width-thickness ratio is small, steel flat plates do not buckle before yielding, and the ultimate limit state is the yielding of the steel flat plates. This leads to large steel consumption and high construction cost of thick steel flat-plate shear walls. Concerning thin steel flat-plate shear walls, the width-thickness ratio is large, and steel flat plates buckle before yielding and possess stable post-buckling strength and deformation capacity due to the tension-field action. This reduces steel consumption, but forwards higher requirements on the strength and stiffness of the boundary elements.

When steel corrugated plates are used as the embedded steel plates, the out-of-plane stiffness of steel plate shear walls increases significantly, and thus the buckling load under in-plane shear force is significantly improved (Dou et al. 2016). Steel corrugated plate shear walls can perform as thick steel flat-plate shear walls to satisfy the requirement that steel plates do not buckle before yielding without large steel consumption. Meanwhile, steel corrugated plates are not easily deformed under external force, which is convenient for transportation and construction. Steel corrugated plates are manufactured by cold rolling of steel flat plates, so the thickness of steel corrugated plates is restricted by the ductility of the steel and the capacity of the equipment. This limits the load-carrying capacity of steel corrugated-plate shear walls.

To improve the load-carrying performance of steel plate shear walls, stiffened steel plates can be used to limit the out-of-plane deformation of steel plates (Chen and Guo 2004; Tong and Guo 2015, 2018), and concrete cover plates can be set to compose buckling-resistant steel plate shear walls (Guo et al. 2009a, b; Zhao and Astaneh-Asl 2004). Moreover, steel corrugated plates can be superimposed to compose double steel corrugated-plate shear walls (Tong 2019; Tong et al. 2018, 2020a, b).

The composite effect of steel and concrete can be effectively utilised by combining concrete and steel plates to compose steel plate concrete composite walls. According to the position of steel plates, steel plate concrete composite walls can be divided into steel plate reinforced concrete composite walls and concrete-infilled double steel plate composite walls. According to the type of steel plates, steel plate concrete composite walls can be divided into steel flat-plate concrete composite walls and steel corrugated-plate concrete composite walls.

When steel plate concrete composite walls are applied in building structures, steel profiles or concrete-filled steel tubes are set at the ends of the walls as boundary elements to improve the load-carrying performance under axial compression and bending moment (Bu 2011; Ding et al. 2011; Hou et al. 2017; Sun et al. 2011). Concerning reinforced concrete shear walls, boundary elements are set at the ends of the walls, and the section height of boundary elements includes the zone where the compression strain exceeds the ultimate compressive strain of plain concrete so that the seismic performance of shear walls is improved (MOHURD 2010). Concerning steel plate concrete composite walls, if embedded steel plates and steel profiles are used, the boundary elements should be strengthened by increasing stirrups as steel-reinforced concrete shear walls (MOHURD 2016); if outer steel plates and concrete-filled steel tubes are used, since outer steel plates and steel tubes can confine infilled concrete, it is not necessary to make strict specification on the width of boundary elements, and generally, the section height of steel tubes should not be less than the wall thickness. In addition, when steel plate concrete composite walls are applied in shield structures (Booth et al. 2015; Bruhl et al. 2015; Kurt et al. 2016; Liu and Li 2018; Sener et al. 2015; Zhang 2015) and marine structures (Yan and Liew 2016; Yan et al. 2016a, b, c, d), steel plate concrete composite walls are mainly subjected to out-of-plane loads, and boundary elements are not set.

Similar to steel plate shear walls, the local buckling of steel plates occurs in steel plate concrete composite walls. To avoid the premature buckling of steel plates and the debonding between steel plates and concrete, it is required that concrete and connectors have sufficient constraining effects on steel plates. Therefore, the minimum thickness of concrete needs to be specified in steel plate reinforced concrete composite walls, and the maximum spacing of connectors needs to be specified in concrete-infilled double steel plate composite walls. The limit values of concrete thickness and connector spacing are influenced by the type of steel plates. Due to the larger out-of-plane stiffness of steel corrugated plates, the corresponding limit values are relaxed compared with steel flat plates.

1.2 Literature Review

For different types of steel plate concrete composite walls, researchers have carried out a large number of experimental investigations and finite element analyses and proposed corresponding design methods. In this section, the studies on four forms of steel plate concrete composite walls are summarised, namely steel flat-plate reinforced concrete composite walls (SFRCWs), steel corrugated-plate reinforced concrete composite walls (SCRCWs), concrete-infilled double steel flat-plate composite walls (CDSFWs), and concrete-infilled double steel corrugated-plate composite walls (CDSCWs).

1.2.1 *Steel Flat-Plate Reinforced Concrete Composite Walls (SFRCWs)*

SFRCWs are composed of reinforced concrete shear walls with embedded steel flat plates and steel profiles (see Fig. 1.1), which are connected by welding or bolts. To ensure the composite effect of steel plates and concrete, headed shear studs are arranged on steel plates. The concrete can effectively constrain the embedded steel plates. To ensure the stability of steel plates, the concrete should not be too thin. It is required that the ratio of concrete thickness to steel flat plate thickness should not be <15 (MOHURD 2016). In boundary elements, to ensure the confining effect on the concrete, stirrups should pass through or be welded with steel plates to form closed stirrups.

The research on SFRCWs focuses on seismic performance and studies the effects of geometric parameters, load forms, construction measures and other factors. Chen et al. (Chen et al. 2011; Jiang et al. 2012; Sun et al. 2008; Zhu et al. 2016) conduct experimental investigations and finite element analyses on the seismic performance

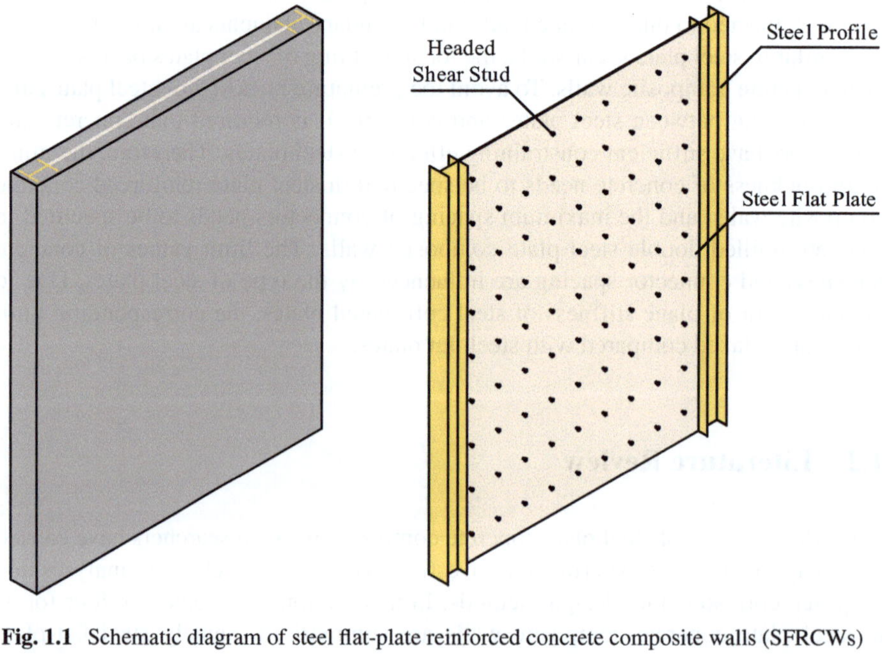

Fig. 1.1 Schematic diagram of steel flat-plate reinforced concrete composite walls (SFRCWs)

of SFRCWs and study the effects of axial compression ratio, shear span ratio, concrete strength, and steel plate contribution. Under cyclic horizontal loads, the concrete at the ends of the SFRCW specimens crack in tension and gradually falls off, and the concrete at the middle part of the SFRCW specimens cracks in tension or shear. Based on the experimental results, the design formulas are proposed to evaluate the bending and shear capacities of SFRCWs by superposing the capacities of steel and concrete. Wang et al. (2018b) conduct experimental investigations on the seismic performance of SFRCWs and study the effects of axial compression ratio, cross-sectional dimensions, shear span ratio, concrete strength, and steel plate thickness. Based on the experimental results, the hysteretic model is proposed to depict the load–displacement relationship of SFRCWs under cyclic horizontal loads. Wang et al. (2017) conduct finite element analyses on the seismic performance of SFRCWs and study the effects of axial compression ratio, steel plate thickness, concrete strength, and reinforcement ratio. The finite element simulation results show that the reinforcement ratio, including vertical and horizontal reinforcement ratios, has negligible effects on the load-carrying capacity of SFRCWs. The main role of reinforcement is to restrain shrinkage and temperature stress. If the composite effect of concrete and steel plates is ensured, there is no need to continue to increase the reinforcement ratio. Wang et al. (2018a) conduct experimental investigations and finite element analyses on the seismic performance of SFRCWs under axial tension and cyclic horizontal loads. Under combined tension-bending-shear loads, tension-bending failure and anchorage failure may occur in SFRCWs. The design formulas are proposed to evaluate the load-carrying capacity of SFRCWs under the tensile-bending-shear load. The anchorage failure can be avoided by increasing the reinforcement and limiting the axial tension ratio and the stud slip.

Some construction measures can improve the seismic performance of SFRCWs. Zhou et al. (2015) conduct experimental investigations on the seismic performance of the SFRCWs with steel plate openings and study the effects of steel plate thickness, connector type and steel plate contribution. By opening the steel plates, tie rebar can be set through steel plates to connect the reinforcement on two sides. The experiment results show that, with the steel plate thickness remaining the same, the tie rebar can improve the seismic performance of SFRCWs compared with studs.

Compared with steel profiles, steel tubes can confine infilled concrete. Nie et al. (Hu et al. 2016; Nie et al. 2013b) conduct experimental investigations on the seismic performance of the SFRCWs with concrete-filled steel tube (CFST) boundary elements at the ends and study the effects of concrete strength and steel plate thickness. Under cyclic horizontal loads, the concrete at the middle part of the SFRCW specimens cracks under tension or shear, and the rectangular steel tubes at the ends buckle under compression. Based on the experimental results, the design formulas are proposed to evaluate the load-carrying capacity of the SFRCWs with CFST boundary elements. Cao et al. (2011a, b, 2014) conduct quasi-static experiments and shaking table experiments of the SFRCWs with CFST boundary elements, and study the effects of shear span ratio, connection type, steel plate thickness, steel plate opening, and concrete contribution. Based on the experimental results, the design formulas are proposed to evaluate the bending and shear capacities of the SFRCWs with CFST boundary elements.

The headed shear studs on steel plates play a role in ensuring the composite effect of steel plates and concrete. Qi et al. (2017) conduct finite element analyses on the sheer force of studs in the SFRCWs and study the effects of stud number, stud diameter, steel plate thickness, concrete panel thickness, and shear span ratio. It is pointed out that the maximum shear force of studs occurs at the boundary of the stud group except for the corners. The formulas are proposed to evaluate the shear force. Zhang et al. (2012) conduct push-out experiments to test the performance of steel plates with headed shear studs and concrete. With the total area of studs remaining the same, the shear capacity of studs can be improved by using studs with a smaller diameter. To ensure enough shear capacity, it is recommended that the ratio of length to diameter should be >4. Based on the experimental results, a simplified formula is proposed to estimate the shear capacity of studs.

The design formulas are specified to evaluate the bending and shear capacities of SFRCWs in JGJ 138 (MOHURD 2016). Compared with reinforced concrete shear walls, the ductility and energy dissipation capacities of SFRCWs are significantly improved. In the evaluation of axial compression ratio, the contribution of the load-carrying capacity of steel plates can be taken into consideration. The requirements of the axial compression ratio are consistent with the requirements for reinforced concrete shear walls specified in GB 50011 (MOHURD and AQSIQ 2016): the axial compression ratio is not appropriate to be >0.4 when the class of seismic measure is class I and the seismic precautionary intensity is 9 degree, 0.5 when and the seismic precautionary intensity is 7 or 8 degree, and 0.6 when is the class of seismic measure is class II or III.

1.2.2 Steel Corrugated-Plate Reinforced Concrete Composite Walls (SCRCWs)

SFRCWs are transformed into SCRCWs by replacing steel flat plates with steel corrugated plates (see Fig. 1.2). The steel corrugated plates in Fig. 1.2 are vertically placed with folding lines upright. Besides, steel corrugated plates can be horizontally placed with folding lines horizontal. With vertical folding lines, steel corrugated plates can sustain vertical load, while with horizontal folding lines, steel corrugated plates can barely sustain vertical loads. Since the stability performance of steel corrugated plates is better than that of steel flat plates, the out-of-plane deformation of steel corrugated plates is relatively small, which improves the problem of concrete falling off caused by the out-of-plane deformation of steel plates (Guo et al. 2015).

Wang et al. (2018c, 2019, 2020) conduct experimental investigations and finite element analyses on the seismic performance of SCRCWs and study the effects of axial compression ratio, shear-span ratio, steel plate placement direction, and stud spacing.

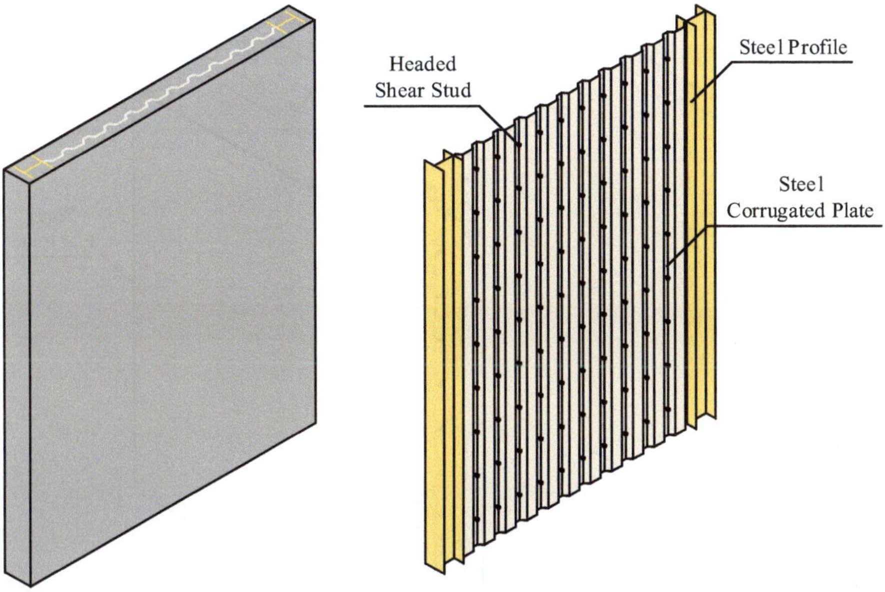

Fig. 1.2 Schematic diagram of steel corrugated-plate reinforced concrete composite walls (SCRCWs)

Based on the experimental and finite element simulation results, the design formulas are proposed to evaluate the shear capacity of SCRCWs with the global and local buckling of steel corrugated plates in consideration.

1.2.3 Concrete-Infilled Double Steel Flat-Plate Composite Walls (CDSFWs)

In steel plate reinforced concrete composite walls, steel plates are constrained by the concrete on two sides, but the concrete is not constrained. By infilling concrete between double steel plates, steel plates can effectively confine infilled concrete. CDSFWs are composed of outer steel plates, infilled concrete, and boundary elements (see Fig. 1.3). Boundary elements are generally rectangular CFSTs, which are connected to steel plates by welding. Shear connectors, such as studs, intermediate bolts, tie bars and panels, are set between steel plates and concrete. Since the out-of-plane stiffness of steel flat plates is small and steel plates are unilaterally constrained by concrete, the steel plate thickness should not be too small, and dense connectors should be set to avoid the premature buckling of steel flat plates. JGJ/T 380 (MOHURD 2015) specifies that the steel plate thickness of CDSFWs should not be <10 mm, and the ratio of wall thickness to steel plate thickness should be between 25 and 100. The upper limit of the connector spacing is related to the

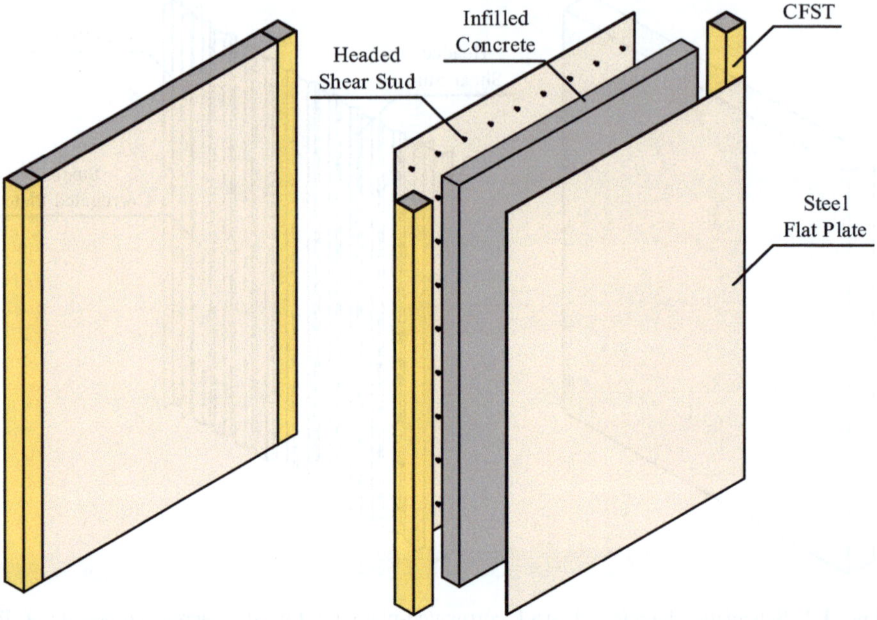

Fig. 1.3 Schematic diagram of concrete-infilled double steel flat-plate composite walls (CDSFWs)

connector type: concerning studs, intermediate bolts, and tie bars, the ratio of connector spacing to steel plate thickness should not be >$40\varepsilon_k$, where ε_k is the correction factor for steel grade; concerning T-shaped stiffeners arranged along the full length of steel plates, the ratio of connector spacing to steel plate thickness should not be >$60\varepsilon_k$, which refers to the requirement for the width-thickness ratio of CFSTs.

Infilled concrete is effectively confined by steel plates, but steel plates are unilaterally constrained by infilled concrete and connectors, and thus the local buckling occurs in steel plates. Liang et al. (2004) conduct finite element analyses on the local buckling of unilaterally constrained steel flat plates under biaxial compression and shear. The finite element model takes into consideration initial imperfection, material nonlinearity, and stud slip effect. The design formulas are proposed to evaluate the buckling and strength of unilaterally constrained steel flat plates under biaxial compression and shear. Nie and Li (2009a, b) conduct finite element analyses on the global stability of double steel flat-plate composite panels and the local stability of steel flat plates under uniaxial compression. Taking into consideration the stud slip effect, the elastic buckling load of double steel flat-plate composite panels is proposed. The upper limits of the stud spacing in two directions and the lower limit of concrete thickness are proposed to ensure that the yielding of steel plates occurs before the local buckling of steel plates and the local buckling of steel plates occurs before the global buckling of composite panels. Choi et al. (2014) conduct experimental investigations on the axial compressive performance of CDSFWs. Based on the experimental results, the design formulas are proposed to evaluate the axial compression capacity of CDSFWs with the local buckling of steel plates in consideration. Zhang et al. (2014) conduct finite element analyses on the stud force of CDSFWs and study the effect of studs on the local buckling of steel plates and the composite effect of CDSFWs. The upper limits of stud spacing are proposed with steel plates under axial compression and tension. Huang and Liew (2016a, b) conduct experimental investigations on the axial compressive performance and out-of-plane eccentric compressive performance of CDSFWs with J-hook connectors. Based on the experimental results, the effective width of steel plates under axial compression and the interaction curve of compression and out-of-plane bending are proposed. Yang et al. (2016) conduct experimental investigations on the axial compressive performance of CDSFWs. Based on the experimental results, the upper limit of the stud spacing to steel plate thickness ratio is proposed. Hao and Cao (2017) conduct experimental investigations on the axial compressive performance of CDSFWs and SFRCWs. Based on the experimental results, the design formulas are proposed to evaluate the axial compression capacity with the local buckling of steel plates and the confined strength of concrete in consideration.

CDSFWs possess good seismic performance. Eom et al. (2009) conduct experimental investigations on the seismic performance of CDSFWs with the rectangular section and T-section. The experimental results show that the damage of CDSFWs is concentrated in the weld fracture of walls and coupling beams, the local buckling of steel plates, and the crush and cracking of infilled concrete. Based on the experimental results, the design formulas are proposed to evaluate the in-plane bending capacity of CDSFWs with the rectangular section and T-section. Nie et al. (Hu et al. 2014; Ma et al. 2013, 2014; Nie et al. 2011, 2013a, 2014; Wang et al. 2015) conduct experimental investigations and finite element analyses on the seismic performance of CDSFWs and study the effects of axial compression ratio, shear-span ratio, material strength, and steel plate thickness. The fibre section method is used to analyse the load-carrying capacity, stiffness, and deformation of CDSFWs under axial compression and bending moment. Ji et al. (2013, 2015) conduct experimental investigations on the seismic performance of CDSFWs with CFST boundary elements and steel tube reinforced CFST boundary elements. Epackachi et al. (2014, 2015, 2017) conduct experimental investigations and finite element analyses on the seismic performance of CDSFWs, study the effects of axial compression ratio, shear-span ratio, steel plate thickness and material strength, and propose the load–displacement relationship under axial compression and in-plane loads. Zhao et al. (2016) conduct theoretical analyses on the load-carrying mechanism of CDSFWs and propose the calculation formulas for bending and shear capacities and the hysteretic model of CDSFWs.

In addition to connecting double steel plates by headed shear studs and other connectors, CDSFWs can be composed by connecting steel profiles. Chen et al. (Chen et al. 2016, 2017; Zhang et al. 2016a, b, 2017) conduct experimental investigations and finite element analyses on the seismic performance of CDSFWs with rectangular section and T-section, which are composed of steel U-channels. Steel U-channels are welded to each other on the flanges, and the webs of U channels, which are inside CDSFWs, play the role of connecting steel plates. Steel U-channels can effectively confine infilled concrete. However, the processing of connecting steel U-channels needs a lot of welding, and the welding position is concentrated in the weak position of cold-formed steel U-channels. Guo et al. (2018a) conduct experimental investigations on the axial compressive and seismic performance of CDSFWs composed of H-section steel. Similar to steel U-channels, H-section steel is welded to each other on the flanges. Based on the experimental results, the design formulas in Eurocode 4 (CEN 2004) are closest to the load-carrying capacity of CDSFWs composed of H-section steel.

CDSFWs are mainly subjected to out-of-plane loads when applied in shell structures. Yang et al. (2013) conduct experimental investigations on the out-of-plane flexural performance of concrete-infilled steel flat-plate composite panels and study the effects of steel plate thickness, shear connection, and constructional reinforcement. The design formulas are proposed to evaluate the out-of-plane flexural capacity. Varma et al. (2014) conduct finite element analyses on the in-plane shear and out-of-plane flexural performance of concrete-infilled steel flat-plate composite panels and propose a simplified design method. Sabouri-Ghomi et al. (2016) conduct finite element analyses on the out-of-plane flexural deformation of concrete-infilled steel flat-plate composite panels, and propose the formulas for the out-of-plane flexural deformation considering slip. Yan et al. (Yan and Liew 2016; Yan et al. 2016a, b, c, d) conduct experimental investigations and finite element analyses on the punching shear performance of concrete-infilled steel flat-plate composite panels under concentrated loads and study the effects of stud spacing, steel plate thickness, concrete strength, and geometric parameters. The design formulas are proposed to evaluate the load-carrying capacity under concentrated loads.

The design formulas are specified in JGJ/T 380 (MOHURD 2015) to evaluate the bending and shear capacities of CDSFWs based on the full-section plastic stress distribution. The design formulas for shear capacity only consider the contribution of steel plates. Like SFRCWs, in the evaluation of axial compression ratio, the contribution of steel plates can be taken into consideration. The requirements of axial compression ratio are consistent with the requirements for reinforced concrete shear walls specified in GB 50011 (MOHURD and AQSIQ 2016): the axial compression ratio is not appropriate to be >0.4 when the class of seismic measure is class I and the seismic precautionary intensity is 9 degree, 0.5 when and the seismic precautionary intensity is 7 or 8 degree, and 0.6 when is the class of seismic measure is class II or III. JGJ/T 380 (MOHURD 2015) specifies that the upper limit of the elastic inter-story drift angle of CDSFWs is 1/400, and the upper limit of the elastic-plastic inter-story drift angle is 1/80.

1.2.4 Concrete-Infilled Double Steel Corrugated-Plate Composite Walls (CDSCWs)

CDSFWs are transformed into CDSCWs by replacing steel flat plates with steel corrugated plates (see Fig. 1.4). Since the out-of-plane stiffness of steel corrugated plates is much larger than that of steel flat plates, the stability performance of steel corrugated plates is improved, and the upper limit of the connector spacing is greater than that of steel flat plates. The thickness of CDSCWs should not be <150 mm so that the spacing between wave valleys is large enough to avoid influencing concrete placement (Guo et al. 2017).

Wright and Gallocher (1995) propose that the application of steel profiled sheeting composite members can be extended from composite floor slabs to vertical load-carrying members. The thickness of steel profiled sheeting is generally not >1 mm. The composite effect of steel and concrete is ensured by embossments on steel profiled sheeting. Wright and Hossain (Wright 1998a, b; Wright and Hossain 1997) conduct experimental investigations on the load-carrying performance of CDSDWs under axial compression, out-of-plane flexural load, and in-plane shear force. Due to the small thickness of steel profiled plates and the sparse arrangement of shear connectors, steel profiled sheeting buckles and CDSDWs cannot achieve full-section yielding. The buckling of steel profiled sheeting needs to be considered in the evaluation of the load-carrying capacity (Wright 1993). Hossain et al. (Hossain et al. 2016a, b; Hossain and Wright 2004; Rafiei et al. 2013, 2015, 2017) conduct

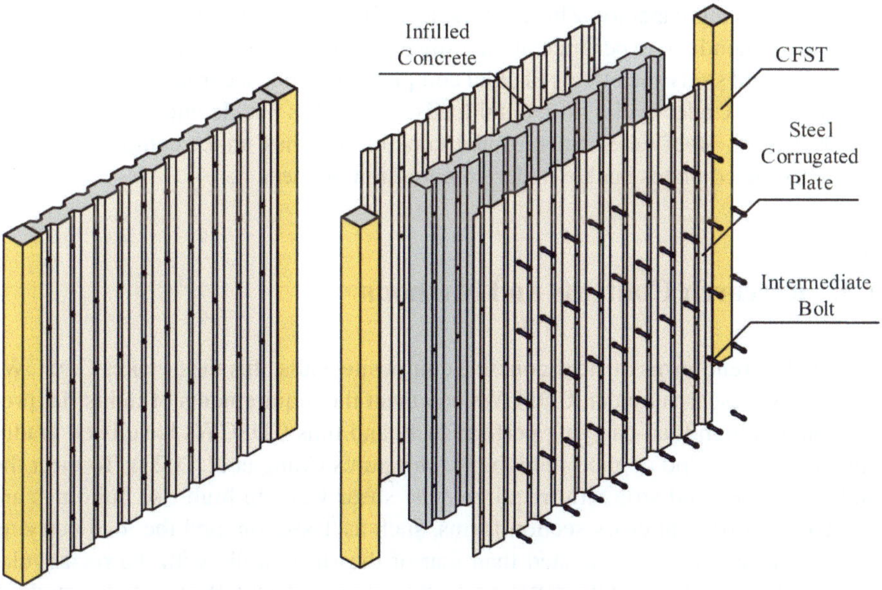

Fig. 1.4 Schematic diagram of concrete-infilled double steel corrugated-plate composite walls (CDSCWs)

experimental investigations and finite element analyses on the in-plane shear performance of CDSCWs, including in-plane horizontal monotonic loading, cyclic loading and impact loading experiments. The effects of intermediate bolts, concrete properties, and geometric parameters of steel profiled sheeting are studied. Design formulas are proposed to evaluate the shear capacity and stiffness of CDSDWs. Compared with the contribution of steel profiled sheeting, the shear capacity contributed by concrete is so small that the contribution of concrete can be neglected. Hossain et al. (2015) conduct experimental investigations on the axial compressive performance of pierced CDSDWs and study the effects of strength enhancement devices. The experimental results show that the compression capacity of pierced CDSDWs can be ensured by proper strength enhancement devices.

To reduce the self-weight of CDSCWs, lightweight concrete can be used as infilled concrete. Othuman Mydin and Wang (2011) conduct experimental investigations on the axial compressive performance of CDSCWs with lightweight concrete and propose that the effective width method can be used to design the compression capacity of CDSCWs. Prabha et al. (2013) conduct experimental investigations on the axial compressive performance of CDSCWs with lightweight concrete and propose the relationship between the failure stress of infilled lightweight concrete and the strength of steel profiled sheeting.

The steel plate thickness can reach 3–8 mm when cold-rolled steel corrugated plates are used. Guo et al. (Guo et al. 2018b; Zhou et al. 2021; Zhu et al. 2019) conduct experimental investigations and finite element analyses on CDSCWs with cold-rolled steel corrugated plates. The design formulas are proposed to evaluate the strength reduction of CDSCWs, the local buckling of steel corrugated plates, and the load-carrying capacity. Zhu et al. (Zhu 2020; Zhu et al. 2020) and Zhao et al. (2020a, b) conduct experimental investigations on the seismic performance of CDSCWs and study the effects of axial compression ratio, shear-span ratio, and bolt arrangement. Under cyclic horizontal loads, steel tubes buckle and concrete cracks under tension, steel tubes fracture and concrete crushes under compression, and steel corrugated plates buckle under compression or shear.

1.3 Research Contents and Methods

For the different forms of steel-concrete composite walls, this thesis takes CDSCWs as the main research object. CDSCWs can meet the requirements of industrial production and high load-carrying performance, and thus CDSCWs have been gradually popularised and applied in high-rise structures (Yang et al. 2021). To meet the building layout and structural requirements, shear walls in high-rise buildings are arranged in different cross-section forms, such as T-section, and the load-carrying mechanism is more complicated than that of the shear walls with the rectangular section. In this thesis, CDSCWs with T-section (T-CDSCWs) will be studied. Experimental investigations, finite element simulations and theoretical analyses will be used to study the cross-sectional capacity, global stability, and seismic performance of T-CDSCWs.

1.3.1 Research Object

According to the building layout and structural requirements, CDSCWs can be arranged in different cross-sectional forms, such as the rectangular section, T-section and L-section. The load-carrying mechanism of CDSCWs is different for different cross-sections. This thesis focuses on T-CDSCWs. T-CDSCWs consist of flange and web wall elements and CFST boundary members (scc Fig. 1.5). The T-section is symmetrical about the web plane. In addition to the boundary elements at the ends of flange and web wall elements, CFST boundary elements are set at the intersection of flange and web wall elements to facilitate the connection between flange and web wall elements.

The load-carrying mechanism of T-CDSCWs is determined by the cross-sectional dimensions, especially the breadth of flange and web walls. JGJ 3 (MOHURD 2010) specifies that when the height-thickness ratio of the cross-section is not >4, the design of shear walls should be carried out according to frame columns; when

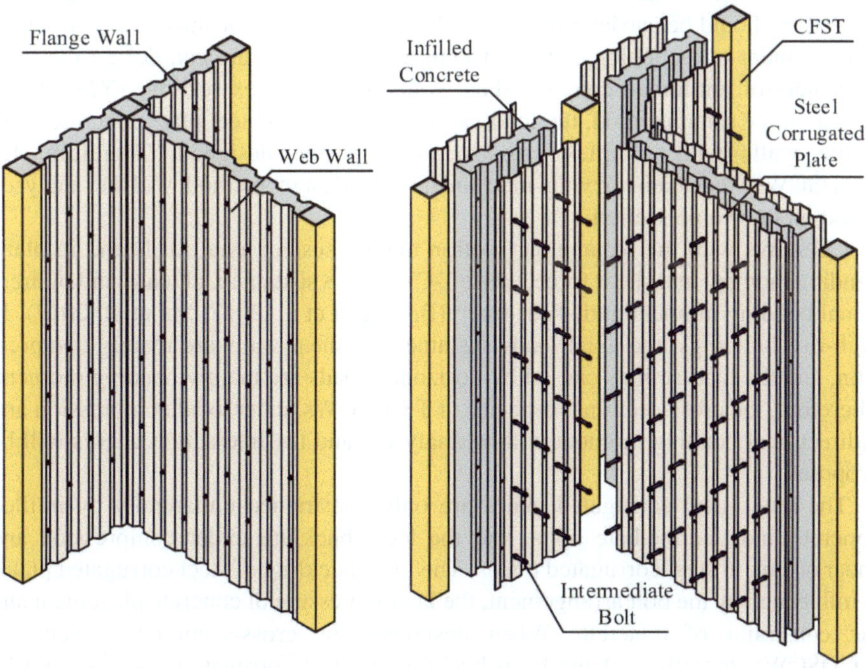

Fig. 1.5 Schematic diagram of concrete-infilled double steel corrugated-plate composite walls with T-section (T-CDSCWs)

the maximum value of the height-thickness ratio is >4 but not >8 and the wall thickness is not >300 mm, the design of shear walls should be carried out according to short-leg shear walls and needs to meet other construction requirements. In GB 50010 (MOHURD and AQSIQ 2010), the effective width of the flange wall is specified in the calculation of the load-carrying capacity. The design of T-CDSCWs should conform to the relevant requirements. However, in the analysis of T-CDSCWs, the cross-sectional dimensions of T-CDSCWs are not strictly limited to meet the relevant requirements, but range in a larger parameter variation to comprehensively reflect the load-carrying performance.

1.3.2 Research Contents

In the structural analysis and design of shear walls, reinforced concrete shear walls with complex cross-section forms, such as T-section, L-section and C-section, are divided into several rectangular segments to design (Fu 2010). This is due to the lack of an easy-to-use method and the support of structural design software for designing the cross-sectional capacity of shear walls with complex sections. However, the design method is inconsistent with the load-carrying mechanism of shear walls and may lead to unsafe or uneconomical design results. To ensure reasonable and reliable design results, internal force analysis and load-carrying capacity design should be carried out on the whole cross-section of shear walls. With the improvement of technology, some structural design software can design the reinforcement of reinforced concrete shear walls with complex sections (YJK 2016). Based on the same method, the load-carrying capacity of steel plate concrete composite walls with complex cross-sections can be designed. Therefore, for T-CDSCWs, the load-carrying mechanism and design method will be analysed based on the whole section.

CDSCWs with the rectangular section mainly sustain axial loads and in-plane bending moment and shear force, while T-CDSCWS sustain axial loads and bidirectional bending moment and shear force (Brueggen et al. 2017; Shi et al. 2017). In high-rise buildings, the axial loads sustained by shear walls are mainly compression, and the load actions caused by horizontal loads are mainly bending moment. Therefore, the cross-sectional capacity of T-CDSCWs under axial compression and bidirectional bending moment will be analysed, and the interaction curves will be proposed.

The outer steel corrugated plates are only constrained unilaterally by infilled concrete and intermediate bolts, and the local buckling under compression and shear occurs in steel corrugated plates. The local buckling of steel corrugated plates is influenced by the bolt arrangement, the lateral pressure of concrete placement and the constraint of concrete. When designing the cross-sectional capacity of T-CDSCWs, the effect of the local buckling of steel corrugated plates should be considered. In this thesis, the design formulas will be proposed to evaluate the local buckling of steel corrugated-plate. When the class of seismic measure is high, steel corrugated plates are not allowed to buckle prematurely before yielding. In this case, it is necessary to limit the bolt spacing. In this thesis, the upper limit of the bolt spacing will be proposed to avoid the local buckling of steel corrugated plates.

Due to the composite effect of steel and concrete, steel plate concrete composite walls have high load-carrying capacity, so steel plate concrete composite walls can be tall and thin. In particular, steel corrugated plates have high buckling performance, and the thickness of CDSCWs may be designed to be smaller. The global buckling of steel plate concrete composite walls may occur under the loads mainly based on axial compression. The global stability design method of reinforced concrete shear walls is specified in JGJ 3 (MOHURD 2010), but the design method is not suitable for steel plate concrete composite walls. It is necessary to propose a global stability design method suitable for steel plate concrete composite walls. Concerning T-CDSCWs, flexural buckling about the asymmetry axis and flexural-torsional buckling about the symmetry axis may occur under axial compression. When the breadth of the web wall is smaller than that of the flange wall and the stiffness of T-CDSCWs about the asymmetry axis is smaller, the first-order buckling mode is flexural buckling about the asymmetry axis; when the breadth of the web wall is larger than that of the flange wall and the stiffness of T-CDSCWs about the asymmetry axis is larger, the first-order buckling mode is flexural-torsional buckling about the symmetry axis. In this thesis, the global stability performance of T-CDSCWs under axial compression will be studied according to different buckling modes, and the corresponding design formulas will be proposed. In-plane instability of T-CDSCWs may occur under axial compression and bending moment in the web plane. In this thesis, the in-plane stability performance of T-CDSCWs will be studied. Combined with the interaction curves of the cross-sectional capacity, the interaction curves of the in-plane stability will be proposed.

Due to the large area of flange walls, flange walls are closer to the neutral axis, the damage of T-CDSCWs under the seismic action is mainly concentrated in the web walls, especially the boundary elements at the end of web walls (Ji et al. 2014, 2017; Jiang et al. 2020). In this thesis, the seismic performance of T-CDSCWs will be studied, and the construction measures to meet seismic requirements will be proposed. Since rectangular steel tubes and outer steel plates can effectively confine infilled concrete, there is no need to specify the section height of constraining boundary elements for concrete-infilled double steel plate composite walls. However, the lower limits of the thickness of rectangular steel tubes and outer steel plates should be specified to ensure the effective confinement on infilled concrete.

1.3.3 Research Methods

Aiming at the cross-sectional capacity, global stability, and seismic performance of T-CDSCWs, experimental investigations, finite element simulations, and theoretical analyses will be used in this thesis to carry out the research.

Three sets of T-CDSCWs tests are designed and conducted with the help of the multi-functional loading device at Tsinghua University, including the cross-sectional capacity test, flexural buckling test, and seismic test. The cross-sectional capacity, global stability, and seismic performance of T-CDSCWs will be investigated, and the failure mode, load–displacement relationship and strain distribution will be studied. Concerning the cross-sectional capacity test, the web wall of the T-CDSCW specimen is designed to be wide since T-CDSCWs mainly bear axial compression and in-plane bending, and the eccentric load is applied to the T-CDSCW specimen. Concerning the flexural buckling test, the web wall of the T-CDSCW specimen is designed to be short to ensure that the flexural buckling occurs. Concerning the seismic test, axial compression and cyclic in-plane horizontal loads are applied to the T-CDSCW specimen. In the experimental investigations, the factors that are difficult to be considered in finite element simulation and theoretical analysis can be observed, which provides the basis for analysis and verification for the subsequent finite element modelling and theoretical derivation.

The finite element models of T-CDCSWs are developed based on the general finite element program Abaqus (Zhuang et al. 2009; Dassault Systemes 2014), and the finite element simulations of T-CDCSWs are carried out to analyse the effects of different parameters on the cross-sectional capacity, global stability, and seismic performance. When analysing the cross-sectional capacity and global stability of T-CDSCWs, the initial imperfection of T-CDSCWs and the interaction between different parts need to be considered. Therefore, the refined solid–shell finite element model is used for parametric analyses. In the refined solid–shell finite element model, the solid element is used to simulate infilled concrete, the shell element is used to simulate steel corrugated plates and rectangular tubes, and the truss element is used to simulate intermediate bolts. At the same time, the contact relationship between steel plates and infilled concrete and the tie relationship between steel plates and intermediate bolts are considered. When analysing the seismic performance of T-CDCSWs, it is necessary to reduce the calculation cost of finite element analyses. Therefore, the simplified shell–beam finite element model is used for parametric analyses and the load–displacement relationship of T-CDSCWs is focused on. In the simplified shell–beam finite element model, the shell element is used to simulate wall elements, and the beam element is used to simulate boundary elements. Since corrugated wall elements are simplified as flat plates by the shell element, the shell–beam finite element model is actually a finite element model of steel plate concrete composite walls with T-section.

Based on reasonable assumptions, the load-carrying mechanism of T-CDSCWs will be analysed. Concerning the cross-sectional capacity, the design formulas will be proposed to evaluate the local buckling of steel corrugated plates and the cross-sectional capacity of T-CDSCWs under axial compression and bidirectional bending moments. Based on the orthotropic plate theory and the energy method (Timoshenko and Gere 2009; Timoshenko and Woinowsky-Krieger 1959), the buckling loads of steel corrugated plates will be analysed with respect to the elastic buckling with intermediate bolt constraints and the contact buckling with infilled concrete and intermediate bolt constraints. Based on the contact buckling load, the normalised slenderness ratio of steel corrugated plates can be derived, and the curve of the strength reduction factor can be established to design the load-carrying capacity of steel corrugated plates. The upper limit of the normalised slenderness ratio will be proposed, below which the strength reduction can be neglected. With the local buckling of steel corrugated plates in consideration, the design formulas for the cross-sectional capacity of T-CDSCWs will be proposed based on the full section plastic stress distribution. The bending capacity of T-CDSCWs can be obtained by reducing the plastic bending capacity, and the interaction curves of axial compression and bending moment in the web plane, bending moment in the flange plane, and bidirectional bending moment will be proposed based on the finite element simulations.

Concerning the global stability, the design formulas will be proposed to evaluate the stability capacity under axial compression and bending moment in the web plane. Based on the elastic stability theory (Timoshenko and Gere 2009), the buckling loads of T-CDSCWs will be analysed with respect to the flexural buckling about the asymmetry axis and the flexural-torsional buckling about the symmetry axis. The critical parameters of buckling modes will be proposed, from which the buckling modes of T-CDSCWs can be directly determined. Based on the elastic buckling loads, the normalised slenderness ratio can be derived, and the curves of the strength reduction factors of flexural and flexural-torsional buckling can be established to design the stability capacity of T-CDSCWs under axial compression. Combined with the interaction curves of the cross-sectional capacity, the interaction curve of T-CDSCWs under axial compression and bending moment in the web plane will be proposed with the second-order effect in consideration.

Concerning the seismic performance, the formulas for ultimate drift ratio and construction requirements will be proposed. Based on existing experimental results, the requirements for drift ratio and axial compression ratio specified in design standards (MOHURD 2015) will be verified, and the formulas will be proposed to evaluate the ultimate drift ratio of steel plate concrete composite walls. Combined with the finite element simulation results, the construction requirements will be proposed to ensure the seismic performance of steel plate concrete composite walls.

1.3.4 Research Significance

In this thesis, comprehensive, systematic and in-depth study is conducted on the load-carrying mechanism and design method of T-CDSCWs. The research results have been adopted by T/CECS 624 "Technical Specification for Structures with Corrugated Steel Plate Components" (CECS 2019). The major contribution of the thesis is as follows.

1. To reveal the cross-sectional mechanism of T-CDSCWs under axial compression and bidirectional bending moment and propose the design method to evaluate the cross-sectional capacity.

In the study of the cross-sectional performance of T-CDSCWs, the excellent cross-sectional capacity of T-CDSCWs is verified by the cross-sectional capacity test, and the effects of different parameters on the cross-sectional capacity are analysed by the refined solid–shell finite element model. Accurate and concise design formulas are proposed to evaluate the local buckling of steel corrugated plates and the cross-sectional capacity of T-CDSCWs. The formulas provide the basis for the engineering application of T-CDSCWs, supplement the research on the cross-sectional capacity of steel plate concrete composite walls with T-section, and provide a reference for the cross-sectional design of similar members.

2. To reveal the global stability mechanism of T-CDSCWs under axial compression and bending moment in the web plane and propose the design method to evaluate the global stability.

   In the study of the global stability performance of T-CDSCWs, the flexural buckling of T-CDSCWs is studied by the global stability test, and the effects of different parameters on the global stability performance are analysed by the refined solid–shell finite element model. Accurate and concise curves of the strength reduction factor are proposed. The curves provide the basis for the engineering application of T-CDSCWs, supplement the research on the global stability of steel plate concrete composite walls with T-section, and provide a reference for the global stability design of similar members.

3. To study the seismic performance of steel plate concrete composite walls with T-section and propose the seismic requirements of steel plate concrete composite walls.

   In the study of the seismic performance of T-CDSCWs, the excellent seismic performance of T-CDSCWs is verified by the seismic test, and the seismic performance of steel plate concrete composite walls with T-section is analysed by the simplified shell–beam finite element model. The seismic requirements of wall and boundary elements are proposed. The research results provide the basis for the seismic design of steel plate concrete composite walls and deepen the understanding of the importance of the boundary elements of steel plate concrete composite walls with T-section.

References

AISC (2016) Seismic provisions for structural steel buildings. American Institute of Steel Construction, Chicago

Booth PN, Varma AH, Sener KC et al (2015) Seismic behavior and design of a primary shield structure consisting of steel-plate composite (SC) walls. Nucl Eng Des 295:829–842

Brueggen BL, French CE, Sritharan S (2017) T-shaped RC structural walls subjected to multidirectional loading: test results and design recommendations. J Struct Eng 143:4017040

Bruhl JC, Varma AH, Johnson WH (2015) Design of composite SC walls to prevent perforation from missile impact. Int J Impact Eng 75:75–87

Bu F (2011) Studies on the mechanical performance of composite shear wall with double steel plates and concrete filled (in Chinese). Ph.D. thesis, Tsinghua University, Beijing

Cao W, Wang Y, Dong H et al (2011a) Shaking table test study on shear walls with concrete-filled steel tube columns and embedded steel-plate (in Chinese). Earthq Eng Eng Vib 31:75–81

Cao W, Xu F, Dong H et al (2014) Shear behavior analysis on low-rise shear wall with CFST columns and concealed steel plate-steel braces (in Chinese). World Earthq Eng 30:1–7

Cao W, Zhang J, Tao J et al (2007) Experimental study on low-rise shear wall with concealed truss (in Chinese). J Southeast Univ (Nat Sci Ed) 37:195–200

Cao W, Zhang W, Zhang J et al (2011b) Seismic research and application on embedded steel plate concrete shear wall with concrete filled steel tube columns (in Chinese). J Civ Eng Manag 28:219–225

CECS (2019) Technical specification for structures with corrugated steel plate components (in Chinese). China Architecture and Building Press, Beijing

CEN (2004) Eurocode 4: design of composite steel and concrete structures, part 1–1: general rules and rules for buildings. European Committee for Standardization, Brussels

Chen G, Guo Y (2004) Ultimate shear-carrying capacity of steel plate shear wall with cross stiffeners (in Chinese). J Build Struct 25:71–78

Chen T, Xiao C, Tian C et al (2011) Experimental study of the compression-bending behavior of composite shear walls of high axial compression ratios (in Chinese). Chin Civ Eng J 44:1–7

Chen Z, Jiang Y, Zhang X et al (2016) Research on deformation property and finite element analysis of T-shaped bundled lipped channel-concrete composite wall (in Chinese). Chin Civil Eng J 49:39–44

Chen Z, Jiang Y, Zhang X et al (2017) Deformation property and finite element analysis of a new type of steel tube bundle composite shear walls (in Chinese). J Vib Shock 36:36–45

Choi B, Kang C, Park H (2014) Strength and behavior of steel plate–concrete wall structures using ordinary and eco-oriented cement concrete under axial compression. Thin Wall Struct 84:313–324

Dan D, Fabian A, Stoian V (2011a) Nonlinear behavior of composite shear walls with vertical steel encased profiles. Eng Struct 33:2794–2804

Dan D, Fabian A, Stoian V (2011b) Theoretical and experimental study on composite steel–concrete shear walls with vertical steel encased profiles. J Constr Steel Res 67:800–813

Dassault Systemes (2014) Abaqus (Version 6.14). Johnston, RI, United States. https://www.3ds.com/products-services/simulia/products/abaqus/abaqusstandard/

Ding Y, Zhang Y, Zhao J (2008) Quasi-static test of chevron-shaped unbonded steel plate brace encased in reinforced concrete panel (in Chinese). Chin Civ Eng J 41:23–30

Ding Z, Jiang H, Zeng J et al (2011) An innovative application of SCS composite wall: structural design of Yancheng TV tower (in Chinese). Build Struct 41:87–91

Dou C, Jiang Z, Pi Y et al (2016) Elastic shear buckling of sinusoidally corrugated steel plate shear wall. Eng Struct 121:136–146

Eom T, Park H, Lee C et al (2009) Behavior of double skin composite wall subjected to in-plane cyclic loading. J Struct Eng 135:1239–1249

Epackachi S, Nguyen NH, Kurt EG et al (2014) In-plane seismic behavior of rectangular steel-plate composite wall piers. J Struct Eng 141:4014176

Epackachi S, Whittaker AS, Aref A (2017) Seismic analysis and design of steel-plate concrete composite shear wall piers. Eng Struct 133:105–123

Epackachi S, Whittaker AS, Huang YN (2015) Analytical modeling of rectangular SC wall panels. J Constr Steel Res 105:49–59

Fu X (2010) Applied structural design of tall buildings (in Chinese), 2nd edn. China Architecture and Building Press, Beijing

Guo L, Wang Y, Zhang S (2018a) Experimental study of rectangular multi-partition steel-concrete composite shear walls. Thin Wall Struct 130:577–592

Guo Y, Dong Q (2005) Research and application of steel plate shear wall in high-rise buildings (in Chinese). Steel Constr 20:1–6

Guo Y, Dong Q, Zhou M (2009a) Tests and analysis on hysteretic behavior of buckling-restrained steel plate shear wall (in Chinese). J Build Struct 30:31–39
Guo Y, Tong J, Jiang Z (2015) Design fundamentals and application of corrugated-web steel structures (in Chinese). Science Press, Beijing
Guo Y, Zhou M (2011) An overview of current state-of-the-art in behaviour and design theory for unstiffened or buckling-restrained steel plate shear walls (in Chinese). J Build Struct 32:1–16
Guo Y, Zhou M, Dong Q (2009b) Hysteretic behavior of buckling-restrained steel plate shear wall (in Chinese). Eng Mech 26:108–114
Guo Y, Zhu J (2020) Research progress of shear walls: types and design methods (in Chinese). Eng Mech 37:19–33
Guo Y, Zhu J, Wang M et al (2017) Research on corrugated-steel-concrete composite wall structure system in high-rise building (in Chinese). Tsinghua University, Beijing
Guo Y, Zhu J, Wang M et al (2018b) Overall instability performance of concrete-infilled double steel corrugated plate wall. Thin Wall Struct 130:372–394
Hao T, Cao W (2017) Study on axial compressive bearing capacity of composite shear wall with double-skin steel plate (in Chinese). Chin J Eng 39:1765–1773
Hossain KMA, Mol LK, Anwar MS (2015) Axial load behaviour of pierced profiled composite walls with strength enhancement devices. J Constr Steel Res 110:48–64
Hossain KMA, Rafiei S, Lachemi M et al (2016a) Finite element modeling of impact shear resistance of double skin composite wall. Thin Wall Struct 107:101–118
Hossain KMA, Rafiei S, Lachemi M et al (2016b) Structural performance of profiled composite wall under in-plane cyclic loading. Eng Struct 110:88–104
Hossain KMA, Wright HD (2004) Performance of double skin-profiled composite shear walls — experiments and design equations. Can J Civ Eng 31:204–217
Hou S, Liu Y, Zhao H (2017) Structural design and construction of composite high strength concrete shear wall with embedded double steel plates in Guangzhou east tower (in Chinese). Build Struct 47:74–79
Hu H, Nie J, Eatherton MR (2014) Deformation capacity of concrete-filled steel plate composite shear walls. J Constr Steel Res 103:148–158
Hu H, Nie J, Fan J et al (2016) Seismic behavior of CFST-enhanced steel plate-reinforced concrete shear walls. J Constr Steel Res 119:176–189
Huang Z, Liew JYR (2016a) Compressive resistance of steel-concrete-steel sandwich composite walls with J-hook connectors. J Constr Steel Res 124:142–162
Huang Z, Liew JYR (2016b) Structural behaviour of steel–concrete–steel sandwich composite wall subjected to compression and end moment. Thin Wall Struct 98:592–606
Ji X, Jia X, Qian J (2015) Experimental study on shear behavior of steel-plate composite shear walls (in Chinese). J Build Struct 36:46–55
Ji X, Jiang F, Qian J (2013) Seismic behavior of steel tube–double steel plate–concrete composite walls: experimental tests. J Constr Steel Res 86:17–30
Ji X, Leong T, Qian J et al (2016) Cyclic shear behavior of composite walls with encased steel braces. Eng Struct 127:117–128
Ji X, Liu D, Qian J (2014) Comparative study on seismic design of RC shear walls between Chinese and U.S. codes (in Chinese). Earthq Eng Eng Dyn 34:424–433
Ji X, Liu D, Qian J (2017) Improved design of special boundary elements for T-shaped reinforced concrete walls. Earthq Eng Eng Vib 16:83–95
Jiang D, Xiao C, Chen T et al (2012) Experimental study of the compression-bending behavior of high-strength concrete steel composite shear walls (in Chinese). Chin Civil Eng J 45:17–25
Jiang Z, Ji X, Wang S et al (2020) Experimental study on design methods of special boundary elements of T-shaped RC walls (in Chinese). Build Struct 50:108–115
Kurt EG, Varma AH, Booth P et al (2016) In-plane behavior and design of rectangular SC wall piers without boundary elements. J Struct Eng 142:4016026
Lan W, Li B, Zhang Z (2018) Seismic performance of steel-concrete composite structural walls with prestressed internal bracing. J Constr Steel Res 140:11–24

Liang QQ, Uy B, Wright HD et al (2004) Local buckling of steel plates in double skin composite panels under biaxial compression and shear. J Struct Eng 130:443–451
Liu J, Li X (2018) Experimental study on the bending shear performance of steel plate reinforced concrete structure for nuclear power plants under out-plane load (in Chinese). Build Struct 48:83–86
Ma X, Nie J, Tao M (2014) Effective stiffness of steel plate-concrete composite shear wall structures in serviceability state (in Chinese). Chin Civil Eng J 47:18–26
Ma X, Nie J, Tao M et al (2013) Numerical model and simplified formula of axial force-moment capacity of composite shear wall with double steel plates and infill concrete (in Chinese). J Build Struct 34:99–106
MOHURD (2010) Technical specification for concrete structures of tall building (in Chinese). China Architecture and Building Press, Beijing
MOHURD (2015) Technical specification for steel plate shear walls (in Chinese). China Architecture and Building Press, Beijing
MOHURD (2016) Code for design of composite structures (in Chinese). China Architecture and Building Press, Beijing
MOHURD, AQSIQ (2010) Code for design of concrete structures (in Chinese). China Architecture and Building Press, Beijing
MOHURD, AQSIQ (2016) Code for seismic design of buildings (in Chinese). China Architecture and Building Press, Beijing
Nie J, Hu H, Fan J et al (2013a) Experimental study on seismic behavior of high-strength concrete filled double-steel-plate composite walls. J Constr Steel Res 88:206–219
Nie J, Hu H, Li S et al (2013b) Experimental study on seismic behavior of steel plate reinforced concrete composite shear walls with square CFST concealed columns (in Chinese). J Build Struct 34:52–60
Nie J, Li F (2009a) Study on the stability of steel-concrete composite plates under uniaxial compressed load (in Chinese). China Railw Sci 30:27–32
Nie J, Li F (2009b) Elastic bending and stability of steel-concrete composite plate (in Chinese). Eng Mech 26:59–66
Nie J, Ma X, Tao M et al (2014) Effective stiffness of composite shear wall with double plates and filled concrete. J Constr Steel Res 99:140–148
Nie J, Tao M, Fan J et al (2011) Research advances of composite shear walls with double steel plates and filled concrete (in Chinese). Build Struct 41:52–60
Othuman Mydin MA, Wang YC (2011) Structural performance of lightweight steel-foamed concrete–steel composite walling system under compression. Thin Wall Struct 49:66–76
Prabha P, Marimuthu V, Saravanan M et al (2013) Effect of confinement on steel-concrete composite light-weight load-bearing wall panels under compression. J Constr Steel Res 81:11–19
Qi Y, Gu Q, Sun G et al (2017) Shear force demand on headed stud for the design of composite steel plate shear wall. Eng Struct 148:780–792
Qian J, Jiang Z, Ji X (2012) Behavior of steel tube-reinforced concrete composite walls subjected to high axial force and cyclic loading. Eng Struct 36:173–184
Rafiei S, Hossain KMA, Lachemi M et al (2013) Finite element modeling of double skin profiled composite shear wall system under in-plane loadings. Eng Struct 56:46–57
Rafiei S, Hossain KMA, Lachemi M et al (2015) Profiled sandwich composite wall with high performance concrete subjected to monotonic shear. J Constr Steel Res 107:124–136
Rafiei S, Hossain KMA, Lachemi M et al (2017) Impact shear resistance of double skin profiled composite wall. Eng Struct 140:267–285
Sabouri-Ghomi S, Jahani Y, Bhowmick AK (2016) Partial interaction theory to analyze composite (steel–concrete) shear wall systems under pure out-of-plane loadings. Thin Wall Struct 104:211–224
Sener KC, Varma AH, Booth PN et al (2015) Seismic behavior of a containment internal structure consisting of composite SC walls. Nucl Eng Des 295:804–816

Shi Q, Wang B, He W et al (2017) Experimental research on seismic behavior of reinforced concrete shear walls with flange (in Chinese). J Build Struct 38:106–115
Sun J, Wang Y, Sun H et al (2011) Application of steel plate concrete composite wall in National Museum of China (in Chinese). Build Struct 41:14–19
Sun J, Xu P, Xiao C et al (2008) Experimental study on shear behavior of steel plate-concrete composite wall (in Chinese). Build Struct
Tao M, Nie J, Fan J et al (2017) Development trends and path for China's civil and structural engineering science and technology to 2035 (in Chinese). Eng Sci 19:73–79
Timoshenko SP, Gere JM (2009) Theory of elastic stability, 2nd edn. Dover Publications, New York
Timoshenko SP, Woinowsky-Krieger S (1959) Theory of plates and shells. McGraw Hill, New York
Tong J (2019) Theoretical and experimental study on design methods for steel corrugated plate shear walls (in Chinese). Ph.D. thesis, Tsinghua University, Beijing
Tong J, Guo Y (2015) Elastic buckling behavior of steel trapezoidal corrugated shear walls with vertical stiffeners. Thin Wall Struct 95:31–39
Tong J, Guo Y (2018) Shear resistance of stiffened steel corrugated shear walls. Thin Wall Struct 127:76–89
Tong J, Guo Y, Pan W (2020a) Ultimate shear resistance and post-ultimate behavior of double-corrugated-plate shear walls. J Constr Steel Res 165:105895
Tong J, Guo Y, Zuo J (2018) Elastic buckling and load-resistant behaviors of double-corrugated-plate shear walls under pure in-plane shear loads. Thin Wall Struct 130:593–612
Tong J, Guo Y, Zuo J et al (2020b) Experimental and numerical study on shear resistant behavior of double-corrugated-plate shear walls. Thin Wall Struct 147:106485
Varma AH, Malushte SR, Sener KC et al (2014) Steel-plate composite (SC) walls for safety related nuclear facilities: design for in-plane forces and out-of-plane moments. Nucl Eng Des 269:240–249
Wang B, Jiang H, Lu X (2017) Seismic performance of steel plate reinforced concrete shear wall and its application in China Mainland. J Constr Steel Res 131:132–143
Wang J, Fan J, Tao M et al (2015) Finite element model on mechanical behavior of composite shear wall with double steel plates and infilling concrete (in Chinese). J Build Struct 36:1–8
Wang J, Tao M, Fan J et al (2018a) Seismic behavior of steel plate reinforced concrete composite shear walls under tension-bending-shear combined cyclic load. J Struct Eng 144:4018075
Wang W, Liu G, Su S et al (2019) Research on the shear bearing capacity of corrugated steel plate shear wall and composite wall (in Chinese). Eng Mech 36:197–206
Wang W, Wang Y, Lu Z (2018b) Experimental study on seismic behavior of steel plate reinforced concrete composite shear wall. Eng Struct 160:281–292
Wang W, Zhang L, Su S et al (2018c) Experimental research on seismic behavior of corrugated steel plate-concrete composite shear walls (in Chinese). J Build Struct 39:75–84
Wang W, Zhao C, Ren T et al (2020) Nonlinear finite element analysis of corrugated steel plate-concrete composite shear wall (in Chinese). Chin J Appl Mech 37:2054–2062
Wright H (1998a) Axial and bending behavior of composite walls. J Struct Eng 124:758–764
Wright H (1998b) The axial load behaviour of composite walling. J Constr Steel Res 45:353–375
Wright HD (1993) Buckling of plates in contact with a rigid medium. Struct Eng 71:209–215
Wright HD, Gallocher SC (1995) The behaviour of composite walling under construction and service loading. J Constr Steel Res 35:257–273
Wright HD, Hossain KMA (1997) In-plane shear behaviour of profiled steel sheeting. Thin Wall Struct 29:79–100
Wu Y, Kang D, Yang Y (2016) Seismic performance of steel and concrete composite shear walls with embedded steel truss for use in high-rise buildings. Eng Struct 125:39–53
Yan J, Liew JYR (2016) Design and behavior of steel–concrete–steel sandwich plates subject to concentrated loads. Compos Struct 150:139–152
Yan J, Liew JYR, Zhang M et al (2016a) Punching shear resistance of steel–concrete–steel sandwich composite shell structure. Eng Struct 117:470–485

Yan J, Qian X, Liew JYR et al (2016b) Damage plasticity based numerical analysis on steel–concrete–steel sandwich shells used in the Arctic offshore structure. Eng Struct 117:542–559

Yan J, Wang J, Liew JYR et al (2016c) Punching shear behavior of steel–concrete–steel sandwich composite plate under patch loads. J Constr Steel Res 121:50–64

Yan J, Wang J, Liew JYR et al (2016d) Ultimate strength behaviour of steel–concrete–steel sandwich plate under concentrated loads. Ocean Eng 118:41–57

Yang H, Li Z, Li X et al (2021) Design and application of corrugated steel plate components structure system in high-rise residential buildings (in Chinese). Paper presented at the 2021 national conference on civil engineering construction technology, Beijing, China

Yang Y, Liu J, Fan J (2016) Buckling behavior of double-skin composite walls: an experimental and modeling study. J Constr Steel Res 121:126–135

Yang Y, Liu J, Fan J et al (2013) Experimental study on flexural capacity of steel plate-concrete composite slabs (in Chinese). J Build Struct 34:24–31

YJK (2016) Reinforcement design of shear walls with complex sections (in Chinese). YJK Building Software, Beijing

Zhang K, Varma AH, Malushte SR et al (2014) Effect of shear connectors on local buckling and composite action in steel concrete composite walls. Nucl Eng Des 269:231–239

Zhang W, Cao W, Dong H et al (2012) Performance of co-work between steel plate with studs and outer reinforced concrete in shear wall (in Chinese). J Beijing Univ Technol 38:828–834

Zhang X, Chen Z, Ren Q et al (2017) Study on finite element analysis and design method of composite shear wall with T-section double steel plates (in Chinese). Build Struct 47:241–248

Zhang X, Qin Y, Chen Z (2016b) Experimental seismic behavior of innovative composite shear walls. J Constr Steel Res 116:218–232

Zhang X, Qin Y, Chen Z et al (2016a) Experimental behavior of innovative T-shaped composite shear walls under in-plane cyclic loading. J Constr Steel Res 120:143–159

Zhang Y (2015) Experimental and numerical analysis study on seismic performance of steel plate reinforced concrete structure for nuclear power plants (in Chinese). Ph.D. thesis, Institute of Engineering Mechanics, China Earthquake Administration, Harbin

Zhao Q, Astaneh-Asl A (2004) Cyclic behavior of traditional and innovative composite shear walls. J Struct Eng 130:271–284

Zhao Q, Li Y, Tian Y (2020b) Cyclic behavior of double-skin composite walls with flat and corrugated faceplates. Eng Struct 220:111013

Zhao Q, Li Y, Tian Y et al (2020a) Cyclic behavior of corrugated double-skin composite walls with different aspect ratios. J Struct Eng 146:4020214

Zhao W, Guo Q, Huang Z et al (2016) Hysteretic model for steel–concrete composite shear walls subjected to in-plane cyclic loading. Eng Struct 106:461–470

Zhou Y, Zhu J, Guo Y et al (2021) Numerical and experimental studies on sectional load capacity of concrete-infilled double steel corrugated-plate walls under combined compression and in-plane bending. Thin Wall Struct 159:107250

Zhou Z, Cao W, Zhang W et al (2015) Seismic performance of composite shear wall with shaped steel concrete boundary columns and embedded steel plate with holes (in Chinese). J Beijing Univ Technol 41:1550–1558

Zhu A, Xiao C, Chen T et al (2016) Experimental study on seismic behavior of embedded steel plate reinforced concrete shear walls with 1.0 shear-span-ratio (in Chinese). Chin Civil Eng J 49:49–56

Zhu J (2020) Theoretical and experimental study on design methods for I-section concrete-infilled double steel corrugated-plate walls (in Chinese). Ph.D. thesis, Tsinghua University, Beijing

Zhu J, Guo Y, Wang M et al (2019) Strength design of concrete-infilled double steel corrugated-plate walls under uniform compressions. Thin Wall Struct 141:153–174

Zhu J, Guo Y, Wang M et al (2020) Seismic performance of concrete-infilled double steel corrugated-plate walls: experimental research. Eng Struct 215:110601

Zhuang Z, You X, Liao J et al (2009) Finite element analyses and applications based on ABAQUS (in Chinese). Tsinghua University Press, Beijing

Chapter 2
Experimental Investigations on T-CDSCWs

To study the cross-sectional capacity, global stability, and seismic performance of T-CDSCWs, three sets of T-CDSCW tests are designed and conducted. This chapter introduces the design, phenomena and results of the tests. Firstly, the cross-sectional capacity test of the T-CDSCW is conducted to study the cross-sectional capacity under axial compression and bending moment in the web plane, to which T-CDSCWs are mainly subjected. The web wall of the T-CDSCW specimen is designed to be wide, and eccentric compression in the web plane is applied to the T-CDSCW specimen during the test. Secondly, the global stability test of the T-CDSCW is conducted to study the global stability under axial compression. Flexural buckling about the asymmetry axis and flexural-torsional buckling about the symmetry axis of T-CDSCWs may occur under axial compression. To ensure that the buckling mode is flexural buckling, the web wall of the T-CDSCW specimen is designed to be short. Thirdly, the seismic performance test of the T-CDSCW is conducted to study the seismic performance under axial compression and cyclic horizontal load. The horizontal load is applied in the web plane and the seismic performances with flange in compression and tension are compared.

2.1 Cross-Sectional Capacity Test

The purpose of the cross-sectional capacity test is to study the cross-sectional capacity of T-CDSCWs under axial compression and bending moment in the web plane. This section introduces the design, phenomena and results of the cross-sectional capacity test, including the specimen design, test setup, instrumentation, material property, failure phenomena, stress distribution, etc.

M.-Z. Wang, *Design Method of Concrete-Infilled Double Steel Corrugated-Plate Walls with T-section*, Springer Theses,
https://doi.org/10.1007/978-981-95-7789-7_2

2.1.1 Experiment Design

A T-CDSCW specimen numbered T-1 is designed (see Figs. 2.1 and 2.2). For the convenience of description, the direction of the flange is defined as the x-axis, the direction of the web is defined as the y-axis, and the positive direction of the y-axis points to the boundary element at the end of the web wall. To better reflect the load-carrying performance of T-CDSCWs in practical applications, the overall and detailed dimensions of specimen T-1 are full-scale as follows.

The height of specimen T-1 H_w is 4500 mm. The breadth of the flange wall b_f is 1077 mm and the overall breadth of the flange wall B_f is 1397 mm. The breadth of the web wall b_w is 1077 mm and the overall breadth of the web wall B_w is 1397 mm. The thickness of wall elements d_w is 150 mm. The cross-sectional dimensions of steel rectangular tubes $b_c \times d_c \times t_c$ are 160 × 160 × 8 mm. For the convenience of

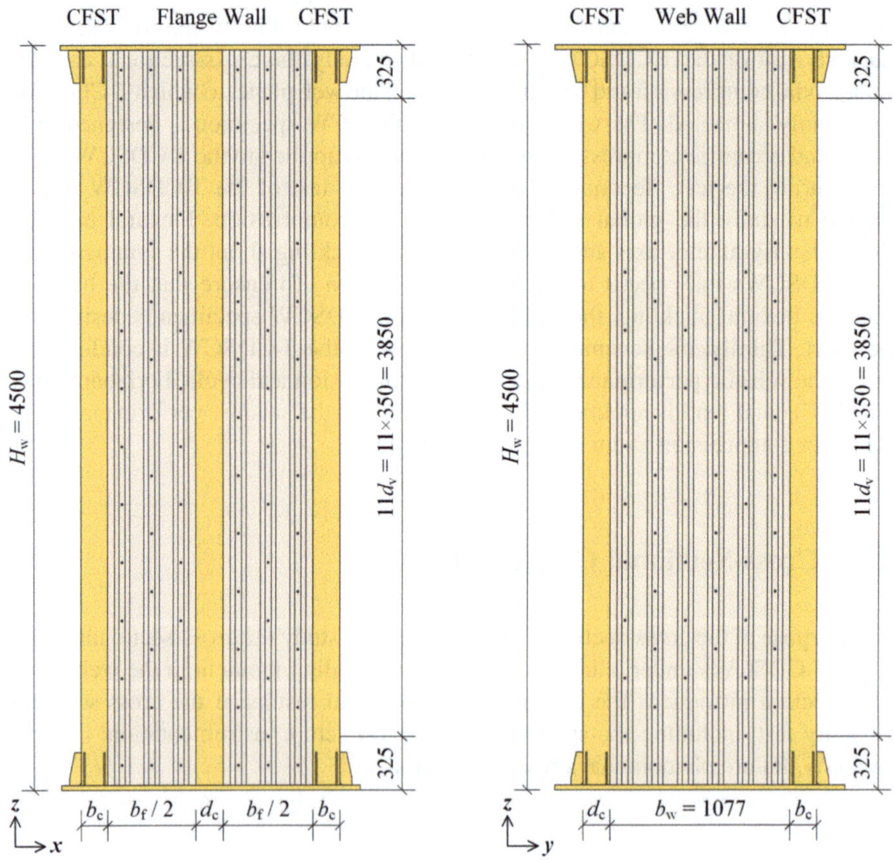

Fig. 2.1 Elevation view of specimen T-1 for the cross-sectional capacity test (unit: mm): (**a**) flange wall; (**b**) web wall

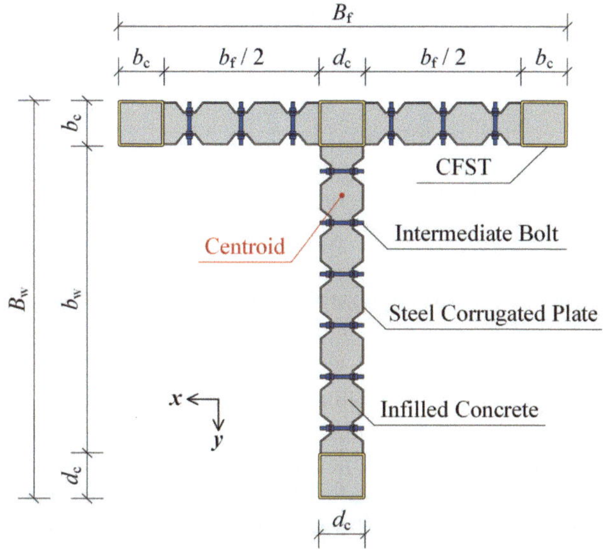

Fig. 2.2 Cross-section view of specimen T-1 for the cross-sectional capacity test

welding operations, the width of CFSTs is larger than the thickness of wall elements d_w. To ensure the load-carrying capacity of CFSTs, the thickness of steel rectangular tubes t_c is larger than the thickness of steel corrugated plates t_s. The width-to-thickness ratio of steel rectangular tubes conforms to the requirements specified in JGJ 138 (MOHURD 2016) so that steel rectangular tubes do not buckle prematurely before yielding. The detailed requirements are shown in Eqs. (2.1) and (2.2).

$$\frac{b_c}{t_c} \leq 60\sqrt{\frac{235}{f_y}} \tag{2.1}$$

$$\frac{d_c}{t_c} \leq 60\sqrt{\frac{235}{f_y}} \tag{2.2}$$

Where f_y denotes the yield strength of steel.

The dimensions of steel corrugated plates are determined by the optimisation of the load-carrying capacity and material consumption (Guo et al. 2017). The wave width of steel corrugated plates q is 180 mm, the crest width q_1 is 85 mm, the transition width q_2 is 27.5 mm, the trough width q_3 is 40 mm, the wave height a is 35 mm, and the thickness t_s is 4 mm (see Fig. 2.3). Intermediate bolts are staggered arranged. The vertical spacing d_v is 350 mm and the horizontal spacing d_h is 180 mm. The bolt specification is M14 with an effective diameter of 12.1 mm.

For the convenience of connection, steel plates are welded at the bottom and top of T-CDSCW specimens, and circular holes are made in the middle of the top steel

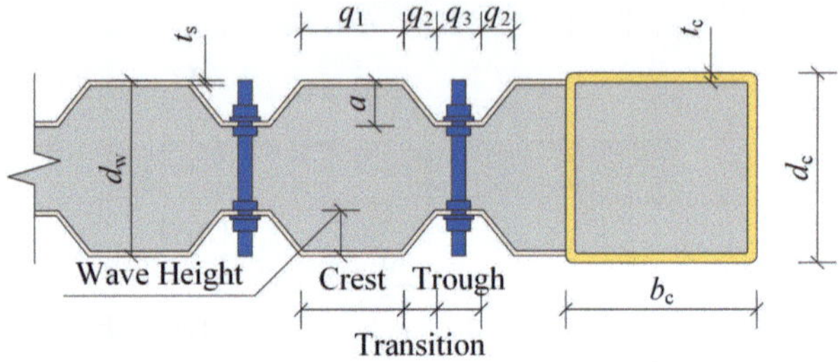

Fig. 2.3 Cross-section view of steel corrugated plates and rectangular tubes

plates for concrete placement. To facilitate the flow of concrete and avoid the formation of voids, rectangular holes are made in steel rectangular tubes on the side connected with wall elements. Screw holes are made around the bottom and top steel plates for the connection with the loading beam and the foundation beam. Stiffening rib plates are set at the bottom and top of the specimens to avoid weld fractures before the failure of the specimens.

Specimen T-1 is made of Q345 steel with the yield strength of 345 MPa standard value, C40 concrete with the cube strength of 40 MPa average value and high-strength bolts with the ultimate strength of 1000 MPa standard value (MOHURD and AQSIQ 2010, 2017). Since most of the steel is used in steel corrugated plates, three steel corrugated plate coupons are prepared for material property tests (AQSIQ and SAC 2010). Meanwhile, three concrete cubes are prepared to measure the compressive strength (MOHURD and AQSIQ 2003a).

The cross-sectional capacity test is conducted with the multifunctional loading device at the Department of Civil Engineering, Tsinghua University (see Fig. 2.4). The loading device includes a vertical hydraulic jack with a loading capacity of 20,000 kN and two horizontal actuators with a loading capacity of 2000 kN. The vertical hydraulic jack is controlled by a hydraulic stabilisation system and connected with a shear force measurement device. The shear force measurement device measures the friction force between the vertical hydraulic jack and the loading beam to correct the horizontal load and obtain the real value of the horizontal load applied to the specimen. With the action of the horizontal actuators, the vertical hydraulic jack can move horizontally together with the top of the specimen. A spherical hinge is arranged inside the vertical hydraulic jack, which can rotate with the specimen, so that the vertical hydraulic jack is always vertical to the top of the specimen.

The vertical hydraulic jack applies compression on the top surface of the loading beam, and the horizontal actuator constrains the sides of the loading beam. The specimen is connected with the loading beam at the top and the foundation beam at the bottom by bolts. The foundation beam is fixed on the ground by lead screws and shear keys. The loading beam is 800 mm high, T-shaped, and the shape is compatible with the specimen. The foundation beam is 800 mm high, 2650 mm long, and 2300 mm wide. The top and bottom plates of the loading beam and the foundation beam are 40 mm thick, the web plates are 30 mm thick, and the stiffening plates are

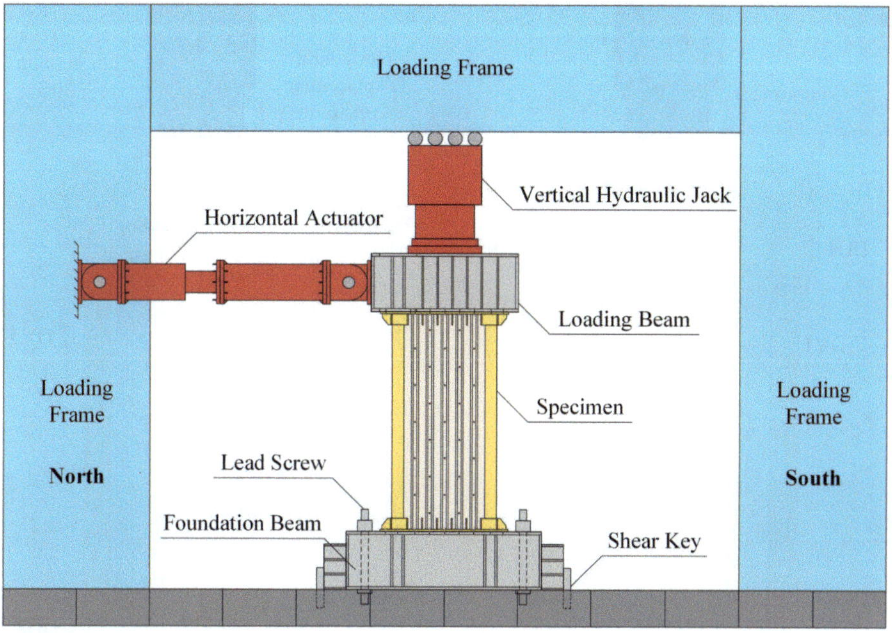

Fig. 2.4 Schematic diagram of the loading device

20 mm thick. Concerning the direction of the T-CDSCW specimen, the positive direction of the x-axis points to the west and the positive direction of the y-axis points to the south. The design of the loading beam and the foundation beam ensures that even under the maximum load that the loading device can apply, the loading beam and the foundation beam still remain in the elastic state and the bolt connection with the specimen does not break.

T-CDSCWs are mainly subjected to axial compression and bending moment in the web plane. To achieve the load action, eccentric compression in the web plane is applied to specimen T-1 (see Fig. 2.5). The loading centre of the vertical hydraulic jack is located on the symmetry axis of specimen T-1, and offset in the positive direction of the y-axis by 300 mm from the centroid.

The position of the centroid y_0 is determined by the weighted average of the axial stiffness of steel and concrete, as shown in Eq. (2.3).

$$y_0 = \frac{E_s A_s y_s + E_c A_c y_c}{E_s A_s + E_c A_c} \tag{2.3}$$

Where E_s, A_s, and y_s denote the elastic modulus, cross-sectional area, and the position of the centroid of the steel parts; E_c, A_c, and y_c denote the elastic modulus, cross-sectional area, and the position of the centroid of the concrete parts.

The definition of the direction of bending moment in the web plane is the same as the x-axis. When the direction of bending moment is along with the positive

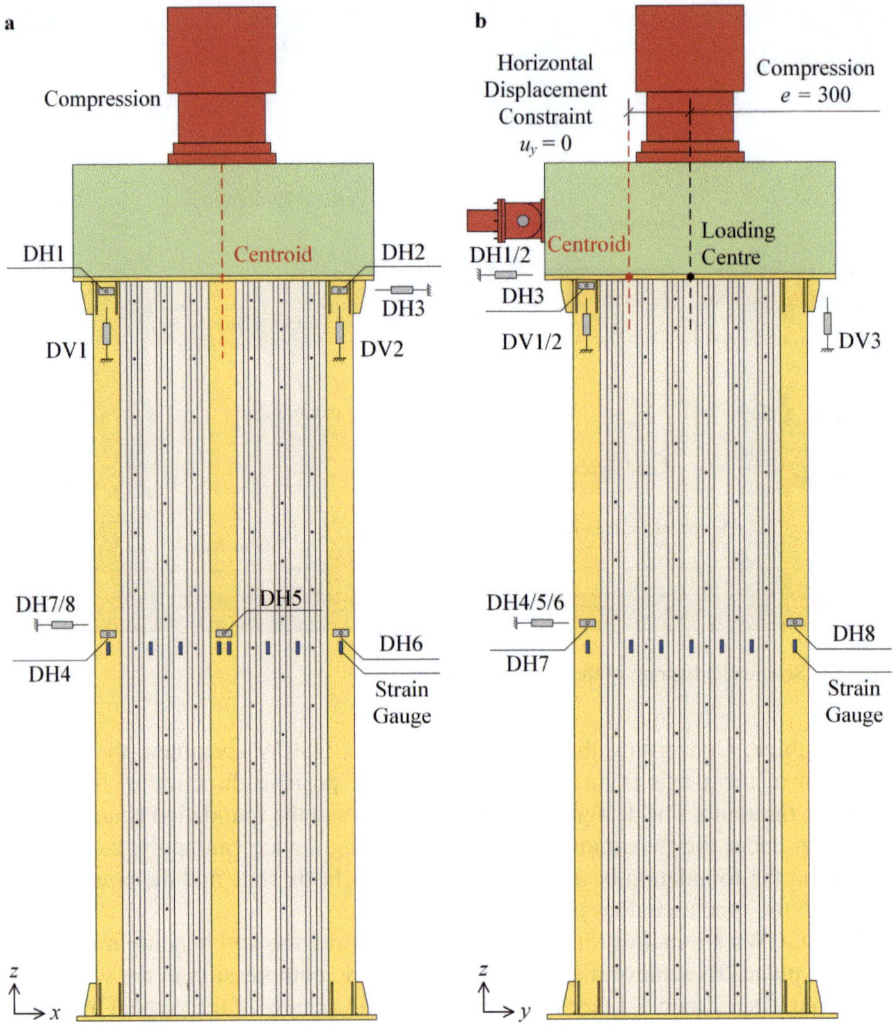

Fig. 2.5 Setup and instrumentation of the cross-sectional capacity test (unit: mm): (**a**) flange wall; (**b**) web wall

direction of the x-axis, the flange is under compression and the web is under tension, and the T-CDSCW is under positive bending moment; when the direction of bending moment is along with the negative direction of the x-axis, the flange is under tension and the web is under compression, and the T-CDSCW is under negative bending moment. The eccentric compression applied by the vertical hydraulic jack is equivalent to the axial compression and negative bending moment applied to the top of specimen T-1. The displacement constraint applied by the horizontal actuator is equivalent to the shear force applied to the top of specimen T-1. In consequence, the bending moment of specimen T-1 decreases linearly from top to bottom, the maximum bending moment is applied to the top and the damage is concentrated in the upper part of specimen T-1.

During the loading process, the horizontal displacement of the horizontal actuator is fixed, and the vertical load is continuously applied by the vertical hydraulic jack. In the initial stage, the loading is force-controlled until the vertical load reaches about 6000 kN. Afterwards, the loading is displacement-controlled until specimen T-1 exhibits noticeable damage, excessive deformation, or a significant decrease in the load-carrying capacity.

During the test, the load–displacement response of specimen T-1 and the strain development on the steel parts are monitored (see Figs. 2.5 and 2.6). The data is collected by the DATA LOGGER TDS-530 data measurement system, which displays the load–displacement, load–strain and other curves in real time for observing and adjusting the loading situation. The failure phenomena of specimen T-1, such as the local buckling of steel corrugated plates, are recorded by manual observation. Concerning the loads applied to specimen T-1, the load cells in the vertical hydraulic jack and the horizontal actuator measure the vertical and horizontal loads automatically. Concerning the displacement of specimen T-1, displacement transducers measure the vertical and horizontal displacement. Vertical displacement transducers DV1, DV2, and DV3 located near the boundary elements measure the vertical displacement of the top section, and the vertical displacement of the centroid can be obtained by the weighted average. Horizontal displacement transducers DH1 and DH2 at the top section measure the horizontal displacement in the y-axis, and

Fig. 2.6 On-site photographs of specimen T-1: (**a**) front view; (**b**) side view

horizontal displacement transducer DH3 at the top section measures the horizontal displacement in the x-axis to monitor whether the horizontal displacement of the top is effectively constrained. Horizontal displacement transducers DH4, DH5, and DH6 at the mid-height section measure the horizontal displacement in the y-axis, and horizontal displacement transducers DH7 and DH8 at the mid-height section measure the horizontal displacement in the x-axis. Besides, the horizontal slippage and the rotation of the foundation beam are measured to eliminate their effects on the displacement measurement of specimen T-1. Concerning the strain development, the normal strain on the steel parts is measured by strain gauges at the mid-height section.

2.1.2 Experimental Phenomena and Results

Before conducting the cross-sectional capacity test, the material property tests of steel and concrete are conducted to determine some data of specimen T-1, such as the centroid position. According to the average value of the material property test results (see Fig. 2.7), the yield strength f_y of the steel is 446 MPa, the ultimate strength f_u is 558 MPa, and the elastic modulus E_s is 1.99×10^5 MPa; the cubic compressive strength f_{cu} of the concrete is 47.6 MPa. The axial compressive strength f_c is 36.1 MPa, the axial tensile strength f_t is 3.22 MPa, and the elastic modulus E_c is 3.41×10^4 MPa, which are determined by Eqs. (2.4), (2.5), and (2.6), respectively (MOHURD and AQSIQ 2010).

$$f_c = 0.76 f_{cu} \tag{2.4}$$

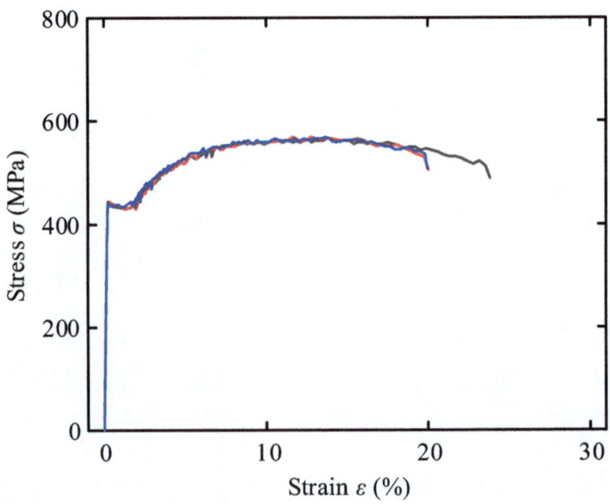

Fig. 2.7 Stress–strain curves of the steel plate coupons of the cross-sectional capacity test

$$f_t = 0.395 f_{cu}^{0.550} \tag{2.5}$$

$$E_c = \frac{10^5}{2.20 + 34.7 / f_{cu}} \tag{2.6}$$

Under the combined action of axial compression, negative bending moment in the web plane, and shear force, specimen T-1 exhibits the failure mode of the cross-section failure, including the local buckling of steel corrugated plates and rectangular tubes (see Fig. 2.8). In the loading process, no significant horizontal displacement in the x-axis occurs.

Before the vertical load reaches 10,000 kN (see Fig. 2.9), the response of the load–displacement curve is almost linear, and no obvious phenomena are observed. After the vertical load reaches 10,000 kN, the slope of the load–displacement curve begins to drop. When the vertical displacement of the centroid reaches 5.6 mm, the vertical load applied to specimen T-1 reaches the maximum of 15,518 kN, the corresponding negative bending moment applied to the top section is 4417 kN·m, and the local buckling occurs in steel corrugated plates.

As the loading continues, the vertical load degrades slowly. The local buckling of steel corrugated plates continues to develop and exhibits the buckling mode under the combined action of compression, bending and shear. The local buckling occurs in the steel rectangular tubes at the end of the web wall. When the vertical load degrades to about 11,000 kN, 70% of the maximum load, the loading ends. It is obviously observed from the experimental phenomena that, although specimen T-1 is 4.5 m tall, which is taller than the general floor-to-floor height, the failure of specimen T-1 is not governed by the global stability, but by the cross-sectional capacity.

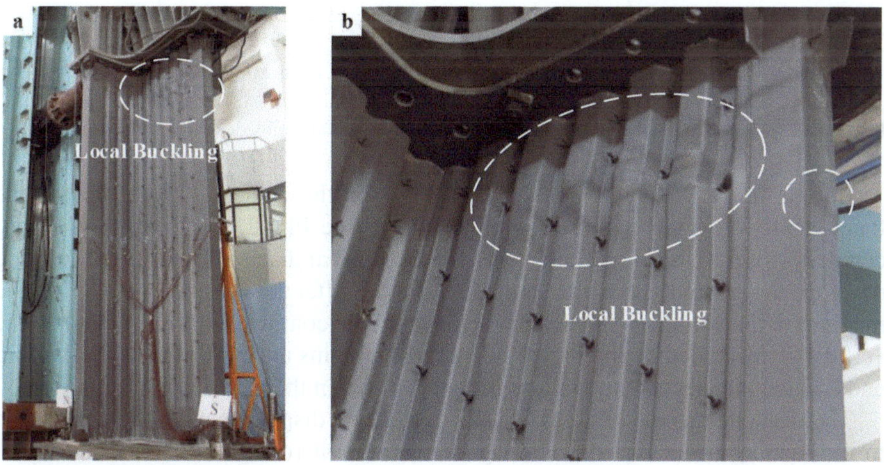

Fig. 2.8 Experimental phenomena of specimen T-1: (**a**) overall view; (**b**) local view

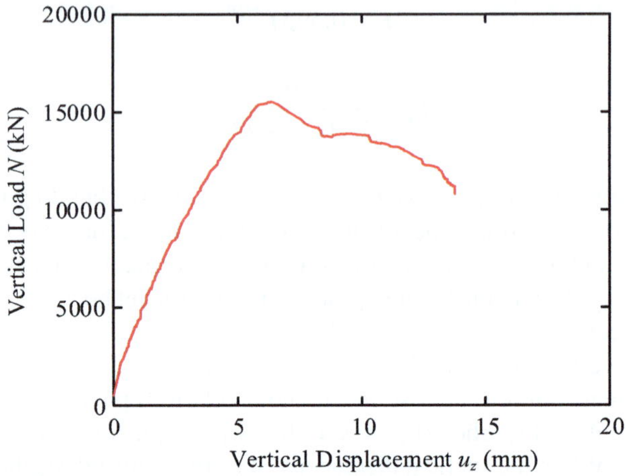

Fig. 2.9 Vertical load–vertical displacement curve of specimen T-1

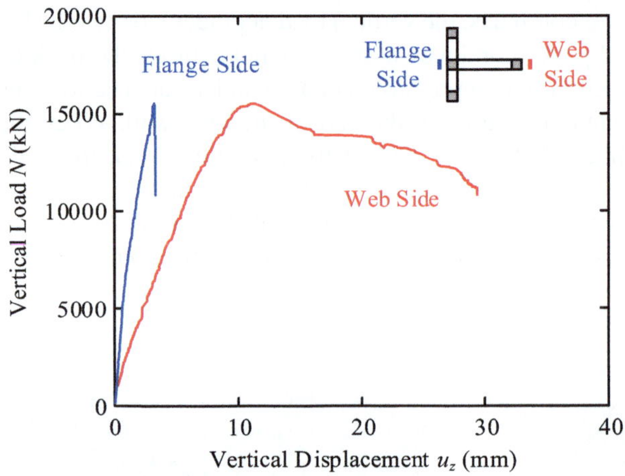

Fig. 2.10 Vertical load–vertical displacement at the flange and web sides curve of specimen T-1

Before the vertical load reaches the maximum, the vertical displacement at the flange and web sides increases simultaneously (see Fig. 2.10). Since the loading centre is closer to the end of the web wall, the vertical displacement at the web side develops more rapidly than that at the flange side. After the vertical load reaches the maximum, the vertical displacement at the web side continues to increase, while the vertical displacement at the flange side almost remains unchanged until the loading ends. This indicates that the failure is concentrated in the web wall.

The horizontal actuator constrains the horizontal displacement at the connection to the loading beam, but the loading beam can still rotate about the connection, resulting in the horizontal displacement at the top section of specimen T-1. When

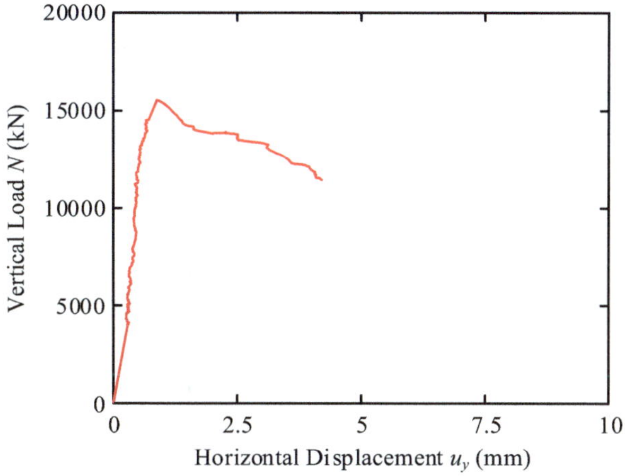

Fig. 2.11 Vertical load–horizontal displacement curve of specimen T-1

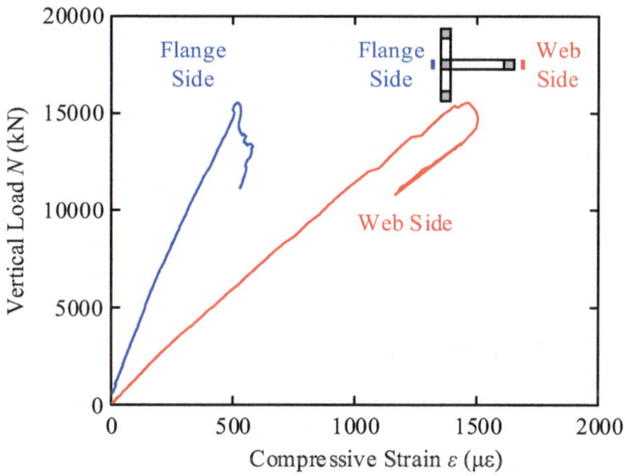

Fig. 2.12 Strain development of the steel parts at the mid-height section of specimen T-1

the vertical load reaches the maximum, the horizontal displacement in the negative direction of the x-axis is 0.9 mm; when the loading ends, the horizontal displacement in the negative direction of the x-axis is 4.4 mm (see Fig. 2.11).

Since the horizontal constraining force reduces the bending moment at the mid-height section, the steel at the mid-height section remains elastic during the loading process (see Fig. 2.12). The strain on the web side develops more rapidly than that on the flange side. After the vertical load reaches the maximum, the strain on the flange side remains basically unchanged, while the strain on the web side decreases due to the degrading load. Before the vertical load reaches the maximum, the strain distribution of steel basically satisfies the plane section assumption (see Fig. 2.13). However, the strain of the steel corrugated plates on the east side (negative x-axis) is smaller than that on the west side (positive x-axis). As the vertical load increases, the difference between the two sides increases. This may result from the eccentricity

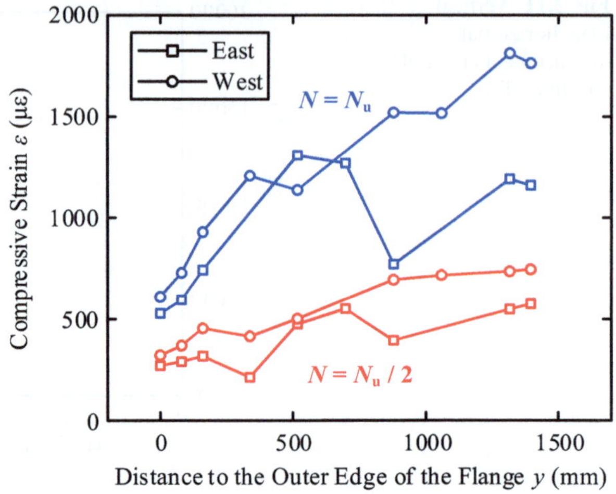

Fig. 2.13 Strain distribution of the steel parts at the mid-height section of the T-1 section

of the loading centre in the positive direction of the *x*-axis, so that the steel corrugated plates on the west side sustain a larger load.

It can be seen from the experimental results that T-CDSCWs exhibit excellent cross-sectional capacity. After T-CDSCWs reach the limit state of the cross-sectional capacity, the local buckling after yielding may occur in steel corrugated plates and rectangular tubes.

2.2 Global Stability Test

The purpose of the global stability test is to study the global stability of T-CDSCWs under axial compression. This section introduces the design, phenomena and results of the global stability test, including the specimen design, test setup, instrumentation, material property, failure phenomena, stress distribution, etc.

2.2.1 Experiment Design

A T-CDSCW specimen numbered T-2 is designed (see Figs. 2.14 and 2.15). Compared with specimen T-1 of the cross-sectional capacity test, the flange wall of specimen T-2 is the same, while the web wall of specimen T-2 is shorter. Under axial compression, two kinds of buckling modes may occur in T-shaped members: when the web is short and the stiffness about the asymmetry axis is smaller, the first-order buckling mode is flexural buckling about the asymmetry axis; when the web is wide

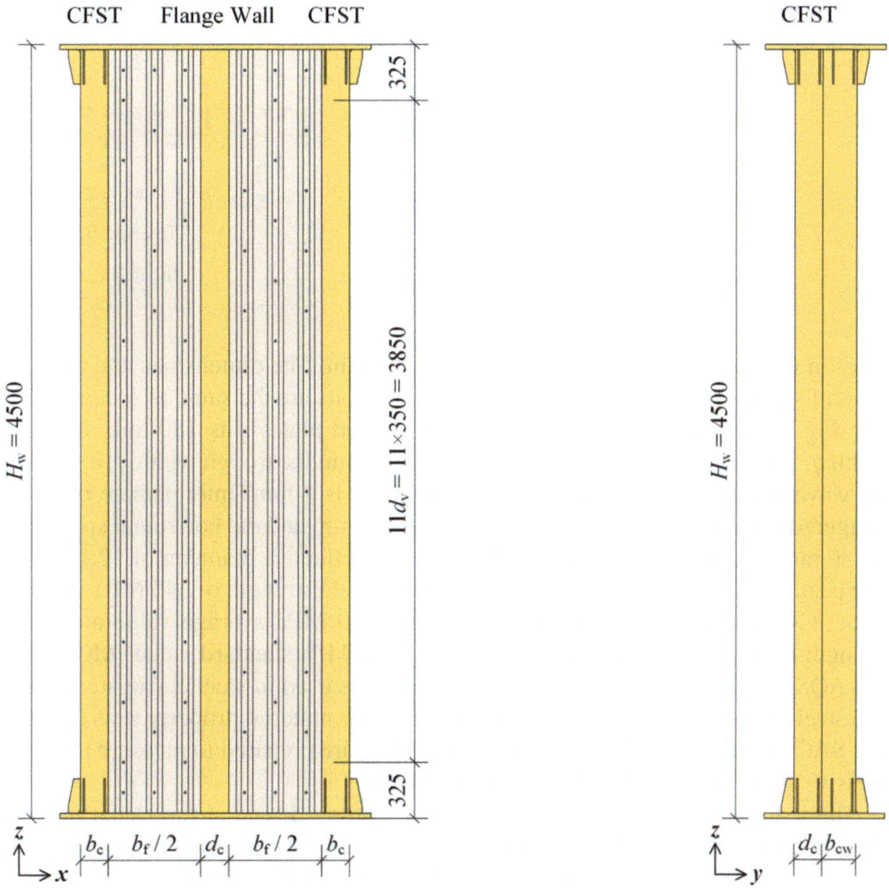

Fig. 2.14 Elevation view of specimen T-2 for the global stability test (unit: mm): (**a**) flange wall; (**b**) web wall

and the stiffness about the symmetry axis is smaller, the first-order buckling mode is flexural-torsional buckling about the symmetry axis. To ensure the flexural buckling occurs, the web wall of specimen T-2 is short and only composed of two boundary elements. To better reflect the load-carrying performance of T-CDSCWs in practical applications, the overall and detailed dimensions of specimen T-2 are full-scale as follows.

The height of specimen T-2 H_w is 4500 mm. The breadth of the flange wall b_f is 1077 mm and the overall breadth of the flange wall B_f is 1397 mm. The overall breadth of the web wall B_w is 360 mm. The thickness of wall elements d_w is 150 mm. The cross-sectional dimensions of the steel rectangular tubes in the flange $b_c \times d_c \times t_c$ are 160 × 160 × 8 mm, and the cross-sectional dimensions of the steel rectangular

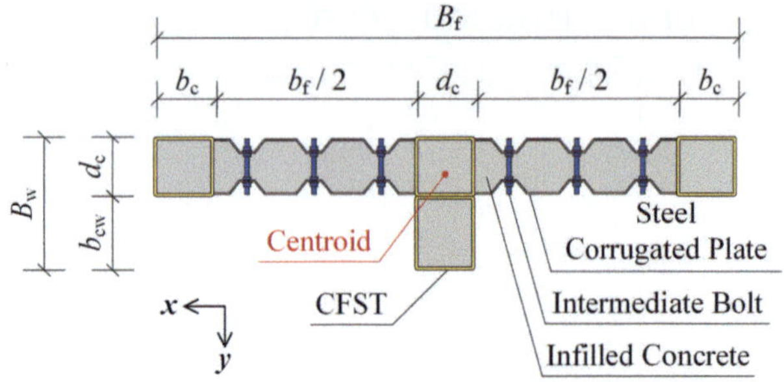

Fig. 2.15 Cross-section view of specimen T-2 for the global stability test

tubes in the web $b_{cw} \times d_c \times t_c$ are 200 × 160 × 8 mm. The dimensions and arrangement of steel corrugated plates and intermediate bolts are the same as specimen T-1 (see Fig. 2.3). The wave width of steel corrugated plates q is 180 mm, the crest width q_1 is 85 mm, the transition width q_2 is 27.5 mm, the trough width q_3 is 40 mm, the wave height a is 35 mm, and the thickness t_s is 4 mm. Intermediate bolts are staggered arranged. The vertical spacing d_v is 350 mm and the horizontal spacing d_h is 180 mm. The bolt specification is M14 with an effective diameter of 12.1 mm.

Specimen T-1 is made of Q345 steel with the yield strength of 345 MPa standard value, C40 concrete with the cube strength of 40 MPa average value and high-strength bolts with the ultimate strength of 1000 MPa standard value (MOHURD and AQSIQ 2010, 2017). Since most of the steel is used in steel corrugated plates, five steel corrugated plate coupons are prepared for material property tests (AQSIQ and SAC 2010). Meanwhile, three concrete cubes are prepared to measure the compressive strength (MOHURD and AQSIQ 2003a).

The loading device is the multifunctional loading device at the Department of Civil Engineering, Tsinghua University (see Fig. 2.4). In the loading process, only the vertical hydraulic jack acts on the loading beam and the loading centre is consistent with the centroid of specimen T-2 (see Fig. 2.16). The loading beam and the foundation beam are the same with those of the cross-sectional capacity test. Since the loading beam needs to be compatible with all the T-CDSCW specimens, the loading beam will be suspended on the web side of specimen T-2 and apply eccentric compression to specimen T-2. The initial loading eccentricity is neglected and the loading is still considered as axial loading since the gravity load of the loading beam is far less than the compressive capacity of specimen T-2. In the following analyses, the initial loading eccentricity will be considered in the initial geometric imperfection.

During the loading process, the vertical load is continuously applied by the vertical hydraulic jack. In the initial stage, the loading is force-controlled until the vertical load reaches about 8000 kN. Afterwards, the loading is displacement-controlled until specimen T-2 exhibits noticeable damage, excessive deformation, or a significant decrease in the load-carrying capacity.

The instrumentation of the global stability test is basically the same as that of the cross-sectional capacity test. During the test, the load–displacement response of specimen T-2 and the strain development on the steel parts are monitored (see

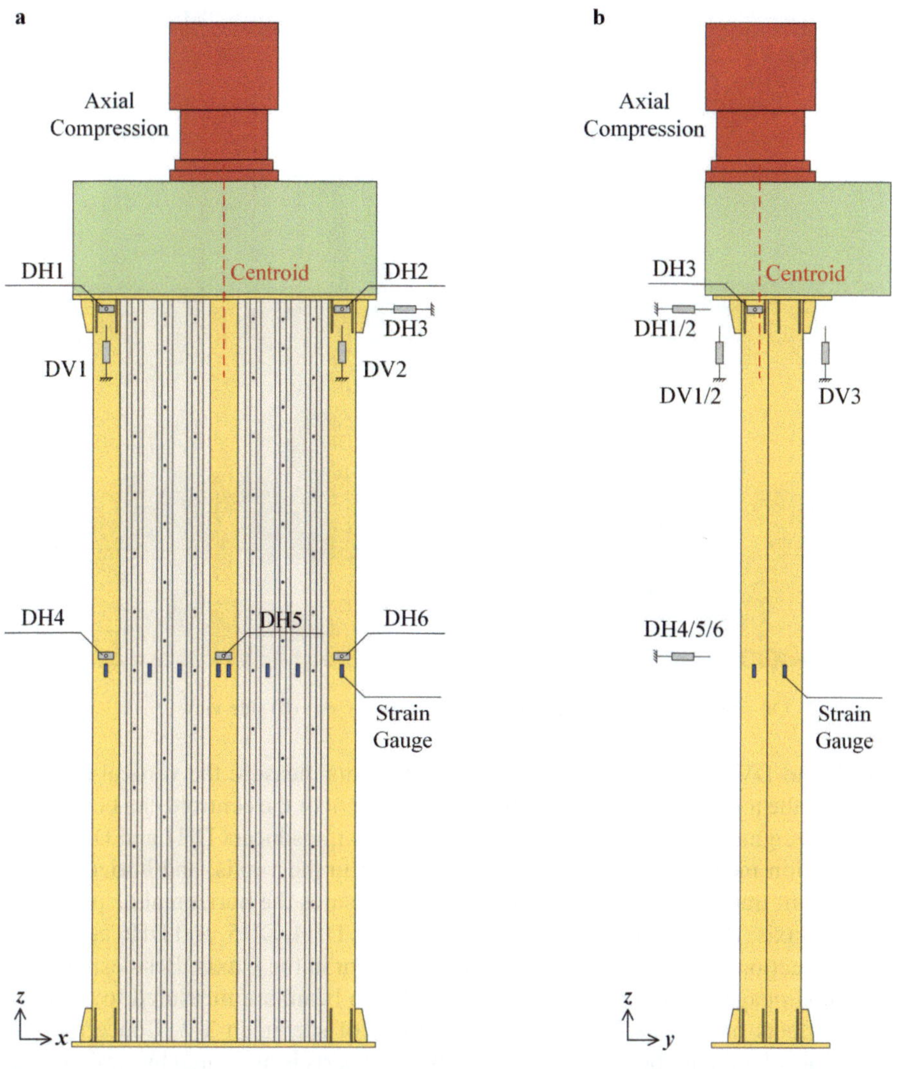

Fig. 2.16 Setup and instrumentation of the global stability capacity test: (**a**) flange wall; (**b**) web wall

Figs. 2.16 and 2.17). The data is collected by the DATA LOGGER TDS-530 data measurement system, which displays the load–displacement, load–strain and other curves in real time for observing and adjusting the loading situation. The failure phenomena of specimen T-2, such as the local buckling of steel corrugated plates, are recorded by manual observation. Concerning the loads applied to specimen T-2, the load cell in the vertical hydraulic jack measures the vertical load automatically. Concerning the displacement of specimen T-2, displacement transducers measure the vertical and horizontal displacement. Vertical displacement transducers DV1,

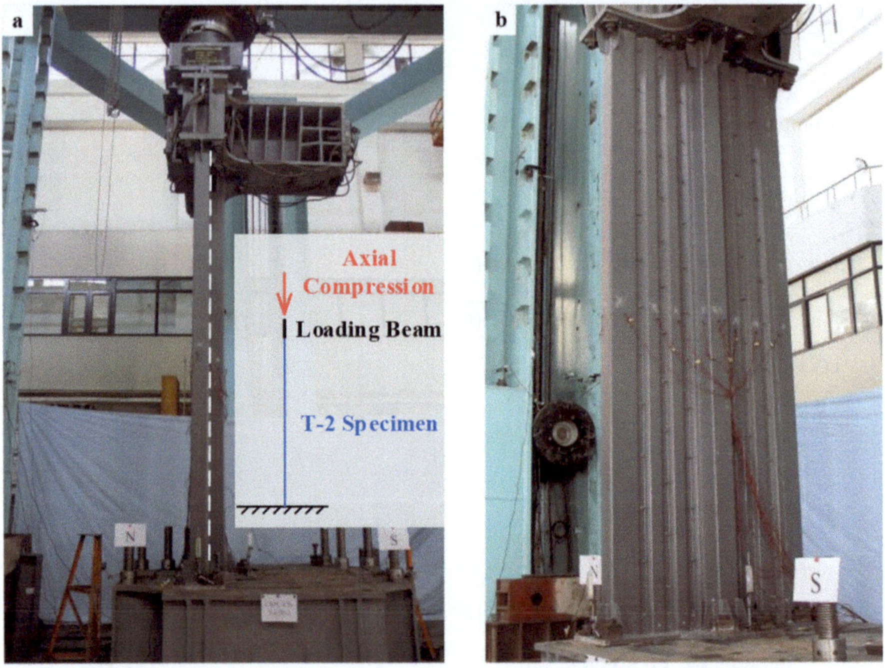

Fig. 2.17 On-site photographs of specimen T-2: (**a**) front view; (**b**) side view

DV2, and DV3 located near the boundary elements measure the vertical displacement of the top section, and the vertical displacement of the centroid can be obtained by the weighted average. Horizontal displacement transducers DH1 and DH2 at the top section measure the horizontal displacement in the y-axis, and horizontal displacement transducer DH3 at the top section measures the horizontal displacement in the x-axis. Horizontal displacement transducers DH4, DH5, and DH6 at the mid-height section measure the horizontal displacement in the y-axis. Besides, the horizontal slippage and the rotation of the foundation beam are measured to eliminate their effects on the displacement measurement of specimen T-2. Concerning the strain development, the normal strain on the steel parts is measured by strain gauges at the mid-height section.

2.2.2 Experimental Phenomena and Results

Before conducting the global stability test, the material property tests of steel and concrete are conducted to determine some data of specimen T-2, such as the centroid position. According to the average value of the material property test results (see Fig. 2.18), the yield strength f_y of the steel is 389 MPa, the ultimate strength f_u is 530 MPa, and the elastic modulus E_s is 1.99×10^5 MPa; the cubic compressive

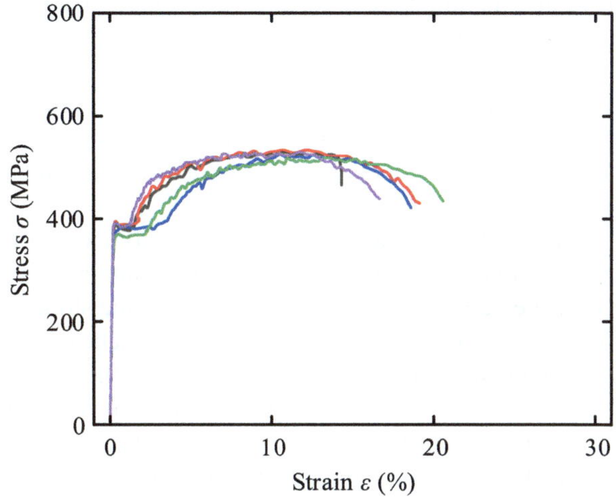

Fig. 2.18 Stress–strain curves of the steel plate coupons of the global stability test

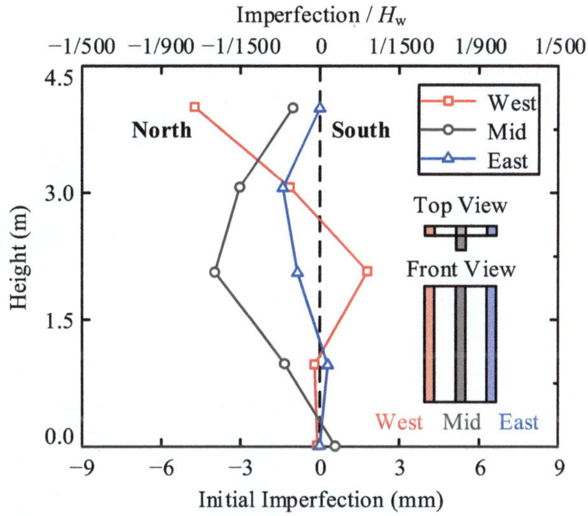

Fig. 2.19 Initial geometric imperfection of specimen T-2

strength f_{cu} of the concrete is 37.1 MPa. The axial compressive strength f_c is 28.2 MPa, the axial tensile strength f_t is 2.88 MPa, and the elastic modulus E_c is 3.19×10^4 MPa, which are determined by Eqs. (2.4), (2.5), and (2.6), respectively (MOHURD and AQSIQ 2010).

The initial geometric imperfection of specimen T-2 is measured to provide the reference for the following analyses (see Fig. 2.19). Specimen T-2 tilts to the north as a whole, that is, in the negative direction of the *y*-axis. The amplitude of the initial geometric imperfection is 6.5 mm, which is 1/692 of the wall height. At the top of the specimen, the west boundary element tilts more to the north, the east one tilts less, and the middle one locates between them. At the mid-height of the specimen, only the middle boundary element tilts to the north.

Under axial compression, specimen T-2 exhibits the failure mode of the flexural buckling about the asymmetry axis and the buckling direction is the direction with the flange on the convex side (see Fig. 2.20). During the loading process, no obvious local buckling of steel corrugated plates nor significant horizontal displacement in the x-axis occur.

Before the vertical load reaches 10,000 kN (see Fig. 2.21), the response of the load–displacement curve is almost linear, and no obvious phenomena are observed. After the vertical load reaches 10,000 kN, the slope of the load–displacement curve begins to drop. After the vertical load reaches 15,000 kN, significant horizontal displacement in the negative direction of the y-axis at the top and mid-sections occurs. When the vertical displacement of the centroid reaches 11.6 mm, the vertical load applied to specimen T-2 reaches the maximum of 16,855 kN, and the corresponding strength reduction factor φ, which is the ratio of the maximum load N_u to the yield load N_y, reaches 0.933.

Fig. 2.20 Experimental phenomena of specimen T-2

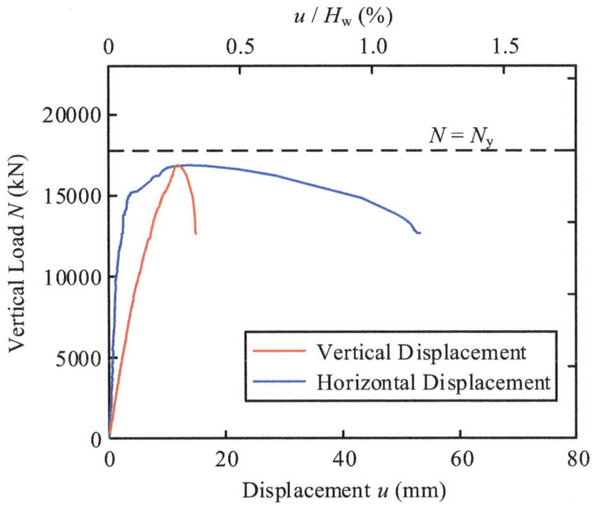

Fig. 2.21 Vertical load–displacement at the top section curve of specimen T-2

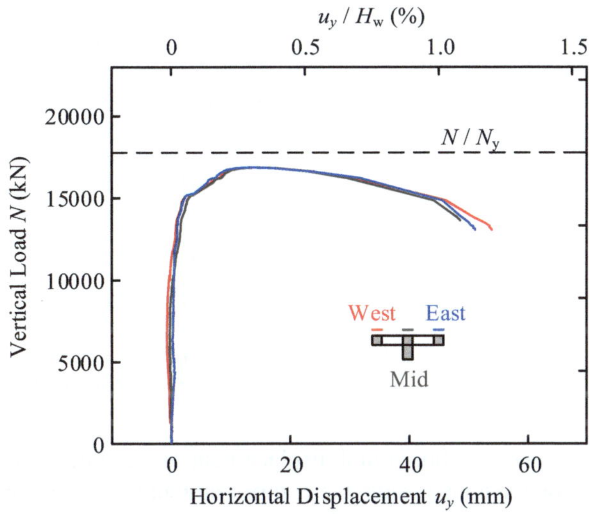

Fig. 2.22 Vertical load–horizontal displacement at the mid-height section curve of specimen T-2

As the loading continues, the vertical load degrades slowly, while the horizontal displacement in the y-axis at the top and mid-height sections develops rapidly. When the vertical load degrades to about 12,000 kN, 70% of the maximum load, the horizontal displacement in the negative direction of the y-axis at the top and mid-height sections reaches 55.4 mm and 55.0 mm respectively and the loading ends (see Fig. 2.22). During the loading process, the horizontal displacement at the mid-height section measured by DH4–DH6 remains roughly the same. This indicates that, despite the distortional imperfection of specimen T-2, the distortional deformation is insignificant, and the deformation mode of specimen T-2 is governed by the global flexural deformation.

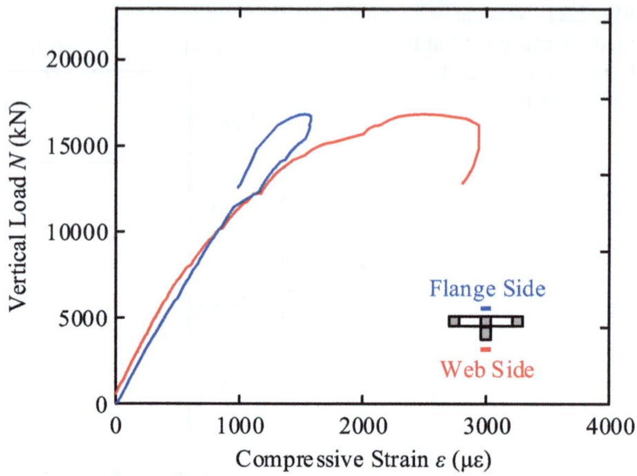

Fig. 2.23 Strain development of the steel parts at the mid-height section of specimen T-2

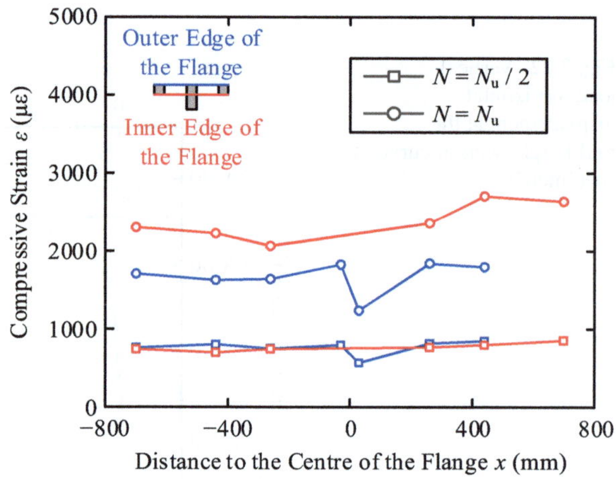

Fig. 2.24 Strain distribution of the steel parts at the mid-height section of specimen T-2

Since the horizontal displacement of the specimen is in the y-axis and the flange is on the convex side, the compression in the flange turns out to be smaller than that in the web, which is reflected in the strain results (see Fig. 2.23). In the initial loading stage, the strain on the flange and web sides develops simultaneously. After the vertical load reaches about 12,000 kN, the strain on the web side develops more rapidly. Such a strain development indicates the second-order effect caused by horizontal displacement.

Concerning the steel corrugated plates in the flange, the strain at the outer and inner edges differs (see Fig. 2.24). When the vertical load reaches the half of the maximum, the strain at the outer edge roughly equals to that at the inner edge. This indicates that specimen T-2 could be treated as axially loaded approximately. When the vertical load reaches the maximum, the strain at the outer edge is obviously less than that at the inner edge.

It can be seen from the experimental results that T-CDSCWs exhibit excellent global stability. After T-CDSCWs reach the limit state of the global stability, the flexural buckling about the asymmetry axis may occur and the buckling direction is the direction with the flange on the convex side.

2.3 Seismic Performance Test

The purpose of the seismic performance test is to study the seismic performance of T-CDSCWs under axial compression and cyclic horizontal load. This section introduces the design, phenomena and results of the seismic performance test, including the specimen design, test setup, instrumentation, material property, failure phenomena, etc.

2.3.1 Experiment Design

A T-CDSCW specimen numbered T-3 is designed (see Figs. 2.25 and 2.26). To better reflect the seismic performance of T-CDSCWs in practical applications, the overall and detailed dimensions of specimen T-3 are full-scale as follows.

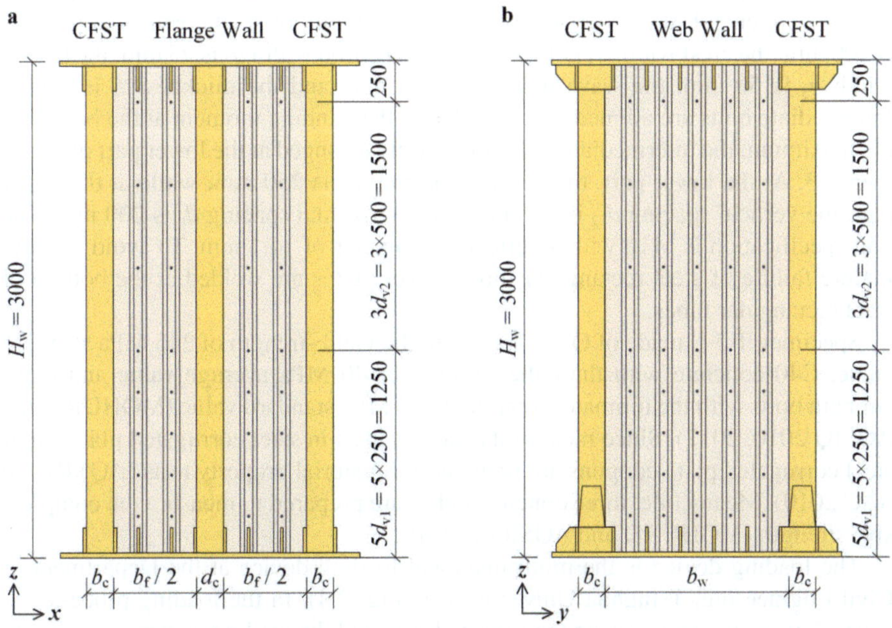

Fig. 2.25 Elevation view of specimen T-3 for the seismic performance test (unit: mm): (**a**) flange wall; (**b**) web wall

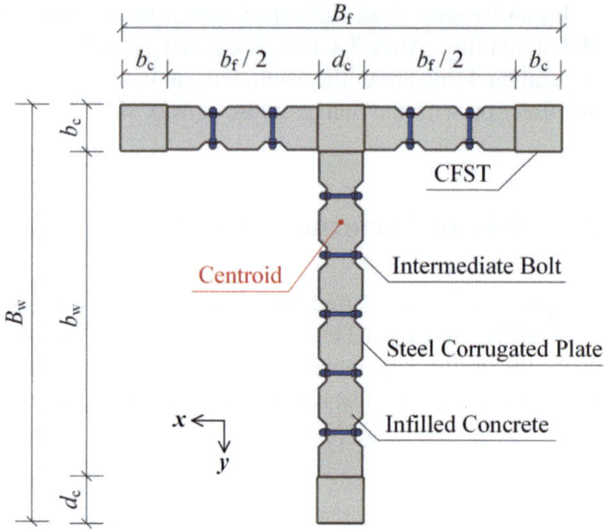

Fig. 2.26 Cross-section view of specimen T-3 for the seismic performance test

The height of specimen T-3 H_w is 3000 mm. The breadth of the flange wall b_f is 1000 mm and the overall breadth of the flange wall B_f is 1480 mm. The breadth of the web wall b_w is 1100 mm and the overall breadth of the web wall B_w is 1420 mm. The thickness of wall elements d_w is 150 mm. The cross-sectional dimensions of the steel rectangular tubes in the flange $b_c \times d_c \times t_c$ are 160 × 160 × 5 mm. The dimensions and arrangement of steel corrugated plates and intermediate bolts are different from specimen T-1 and specimen T-2. The wave width of steel corrugated plates q is 200 mm, the crest width q_1 is 100 mm, the transition width q_2 is 25 mm, the trough width q_3 is 50 mm, the wave height a is 25 mm, and the thickness t_s is 4 mm. Intermediate bolts are aligned arranged. Since the bending moment at the bottom is the maximum, the intermediate bolts are densely arranged at the lower part of specimen T-3. At the lower part, the vertical spacing d_{v1} is 250 mm, while at the upper part, the vertical spacing d_{v2} is 500 mm. The horizontal spacing d_h is 200 mm. The bolt specification is M16 with an effective diameter of 14.1 mm. To avoid the premature failure of steel rectangular tubes, cover plates are welded at the bottom of steel rectangular tubes.

Specimen T-3 is made of Q345 steel with the yield strength of 345 MPa standard value, C40 concrete with the cube strength of 40 MPa average value and high-strength bolts with the ultimate strength of 470 MPa standard value (MOHURD and AQSIQ 2010, 2017). Since most of the steel is used in steel corrugated plates, four steel corrugated plate coupons are prepared for material property tests (AQSIQ and SAC 2010). Meanwhile, three concrete cubes are prepared to measure the compressive strength (MOHURD and AQSIQ 2003a).

The loading device is the multifunctional loading device at the Department of Civil Engineering, Tsinghua University (see Fig. 2.4). In the loading process, the vertical hydraulic jack acts on the loading beam and the loading centre is consistent with the centroid of specimen T-3, and the horizontal actuators act on the two sides of the loading beams (see Fig. 2.27). The loading beam and the foundation beam are the same with those of the cross-sectional capacity test.

The vertical hydraulic jack applies the constant axial compression to specimen T-3 according to the axial compression ratio n. The design value n_d and the test value n_t of the axial compression ratio, which consider the contribution of steel and concrete (MOHURD 2015a, 2016; MOHURD and AQSIQ 2016), are evaluated by Eqs. (2.7) and (2.8).

$$n_d = \frac{1.20N}{f_y A_s / 1.11 + f_c A_c / 1.40} \tag{2.7}$$

$$n_t = \frac{N}{f_y A_s + f_c A_c} \tag{2.8}$$

Where N denotes the axial compression; 1.20 is the partial coefficient of load combination (MOHURD and AQSIQ 2012), 1.11 is the partial coefficient for the resistance of steel (MOHURD and AQSIQ 2003b), and 1.40 is the partial coefficient for the resistance of concrete (MOHURD and AQSIQ 2010). In the test, the design

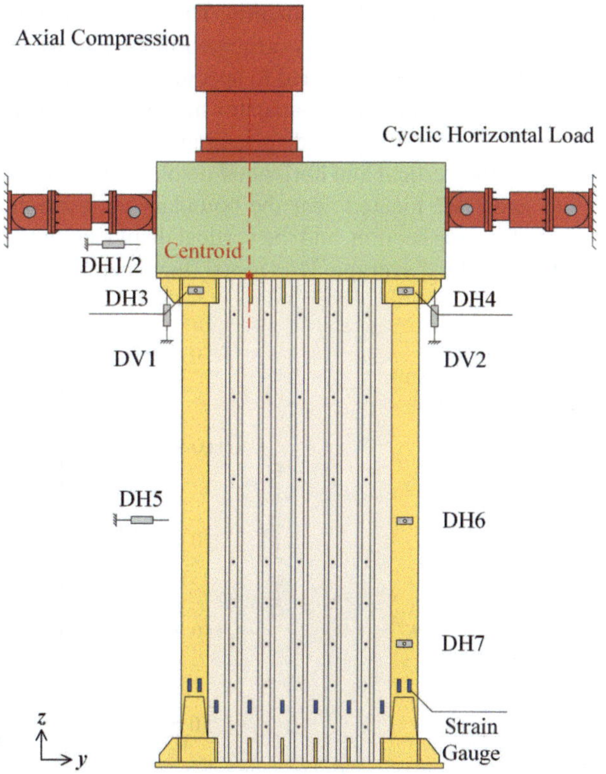

Fig. 2.27 Setup and instrumentation of the seismic performance test

value of the axial compression ratio n_d is 0.56, the test value of the axial compression ratio n_t is 0.39, and the axial compression N is 8443 kN.

After the vertical hydraulic jack applies the axial compression, the horizontal actuators apply cyclic horizontal load to specimen T-3. The loading protocol refers to JGJ/T 101 (MOHURD 2015b) (see Fig. 2.28). In the early stage, the loading is force-controlled, while in the later stage, the loading is displacement-controlled. In the force-controlled loading stage, the horizontal force is in three levels, 500 kN, 1000 kN, and 1500 kN, and each level corresponds to 1 cycle. In the displacement-controlled stage, the horizontal displacement is in four levels, 15 mm, 22.5 mm, 30 mm, and 45 mm, of which the drift angles θ are 1/200, 1/133, 1/100, and 1/67, and each level corresponds to 2 cycles. The loading starts along the negative direction of the y-axis, and specimen T-3 is subjected to negative bending moment. Then the loading acts along the positive direction of the y-axis, and specimen T-3 is subjected to positive bending moment. The loading continues until specimen T-3 exhibits noticeable damage, excessive deformation, or a significant decrease in the load-carrying capacity.

During the test, the load–displacement response of specimen T-3 and the strain development on the steel parts are monitored (see Figs. 2.27 and 2.29). The data is collected by the DATA LOGGER TDS-530 data measurement system, which displays the load–displacement, load–strain and other curves in real time for observing and adjusting the loading situation. The failure phenomena of specimen T-3, such as the local buckling of steel corrugated plates, are recorded by manual observation. Concerning the loads applied to specimen T-3, the load cells in the vertical hydraulic jack and the horizontal actuator measure the vertical and horizontal loads automatically. Concerning the displacement of specimen T-3, displacement transducers measure the vertical and horizontal displacement. Vertical displacement transducers DV1 and DV2 located near the boundary elements measure the vertical displacement of the top section, and the vertical displacement of the centroid can be obtained by the weighted average. Horizontal displacement transducers DH1 and DH2 at the

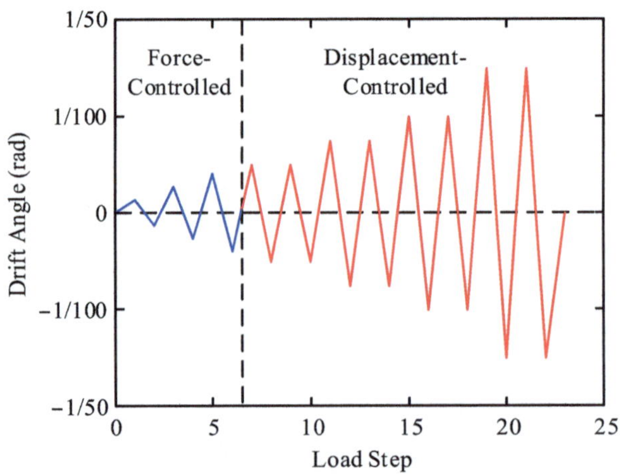

Fig. 2.28 Loading protocol of the seismic performance test

Fig. 2.29 On-site photograph of specimen T-3

top section measure the horizontal displacement in the y-axis, and horizontal displacement transducer DH3 and DH4 at the top section measure the horizontal displacement in the x-axis. Horizontal displacement transducer DH5 at the mid-height section measure the horizontal displacement in the y-axis, and horizontal displacement transducer DH6 at the mid-height section measure the horizontal displacement in the x-axis. Horizontal displacement transducer DH7 at the 1/4-height section measure the horizontal displacement in the x-axis. Besides, the horizontal slippage and the rotation of the foundation beam are measured to eliminate their effects on the displacement measurement of specimen T-3. Concerning the strain development, the normal strain on the steel parts is measured by strain gauges at the section in the middle of the bottom two lines of intermediate bolts.

2.3.2 Experimental Phenomena and Results

Before conducting the seismic performance test, the material property tests of steel and concrete are conducted to determine some data of specimen T-3, such as the centroid position. According to the average value of the material property test results (see Fig. 2.30), the yield strength f_y of the steel is 440 MPa, the ultimate strength f_u is 515 MPa, and the elastic modulus E_s is 2.08×10^5 MPa; the cubic compressive strength f_{cu} of the concrete is 34.5 MPa. The axial compressive strength f_c is 26.2 MPa, the axial tensile strength f_t is 2.77 MPa, and the elastic modulus E_c is 3.12×10^4 MPa, which are determined by Eqs. (2.4), (2.5), and (2.6), respectively (MOHURD and AQSIQ 2010).

Under axial compression and cyclic horizontal load, specimen T-3 exhibits the failure mode of the flexural failure (see Figs. 2.31 and 2.32), including the local buckling of steel corrugated plates and rectangular tubes, the fracture of the steel

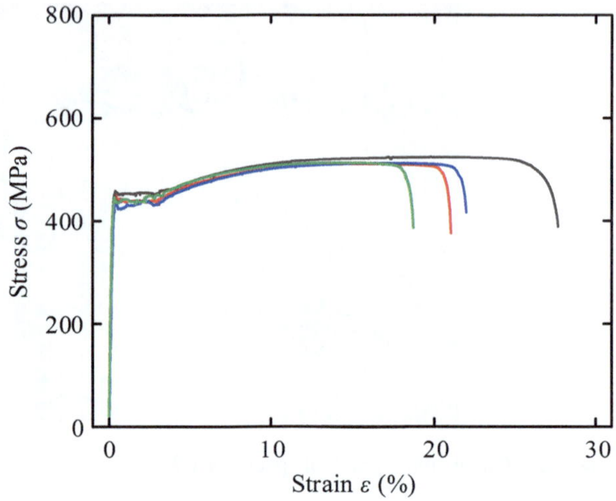

Fig. 2.30 Stress–strain curves of the steel plate coupons of the seismic performance test

Fig. 2.31 Experimental phenomena of specimen T-3

Fig. 2.32 Local damage of specimen T-3: (**a**) local buckling of steel corrugated plates; (**b**) tensile fracture of steel rectangular tubes and corrugated plates

rectangular tubes at the end of the web wall, and the cracking and crush of infilled concrete.

In the initial stage, the axial compression is applied and no significant phenomena occur. Afterwards, the cyclic horizontal load is applied. Under the negative loading at the first cycle of the 15 mm displacement level (drift angle $\theta = 1/200$), the horizontal load reaches the maximum of 1498 kN in the negative direction, and slight local buckling under compression occurs in the steel corrugated plates between the bottom two lines of intermediate bolts. Under the positive loading at the first cycle of the 15 mm displacement level (drift angle $\theta = 1/200$), the horizontal load reaches 70% of the maximum in the positive direction. The local buckling of the steel corrugated plates in the web wall develops as the loading continues.

Under the negative loading at the first cycle of the 22.5 mm displacement level (drift angle $\theta = 1/133$), the horizontal load drops to 89% of the maximum in the negative direction, and the local buckling develops to the steel rectangular tube above the cover plates and the steel corrugated plates near the third line of intermediate bolts.

Under the positive loading at the first cycle of the 22.5 mm displacement level (drift angle $\theta = 1/133$), the horizontal load reaches 88% of the maximum in the positive direction, and the local buckling under compression occurs in the steel corrugated plates at the outer edge of the flange wall.

Under the positive loading at the first cycle of the 45 mm displacement level (drift angle $\theta = 1/67$), the horizontal load reaches the maximum of 2186 kN in the positive direction. Under the positive loading at the second cycle of the 45 mm displacement level (drift angle $\theta = 1/67$), tensile fractures occur at the steel rectangular tube and corrugated plates above the cover plates, and the cracks of infilled concrete are observed. After the fracture of the steel rectangular tube at the end of the web wall, the horizontal load degrades significantly and the loading ends. In the loading process, no significant horizontal displacement in the x-axis occurs.

The horizontal load–horizontal displacement hysteretic curve of specimen T-3 exhibits asymmetry (see Fig. 2.33): with specimen T-3 under positive loading, the hysteretic curve is plump; with specimen T-3 under negative loading, the curve drops significantly and the load degrades in each level.

From the hysteresis curve of specimen T-3, the peak values of each level in the same direction are extracted and connected in sequence to obtain the horizontal load–horizontal displacement skeleton curve (see Fig. 2.34). According to the energy equivalence method (Guo 2014) (see Fig. 2.35), a straight line is drawn in the skeleton curve to satisfy equal quantity of area A and area B, and the yield point of specimen T-3 can be determined. The failure point of specimen T-3 can be determined based on the decrease in the horizontal load to 85% of the maximum load.

The maximum load and failure displacement of specimen T-3 under positive loading are significantly greater than those under negative loading (see Table 2.1).

The stiffness of specimen T-3 under cyclic loading can be evaluated by the cyclic stiffness Kj (MOHURD 2015b), which is defined as Eq. (2.9).

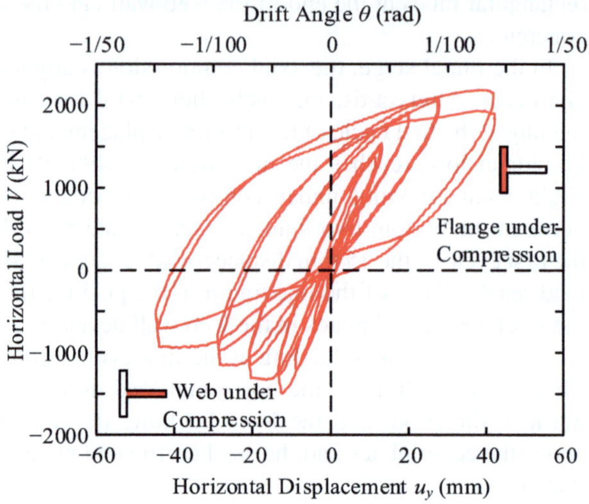

Fig. 2.33 Horizontal load–horizontal displacement hysteretic curve of specimen T-3

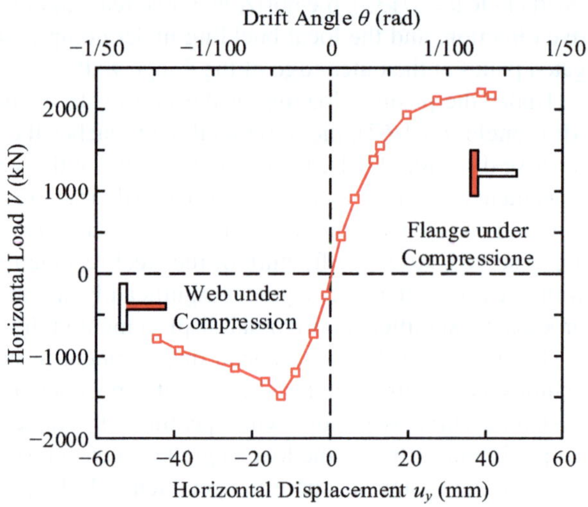

Fig. 2.34 Horizontal load–horizontal displacement skeleton curve of specimen T-3

$$K_j = \frac{\sum_{i=1}^{2} V_j^i}{\sum_{i=1}^{2} \delta_j^i} \tag{2.9}$$

When δ_j^i denote the maximum horizontal displacement at the i cycle of the j level load, and V_j^i denotes the corresponding horizontal load.

In the loading process, the cyclic stiffness degrades continuously (see Fig. 2.36). This results from the crackling and crush of infilled concrete and the yielding of steel.

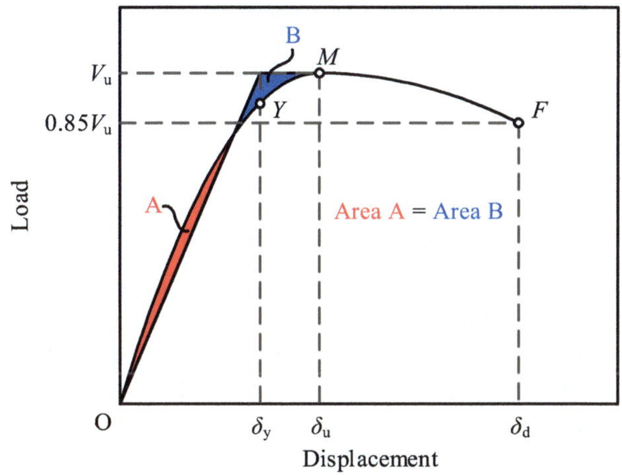

Fig. 2.35 Schematic diagram of the energy equivalence method

Table 2.1 Yield point, maximum point, and failure point of specimen T-3

Direction	Yield			Maximum			Failure		
	δ (mm)	θ (rad)	V (kN)	δ (mm)	θ (rad)	V (kN)	δ (mm)	θ (rad)	V (kN)
Positive	20.6	1/146	1930	39.1	1/77	2186	/	/	/
Negative	10.2	1/294	1297	12.9	1/233	1498	19.3	1/155	1273

δ denotes the horizontal displacement, θ denotes the drift angle, and V denotes the horizontal load.

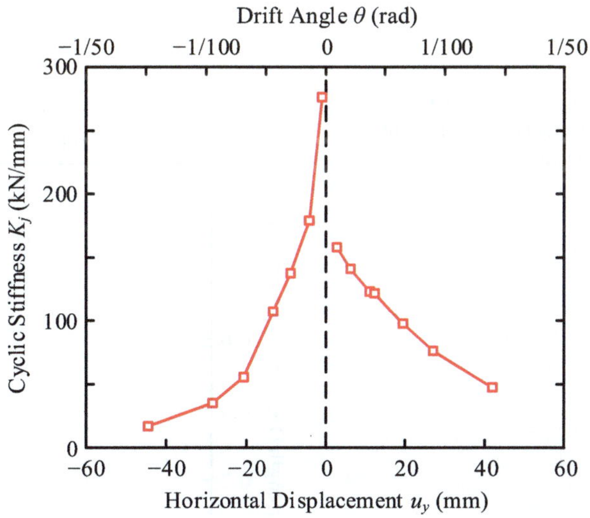

Fig. 2.36 Cyclic stiffness of specimen T-3

Since the damage of specimen T-3 is continuously cumulated under cyclic horizontal load, the load of the same level decreases continuously with the increase of the cycles. The strength degradation coefficient η_j is defined as Eq. (2.10) (MOHURD 2015b).

$$\eta_j = \frac{V_j^2}{V_j^1} \tag{2.10}$$

When V_j^1 and V_j^2 denote the maximum load of the first cycle and the second cycle of the j level load.

The strength degradation coefficient η_j ranges from 0.8 to 1.0 (see Fig. 2.37), which implies that specimen T-3 has relatively stable load-carrying performance.

The cumulative dissipated energy of specimen T-3 is the area enclosed by the horizontal load–horizontal displacement hysteretic curve (see Fig. 2.38). The dissipated energy of each level of specimen T-3 increases continuously with the loading. In each load level, the half-cycle dissipated energy decreases gradually (see Fig. 2.39).

The equivalent viscous ratio ζ_{eq} is defined as Eq. (2.11) (MOHURD 2015b).

$$\zeta_{eq} = \frac{E_D}{4\pi E_{S0}} \tag{2.11}$$

Where E_D denotes the dissipated energy of a cycle, which is the enclosed area of the hysteretic curve, and E_{S0} denotes the maximum elastic strain energy of a cycle.

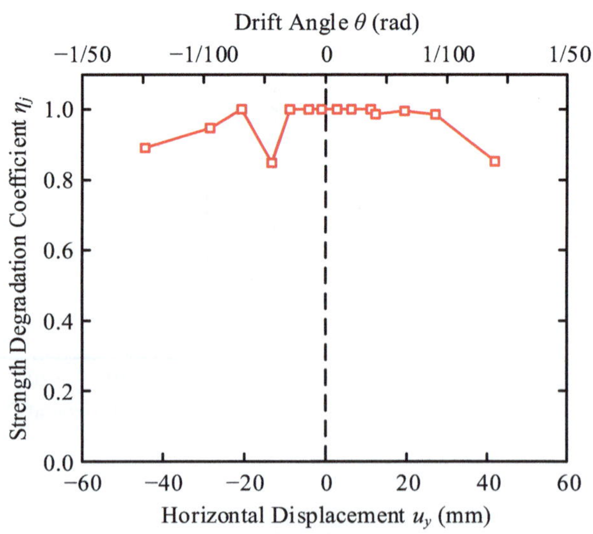

Fig. 2.37 Strength degradation coefficient of specimen T-3

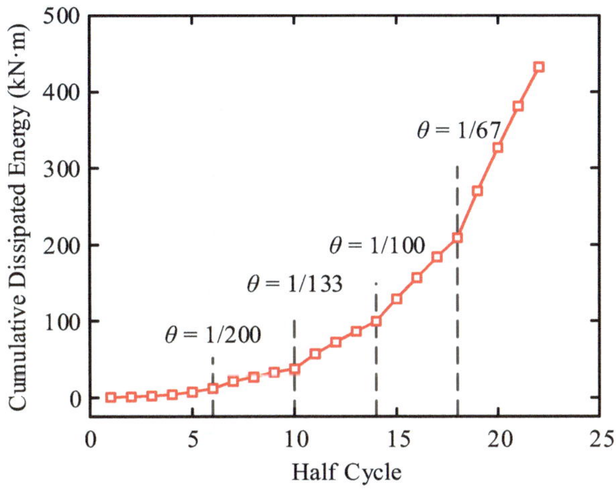

Fig. 2.38 Cumulative dissipated energy of specimen T-3

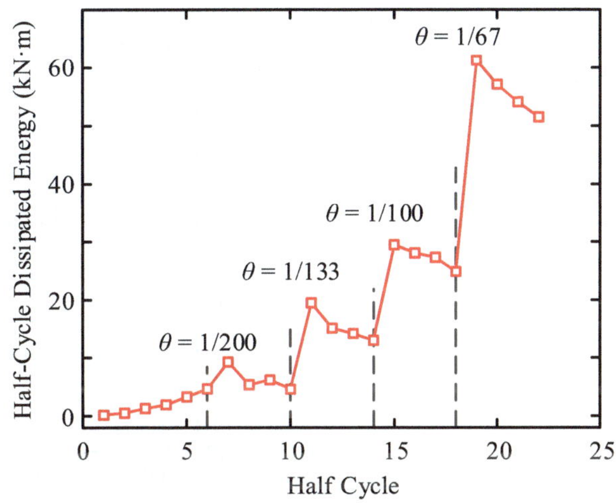

Fig. 2.39 Half-cycle dissipated energy of specimen T-3

When the drift angle θ is greater than 1/100, the equivalent viscous ratio ζ_{eq} ranges from 0.2 to 0.3 (see Fig. 2.40), which implies that specimen T-3 has good energy consumption performance.

The strain distribution of specimen T-3 basically satisfies the plane section assumption (see Figs. 2.41 and 2.42). The dashed line in Figs. 2.41 and 2.42 denotes the strain under the axial compression calculated according to the plane section assumption. Under axial compression and the horizontal load V of 1500 kN, the strain of steel corrugated plates and rectangular tubes is basically linearly distributed. When the drift angle θ reaches 1/133, the strain at the end of the web wall deviates from the straight line to a certain extent, which is affected by the local buckling of steel corrugated plates and rectangular tubes.

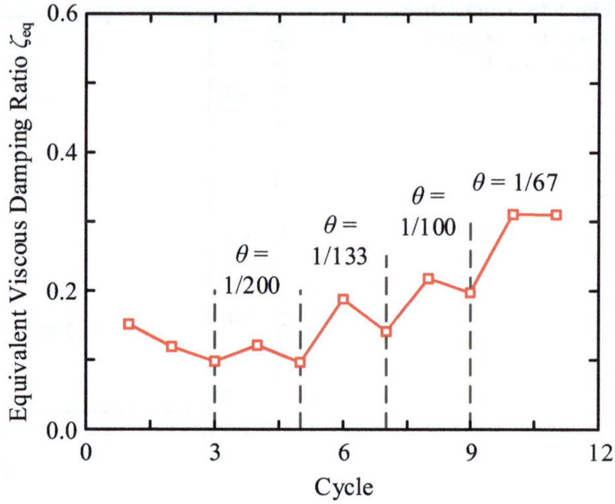

Fig. 2.40 Equivalent viscous damping ratio of specimen T-3

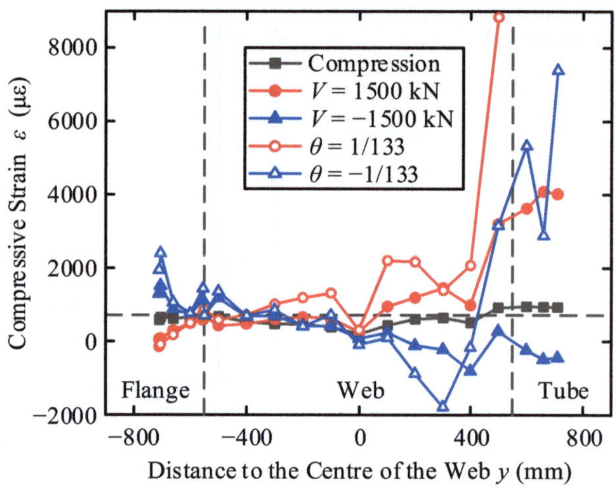

Fig. 2.41 Stress distribution along the web wall of specimen T-3

It can be seen from the experimental results that T-CDSCWs exhibit excellent seismic performance. The failure phenomena of T-CDSCWs are the flexural failure, including the local buckling of steel corrugated plates and rectangular tubes, the fracture of the steel rectangular tubes at the end of the web wall, and the cracking and crush of infilled concrete.

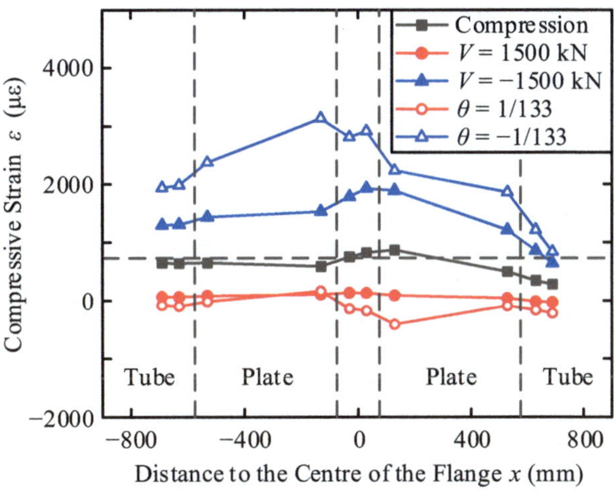

Fig. 2.42 Stress distribution along the flange wall of specimen T-3

2.4 Chapter Summary

Three sets of T-CDSCW tests are conducted, namely the cross-sectional capacity test, the global stability test, and the seismic performance test. This chapter introduces the design, phenomena and results of the tests. Based on the experimental results, the main conclusions are as follows.

1. T-CDSCWs exhibit excellent cross-sectional capacity. To investigate the cross-sectional capacity under axial compression and bending moment in the web plane, specimen T-1 with a wide web is designed. Eccentric compression in the web plane is applied to specimen T-1 to achieve the effect of axial compression and negative bending moment. Under axial compression and bending moment, specimen T-1 fails by flexure and steel corrugated plates and rectangular tubes fail by local buckling.
2. T-CDSCWs exhibit excellent global stability. To investigate the flexural buckling about the asymmetry axis under axial compression, specimen T-2 with a short web is designed. Under axial compression, specimen T-2 fails by the flexural buckling and the buckling direction is the direction with the flange on the convex side. The local buckling does not occur in steel corrugated plates and rectangular tubes. The strength reduction factor of specimen T-2 is 0.933.
3. T-CDSCWs exhibit excellent seismic performance. To investigate the seismic performance, specimen T-3 is designed. Under constant axial compression and cyclic horizontal load, specimen T-3 fails by flexure. The failure phenomena include the local buckling of steel corrugated plates and rectangular tubes, the fracture of the steel rectangular tubes at the end of the web wall, and the cracking and crush of infilled concrete. The horizontal load–horizontal displacement hysteretic curve of specimen T-3 exhibits asymmetry: with specimen T-3 under positive loading, the hysteretic curve is plump; with specimen T-3 under negative loading, the curve drops significantly and the load degrades in each level.

However, the number of the T-CDSCW specimens is not enough to comprehensively analyse the cross-sectional capacity, global stability, and seismic performance. The cross-sectional capacity test investigates the cross-sectional capacity under axial compression and negative bending moment in the web plane, while the cross-sectional capacities under positive bending moment in the web plane, bending moment in the flange plane, and bidirectional bending moment still need further investigation. The global stability test investigates the flexural buckling about the asymmetry axis under axial compression, while the flexural-torsional buckling about the symmetry axis and the instability under axial compression and bending moment still need further investigation. The seismic performance of T-CDSCWs also needs more data to support the study of the design method.

As a supplement to the experimental investigations, the finite element model of T-CDSCWs will be established and validated by the experimental results in next chapter. Through finite element simulations, the cross-sectional capacity, global stability, and seismic performance of T-CDSCWs will be further analysed.

References

AQSIQ, SAC (2010) Metallic materials—tensile testing—part 1: method of test at room temperature (in Chinese). Standards Press of China

Guo Y, Zhu J, Wang M et al (2017) Research on corrugated-steel-concrete composite wall structure system in high-rise building (in Chinese). Tsinghua University, Beijing

Guo Z (2014) Principle of reinforced concrete. Tsinghua University Press, Beijing

MOHURD (2015a) Technical specification for steel plate shear walls (in Chinese). China Architecture and Building Press, Beijing

MOHURD (2015b) Specification for seismic test of buildings (in Chinese). China Architecture and Building Press, Beijing

MOHURD (2016) Code for design of composite structures (in Chinese). China Architecture and Building Press, Beijing

MOHURD, AQSIQ (2003a) Standard for test method of mechanical properties on ordinary concrete (in Chinese). China Architecture and Building Press, Beijing

MOHURD, AQSIQ (2003b) Code for design of steel structures (in Chinese). China Planning Press, Beijing

MOHURD, AQSIQ (2010) Code for design of concrete structures (in Chinese). China Architecture and Building Press, Beijing

MOHURD, AQSIQ (2012) Load code for the design of building structures (in Chinese). China Architecture and Building Press, Beijing

MOHURD, AQSIQ (2016) Code for seismic design of buildings (in Chinese). China Architecture and Building Press, Beijing

MOHURD, AQSIQ (2017) Standard for design of steel structures (in Chinese). China Architecture and Building Press, Beijing

Chapter 3
Finite Element Simulations on T-CDSCWs

To further analyse the load-carrying capacity of T-CDSCWs, in this chapter, the finite element models of T-CDSCWs are developed based on Abaqus (Zhuang et al. 2009; Dassault Systèmes 2014d) and validated by the comparison between the finite element simulation results and the results of T-CDSCW experiments and other existing experiments.

Firstly, two kinds of finite element models are developed: the solid–shell finite element model and the shell–beam finite element model. The solid–shell finite element model is a refined model, which can effectively simulate the interactions among infilled concrete, steel corrugated plates and rectangular tubes, and intermediate bolts. In the solid–shell finite element model, infilled concrete is simulated by the solid element, steel corrugated plates and rectangular tubes are simulated by the shell element, and intermediate bolts are simulated by the truss element. The shell–beam finite element model is a simplified model, which can efficiently analyse the load–displacement relationship of T-CDSCWs. In the shell–beam finite element model, wall elements are simulated by the shell element and boundary elements are simulated by the beam element.

Secondly, the finite element simulation results are compared with the results of T-CDSCW experiments and other existing experiments to validate the finite element models. The finite element simulation results of the solid–shell finite element model are compared with the results of the cross-section capacity test, the global stability test, and other existing tests regarding the failure mode, load–displacement relationship, load–strain response, etc. The finite element simulation results of the shell–beam finite element model are compared with the results of the seismic performance test and other existing tests regarding the load–displacement hysteretic curves.

M.-Z. Wang, *Design Method of Concrete-Infilled Double Steel Corrugated-Plate Walls with T-section*, Springer Theses,
https://doi.org/10.1007/978-981-95-7789-7_3

3.1 Finite Element Models

The refined solid–shell finite element model and the simplified shell–beam finite element model are developed. This section introduces the element types, material properties, interactions, boundary conditions, and load actions of the finite element models.

3.1.1 Solid–Shell Finite Element Model

The interactions among infilled concrete, steel corrugated plates and rectangular tubes, and intermediate bolts can be effectively simulated by the solid–shell finite element model. The following assumptions are made to develop the solid–shell finite element model.

1. Since the failure of intermediate bolts is not observed in the experiments, the material property of intermediate bolts is set to be elastic and the failure is not considered.
2. Since bolt holes have no significant effect on the load-carrying capacity of steel corrugated plates (Zhou et al. 2021; Zhu et al. 2019), bolt holes are not considered.

The solid–shell finite element model is composed of the solid element, shell element, and truss element (see Fig. 3.1). Infilled concrete is simulated by the C3D8R solid element, steel corrugated plates and rectangular tubes are simulated by the S4R shell element, and intermediate bolts are simulated by the T3D2 truss element (Dassault Systèmes 2014a). In the finite element modelling, steel corrugated plates and rectangular tubes are one part, where they have common nodes at the connection to simulate the welding connection. The C3D8R solid element is an eight-node brick element with reduced integration. The S4R shell element is a four-node shell element with reduced integration. The T3D2 truss element is a two-node 3D truss element. Linear elements are used because they are suitable for the contact analysis between infilled concrete and steel plates. The elements with reduced integration are used to avoid the possible locking phenomenon. At the same time, the elements with reduced integration need to refine the mesh to avoid the possible hourglass numerical problem. The default number of integration points for shell elements along the thickness is 5. To increase the calculation accuracy, the number of integration points is increased to 9. Through the trial calculation of finite element models with different mesh sizes, the mesh size of each part is set to 50 mm, which can balance the requirements of accuracy and efficiency.

The solid–shell finite element model simulates the interactions among different parts of T-CDSCWs (Hossain et al. 2016), including the composite effect of infilled concrete and steel plates and the connection between intermediate bolts and steel corrugated plates. The outer surfaces of infilled concrete and the inner surfaces of

Fig. 3.1 Scheme diagram of the solid–shell finite element model

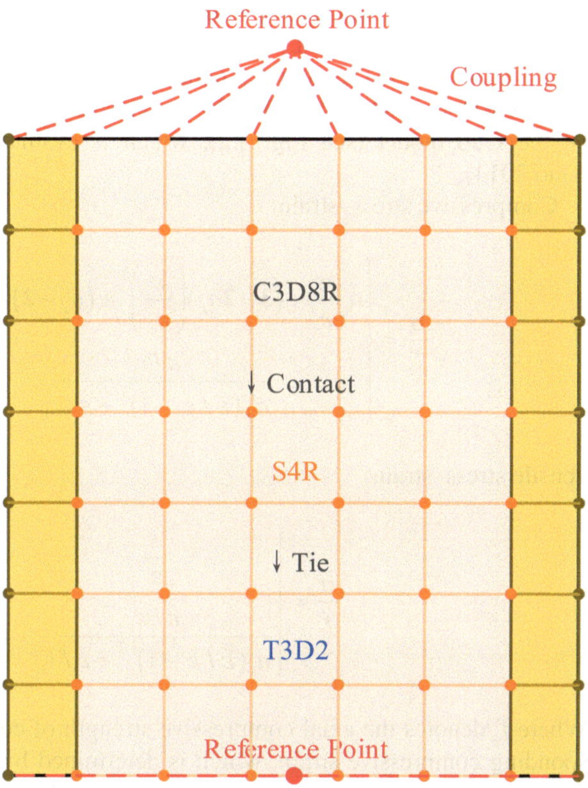

steel plates are in contact with each other, and the surfaces are set in a contact pair. The normal contact behaviour is hard contact and the tangential contact behaviour is frictionless. The outer surface of infilled concrete is the master surface and the inner surface of steel plates is the slave surface. The master surface can penetrate the slave surface between slave nodes, while nodes on the slave surface cannot penetrate the master surface (Dassault Systèmes 2014a). Tangential friction is not considered because the interfacial behaviour between steel and concrete is constrained by intermediate bolts and corrugation. The nodes at the ends of intermediate bolts are tied to steel corrugated plates. The nodes on steel corrugated plates are the master nodes and the nodes of intermediate bolts are the slave nodes.

For the application of boundary conditions and load actions, the bottom and top sections of the T-CDSCW are coupled with reference points, respectively, and the bottom and top sections move and rotate as rigid surfaces about the corresponding reference points. Restraints and loads on the T-CDSCW can be achieved by directly applying restraints and loads at the reference points.

The ideal elastic-plastic model is adopted for steel and the yield criterion is the von Mises criterion. The concrete damaged plasticity model is adopted for concrete (Dassault Systèmes 2014a). The Poisson's ratio of steel υ_s is 0.3 (MOHURD and AQSIQ 2017). The parameters of concrete are as follows: dilation angle $\psi = 30°$, eccentricity $\epsilon = 0.1$, the ratio of initial equibiaxial compressive yield stress to initial uniaxial compressive yield stress f_{b0}/f_{c0} = 1.16, invariant stress ratio K_c = 0.6667 (Jiang et al. 2013; Dassault Systèmes 2014b), and the Poisson's ratio υ_c is 0.2 (MOHURD and AQSIQ 2010). The stress–strain relationship of concrete is based on the Guo model (see Fig. 3.2), which is evaluated by Eqs. (3.1) and (3.2) (Guo 2014).

Compressive stress–strain:

$$\frac{\sigma}{f_c} = \begin{cases} a_a \dfrac{\varepsilon}{\varepsilon_c} + (3 - 2a_a)\left(\dfrac{\varepsilon}{\varepsilon_c}\right)^2 + (a_a - 2)\left(\dfrac{\varepsilon}{\varepsilon_c}\right)^3 & \varepsilon \le \varepsilon_c \\ \dfrac{\varepsilon / \varepsilon_c}{a_d(\varepsilon / \varepsilon_c - 1)^2 + \varepsilon / \varepsilon_c} & \varepsilon > \varepsilon_c \end{cases} \tag{3.1}$$

Tensile stress–strain:

$$\frac{\sigma}{f_t} = \begin{cases} \dfrac{\varepsilon}{\varepsilon_t} & \varepsilon \le \varepsilon_t \\ \dfrac{\varepsilon / \varepsilon_t}{a_t(\varepsilon / \varepsilon_t - 1)^{1.7} + \varepsilon / \varepsilon_t} & \varepsilon > \varepsilon_t \end{cases} \tag{3.2}$$

Where f_c denotes the axial compressive strength of concrete; ε_c denotes the corresponding compressive strain, which is determined by Eq. (3.3); a_a and a_d are the ascending and descending parameters, which are determined by Eqs. (3.4) and (3.5).

$$\varepsilon_c = \left(700 + 172\sqrt{f_c}\right) \times 10^{-6} \tag{3.3}$$

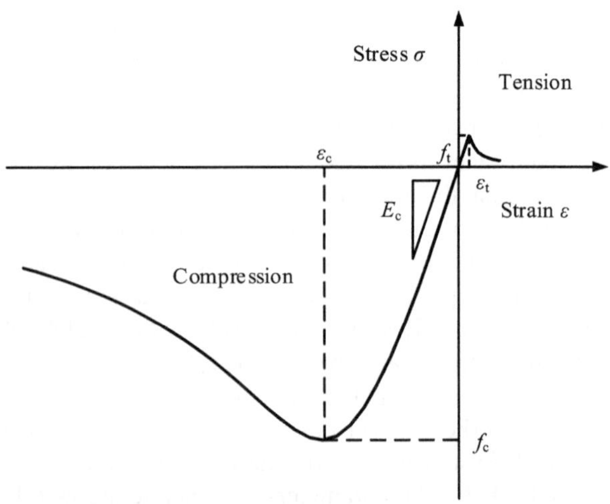

Fig. 3.2 Stress–strain relationship of concrete

$$a_a = 2.4 - 0.0125 f_c \tag{3.4}$$

$$a_d = 0.157 f_c^{0.785} - 0.905 \tag{3.5}$$

Where f_t denotes the axial tensile strength of concrete; ε_t denotes the corresponding tensile strain, which is determined by Eq. (3.6); a_t is the descending parameter, which is determined by Eq. (3.7).

$$\varepsilon_t = f_t^{0.54} \times 65 \times 10^{-6} \tag{3.6}$$

$$a_t = 0.312 f_t^2 \tag{3.7}$$

After the solid–shell finite element model is developed, the elastic buckling behaviour of T-CDSCWs can be analysed by the eigenvalue buckling analysis method, and the load-carrying performance of T-CDSCWs can be analysed by the static analysis method.

3.1.2 Shell–Beam Finite Element Model

The shell–beam finite element model has high calculation efficiency. When using the shell–beam finite element model to simulate the seismic performance of T-CDSCWs, the following assumptions are implied.

1. The slenderness ratio of composite walls is small enough to ensure that the global buckling of composite walls does not occur.
2. The width-thickness ratio of steel rectangular tubes is small enough to ensure the local buckling of steel rectangular tubes does not occur; the bolt spacing is small enough to ensure the local buckling of steel corrugated plates does not occur.
3. The bond slip is not considered, that is, there is no bond slip between infilled concrete and steel plate and the effect of intermediate bolts is not considered.
4. Since composite walls are mainly subjected to bending moment, the thickness of steel corrugated plates and infilled concrete in wall elements is set according to the principle of equivalent areas, that is, the equivalent thickness of concrete t_{ce} equals the ratio of the concrete area to the breadth of the wall element A_{cw}/b_w, and the equivalent thickness of steel corrugated plates t_{se} equals the ratio of the steel area to the breadth of the wall element A_{sw}/b_w.

Assumptions (1)–(3) result from that it is difficult to simulate the global buckling of composite walls, the local buckling of steel corrugated plates and rectangular tubes, and the bond slip between steel and concrete by using the shell element to simulate wall elements. Assumption (4) is for the convenience of using the shell element to simulate corrugated wall elements. In the seismic design, especially when the class of seismic measure is high, it is necessary to avoid the global buckling of composite walls and the local buckling of steel corrugated plates and rectangular tubes which cause adverse effects on the seismic performance of composite walls. The global instability of composite walls can be avoided by limiting the normalised slenderness ratio of composite walls, and the local buckling of steel corrugated plates and rectangular tubes can be avoided by limiting the aspect ratio of steel plates. Although local buckling occurs in the steel corrugated plates and rectangular tubes of specimen T-3, the local buckling occurs after yielding and has no significant effect on the seismic performance of specimen T-3. When the local buckling of steel corrugated plates does not occur before yielding and intermediate bolts are arranged densely enough, steel corrugated plates can maintain a good composite effect with infilled concrete.

The shell–beam finite element model is composed of the shell element and the beam element (see Fig. 3.3). Wall elements are simulated by the S8R shell element and boundary elements are simulated by the B33 beam element (Dassault Systèmes 2014a). The S8R shell element is an eight-node shell element with reduced integration. The edges of the shell elements and the beam elements are tied to simulate the welding connection. Steel plates and infilled concrete in the wall elements are simulated by multilayer shell model, which is the composite shell in Abaqus. The default

Fig. 3.3 Scheme diagram of the solid–shell finite element model

number of integration points for shell elements along the thickness is 5. To increase the calculation accuracy, the number of integration points is increased to 9. The B33 beam element is a three-node 3D Euler beam element. For box and rectangular sections, the default number of integration points for beam sections along the section height is 5. To increase the calculation accuracy, the number of integration points is increased to 25. Through the trial calculation of finite element models with different mesh sizes, the mesh size of each part is set in the range of 100–150 mm, which can balance the requirements of accuracy and efficiency.

For the application of boundary conditions and load actions, the bottom and top sections of the T-CDSCW are coupled with reference points, respectively, and the bottom and top sections move and rotate as rigid surfaces about the corresponding reference points. Restraints and loads on the T-CDSCW can be achieved by directly applying restraints and loads at the reference points.

In the shell element, the material properties of steel and concrete are similar to those in the solid–shell finite element model. The kinematic hardening model is adopted for steel. The concrete damaged plasticity model is adopted for concrete (Dassault Systèmes 2014a). Since the kinematic hardening model and the concrete damaged plasticity model are not supported in the B33 beam element, the uniaxial stress–strain relationship of steel and concrete is defined by user-defined UMAT (Qu 2004; Dassault Systèmes 2014c). Since the shell–beam finite element model is used to analyse the seismic performance of composite walls, the hardening parameters of steel and the damage parameters of concrete are set.

In the shell element, the rate-independent kinematic hardening model is adopted for steel (Chaboche 1986), and the back stress α is defined as Eq. (3.8).

$$\alpha = \sum_k \alpha_k = \sum_k \left[\frac{C_k}{\gamma_k} \left(1 - e^{-\gamma_k \varepsilon_p} \right) \right] \tag{3.8}$$

Where ε_p denotes the equivalent plastc strain; the parameters are set according to the recommendation values for Q345 steel (Shi et al. 2012): $C_1 = 7993$ MPa, $\gamma_1 = 175$, $C_2 = 6773$ MPa, $\gamma_2 = 116$, $C_3 = 2854$ MPa, $\gamma_3 = 34$, $C_4 = 1450$ MPa, $\gamma_4 = 29$.

In the beam element, the uniaxial stress–strain relationship is defined by UMAT (Qu 2004; Dassault Systèmes 2014c). The stress–strain curve adopts a three-line type: when the strain exceeds the yield strain ε_y, the tangent modulus turns to E_t; when the strain exceeds the failure strain ε_f, the tangent modulus turns to E_f. Under cyclic loading, the steel is unloaded according to the elastic modulus E_s at first, and when the direction of the stress changes and the reverse loading starts, the loading path points to the yield point or the maximum point previously loaded.

In the experiments, it is observed that when the composite walls are subjected to cyclic in-plane loading, the failure is concentrated in the boundary elements, where the maximum strain occurs. In the beam element, the strength degradation under cyclic loading is set in the material property: when the strain exceeds the failure strain ε_f, the strength of steel starts to degrades and the degraded strength is evaluated by Eqs. (3.9) (Qu and Ye 2011).

$$f_{yi} = f_{y}\left[1-\frac{E_{\mathrm{eff},i}}{3f_{y}\varepsilon_{f}\left(1-E_{t}/E_{s}\right)}\right] \tag{3.9}$$

Where $E_{\mathrm{eff},i}$ denotes the effective cumulative dissipated energy of the i cycle, which is defined as (3.10).

$$E_{\mathrm{eff},i} = \sum_{j=1}^{i}\left[E_{j}\left(\frac{\varepsilon_{j}}{\varepsilon_{f}}\right)^{2}\right] \tag{3.10}$$

Where E_j denotes the dissipated energy of the j cycle, and ε_j denotes the maximum strain of the j cycle.

Under cyclic loading, damage occurs in the concrete, and the stiffness degrades. The concrete is unloaded according to the degraded stiffness and when the direction of stress changes, the loading continues on the skeleton curve. The compressive damage parameter of concrete d_c is defined as Eq. (3.11) (Mander et al. 1988; MOHURD and AQSIQ 2010).

$$d_{c} = 1-\frac{E_{un}}{E_{c}} \tag{3.11}$$

Where E_{un} denotes the unloading stiffness, which is evaluate by Eq. (3.12).

$$E_{un} = \frac{\sigma_{un}}{\varepsilon_{un}-\varepsilon_{z}} \tag{3.12}$$

Where σ_{un} denotes the unloading stress, ε_{un} denotes the unloading strain, and ε_z is evaluated by Eqs. (3.13) and (3.14).

$$\varepsilon_{z} = \varepsilon_{un} - \frac{\varepsilon_{un}+\varepsilon_{ca}}{\sigma_{un}+E_{c}\varepsilon_{ca}}\sigma_{un} \tag{3.13}$$

$$\varepsilon_{ca} = \max\left(\frac{\varepsilon_{c}}{\varepsilon_{c}+\varepsilon_{un}}, \frac{0.09\varepsilon_{un}}{\varepsilon_{c}}\right)\sqrt{\varepsilon_{c}\varepsilon_{un}} \tag{3.14}$$

In the setting of concrete damage plasticity, the unloading path cannot point to the original point. The unloading is set to point to a focus point and the tensile damage parameter of concrete d_t is defined as Eq. (3.15) (Nguyen and Whittaker 2017).

$$d_{t} = 1-\frac{E_{un}}{E_{c}} = 1-\frac{\sigma_{un}+f_{t}}{E_{c}\left(\varepsilon_{un}+\varepsilon_{t}\right)} \tag{3.15}$$

In the shell and beam elements, the tensile stress–strain curve adopts Eq. (3.2) based on the Guo model. Since the composite effect of steel and concrete is not considered in the shell and beam elements, the improvement of strength and ductility caused by the composite effect needs to be considered in the uniaxial compressive stress–strain curve. The compressive stress–strain curve adopts the Han model proposed for the confined concrete in CFSTs, which is defined as Eq. (3.16) (Tao and Nie 2015; Han et al. 2001).

$$\frac{\sigma}{\sigma_c}=\begin{cases} 1+q\left[\left(\frac{\varepsilon}{\varepsilon_c}\right)^{0.1\xi}-1\right] & \varepsilon\le\varepsilon_c \\ \frac{\varepsilon/\varepsilon_c}{\beta\left(\varepsilon/\varepsilon_c-1\right)^{\eta}+\varepsilon/\varepsilon_c} & \varepsilon>\varepsilon_c \end{cases} \tag{3.16}$$

Where ξ denotes the confinement factor of the confined concrete, the ratio of the cross-sectional strength of steel to the cross-sectional strength of concrete, which is defined as Eq. (3.17).

$$\xi=\frac{f_y A_{sc}}{f_c A_{cc}} \tag{3.17}$$

Where A_{sc} and A_{cc} denote the cross-sectional areas of steel and concrete, respectively. That is, for boundary elements, A_{sc} and A_{cc} denote the cross-sectional areas of steel rectangular tubes and infilled concrete; for wall elements, A_{sc} and A_{cc} denote the cross-sectional areas of steel corrugated plates and infilled concrete.

In Eq. (3.16), σ_c denotes the axial compressive strength of the confined concrete, which is defined as Eq. (3.18), and ε_c denotes the corresponding compressive strain, which is defined as Eq. (3.19).

$$\sigma_c=\left[1+\left(0.1\xi-0.0135\xi^2\right)\left(\frac{24}{f_c'}\right)^{0.45}\right]f_c' \tag{3.18}$$

$$\varepsilon_c=1300+12.5f_c'+\left[1330+760\left(\frac{f_c'}{24}-1\right)\right]\xi^{0.2} \tag{3.19}$$

Where the parameters q, η, and β are defined as Eqs. (3.20), (3.21), and (3.22), respectively; f denotes the cylinder strength of concrete, which is evaluated by Eq. (3.23) (Chen et al. 1992).

$$q=\frac{\xi^{0.745}}{2+\xi} \tag{3.20}$$

$$\eta = 1.6 + \frac{1.5}{\varepsilon / \varepsilon_c} \tag{3.21}$$

$$\beta = \begin{cases} \dfrac{(f_c')^{0.1}}{1.35\sqrt{1+\xi}} & \xi \le 3.0 \\ \dfrac{(f_c')^{0.1}}{1.35\sqrt{1+\xi(\xi-2)^2}} & \xi > 3.0 \end{cases} \tag{3.22}$$

$$f_c' = \begin{cases} 0.8 f_{cu} & f_{cu} \le 50\,\mathrm{M}Pa \\ f_{cu} - 10 & f_{cu} > 50\,\mathrm{M}Pa \end{cases} \tag{3.23}$$

After the shell–beam finite element model is developed, the seismic performance of T-CDSCWs can be analysed by the static analysis method.

3.2 Finite Element Simulation Results

The solid–shell finite element model and the shell–beam finite element model are validated by the comparison between the finite element simulation results and the experimental results. Zhu et al. (Guo et al. 2018; Zhou et al. 2021; Zhu 2020; Zhu et al. 2019) has developed the finite element model similar to the solid–shell finite element model and validated the finite element model by the comparison with the experiment results with CDSCWs, which validates the solid–shell finite element model in some extent. The section introduces the validity of the finite element models to simulate the cross-sectional capacity, global stability, and seismic performance of T-CDSCWs. The simulation results of the solid–shell finite element model are compared with the results of the cross-sectional capacity test and the global stability test of T-CDSCWs and other existing tests regarding the failure mode, load–displacement relationship, and load–strain relationship. The simulation results of the shell–beam finite element model are compared with the results of the seismic performance test of the T-CDSCW and other existing tests regarding the load–displacement hysteretic curve.

3.2.1 Finite Element Simulations on the Cross-Sectional Capacity

The finite element simulation results of the cross-section capacity test of specimen T-1 are compared with the experimental results. The material properties of steel and concrete are set according to the material property test results. In the finite element model (see Fig. 3.4), the loading beam is developed. The bottom section of specimen T-1 is coupled with reference point A, which is located at the centroid of the bottom section. To achieve the fixed boundary condition, all degrees of freedom of reference point A are restrained. Moreover, to simulate the action of the vertical hydraulic jack and the horizontal actuator, the top surface of the loading beam is coupled with reference point B, which is right above the centroid of the top section, and the north surface of the loading beam is coupled with reference point C. Reference point B acts as the role of the spherical joint, and the vertical distance between reference point B and the top section of the specimen is 1150 mm. Reference point C acts as the role of the horizontal actuator, the horizontal distance between reference point C and the north surface of the loading beam is 644 mm and the height from the top section of the specimen is 400 mm. The translational degree of freedom of reference point B is restrained in the direction of x-axis. The vertical load is applied at reference point B. The translational degree of freedom of reference point C is restrained in the direction of y-axis.

Since only the local buckling of steel corrugated plates and rectangular tubes is observed in the cross-sectional capacity test, the local initial geometric imperfection is introduced in the finite element model. It is found in the previous study that the lateral pressure caused by concrete placement leads to the out-of-plane deformation

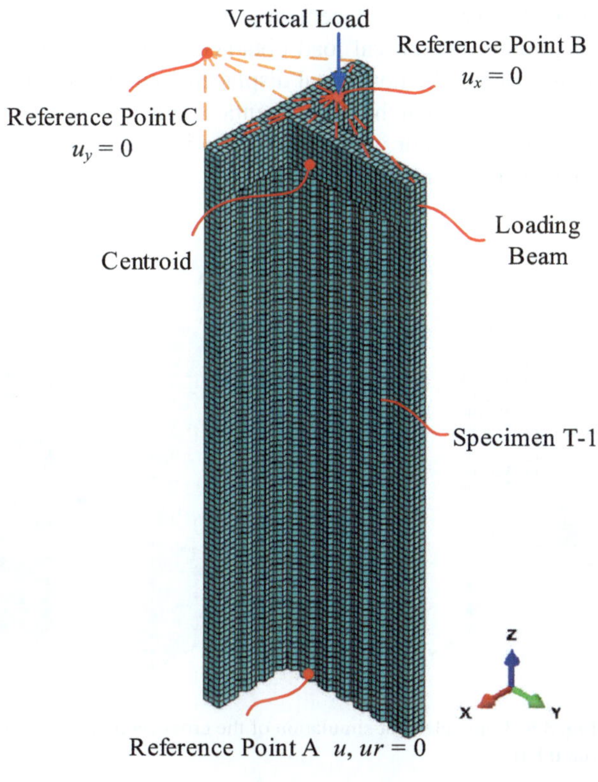

Fig. 3.4 Finite element model of the cross-sectional capacity test

of steel corrugated plates, and for the bolt spacing of 500 mm, the deformation is 0.3–0.5 mm outward (Guo et al. 2017). To consider the effects of residual stress and manufacturing deviation, the local initial geometric imperfection follows the distribution of the out-of-plane deformation caused by lateral pressure and the amplitude of the local initial imperfection is enlarged to $d_v/750$.

After introducing the initial geometric imperfection, the load-carrying capacity of specimen T-1 under axial compression and bending moment in the web plane is analysed by the static analysis method. The displacement and stress development (see Figs. 3.5 and 3.6) corresponds to points A, B, and C in Fig. 3.8. In the initial stage, the steel remains elastic. As the loading continues, the steel in the compressive zone yields and the concrete reaches the axial compressive strength. Then, the plastic zone of the steel expands to all the web wall. The local buckling occurs in the steel corrugated plates and rectangular tubes and the amplitude of the local buckling is 9.8 mm (see Fig. 3.7).

The load–displacement responses obtained from the finite element simulation agree well with the experimental results (see Figs. 3.8, 3.9, and 3.10). The maximum vertical load obtained from the finite element simulation is 15,236 kN, 1.8% less than the experimental result, and the corresponding vertical displacement of the finite element simulation is 5.5 mm, 13% less than the experimental result. Before the vertical load reaches the maximum, the response of vertical load and vertical displacement at the top section is almost the same as the experimental result, especially the vertical displacement at the flange side. After the vertical load reaches the maximum, the vertical load obtained from the finite element simulation degrades more slowly. The horizontal displacement obtained from the finite element simulation is greater than the experimental result. This results from that the restraint of the horizontal actuator is not ideal and with the increase in the vertical load, the horizontal actuator moves.

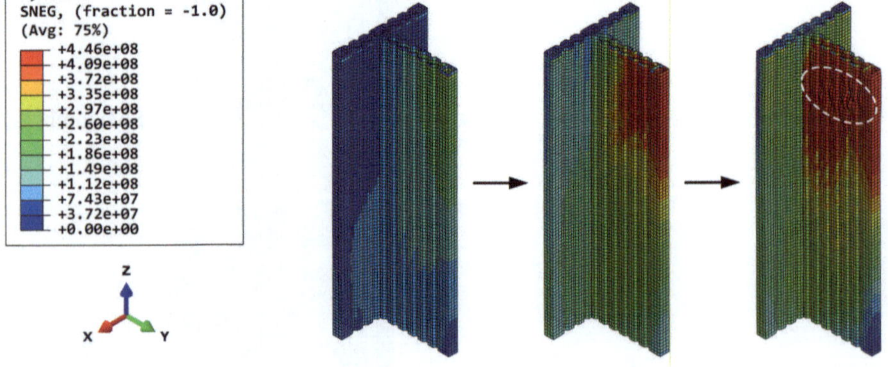

Fig. 3.5 Finite element simulation of the cross-sectional capacity test: strain distribution of steel (unit: Pa)

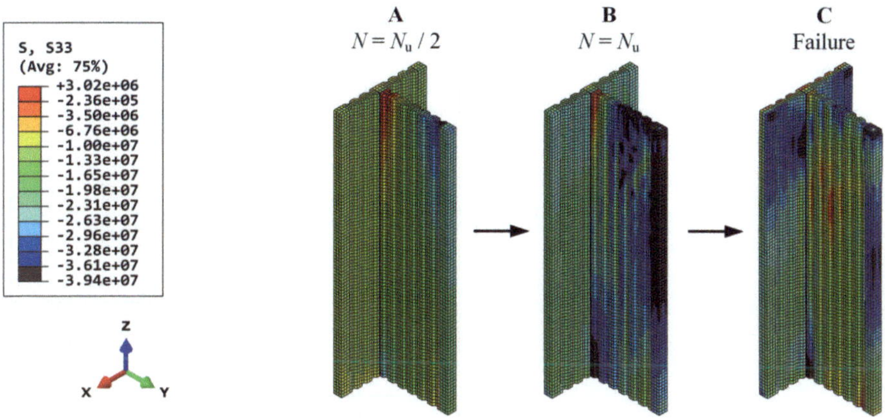

Fig. 3.6 Finite element simulation of the cross-sectional capacity test: strain distribution of concrete (unit: Pa)

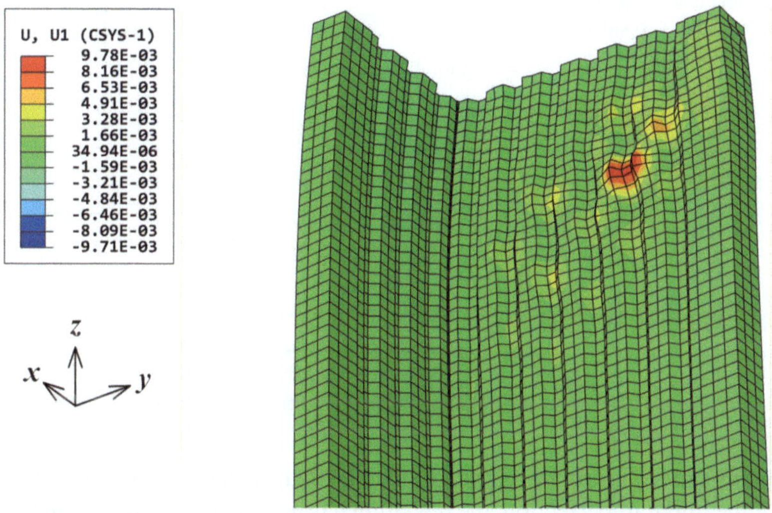

Fig. 3.7 Finite element simulation of the cross-sectional capacity test: local buckling of steel corrugated plates (the deformation is magnified 5 times, unit: m)

The strain development and distribution obtained from finite element simulation agree well with the experimental results (see Figs. 3.11 and 3.12). The strain development on the flange and web sides is almost the same as the experimental results. The strain distribution obtained from the finite element simulation basically lies in the middle of the experimental results. The deviation results from unpredictable imperfection that affects stress and strain distribution sensitively.

Although the load–displacement curves obtained from the finite element simulation are different from the experimental results after the vertical load reaches the

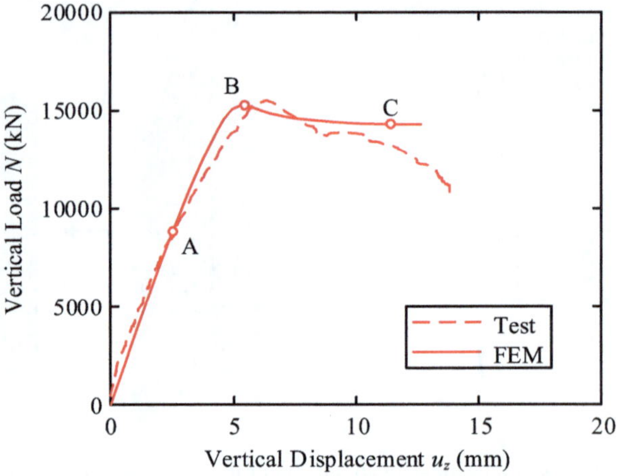

Fig. 3.8 Vertical load–vertical displacement curve of specimen T-1

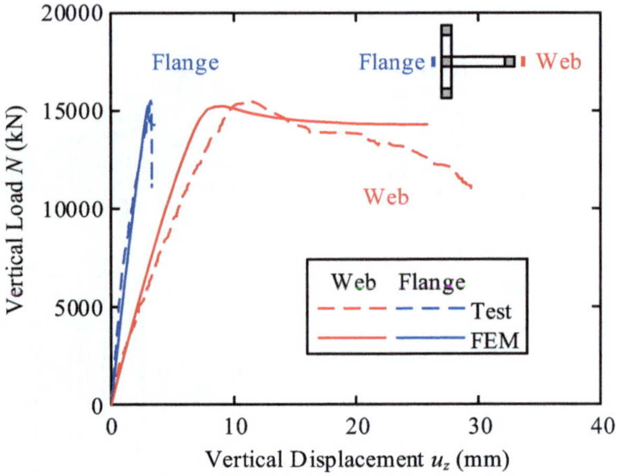

Fig. 3.9 Vertical load–vertical displacement at the flange and web sides curve of specimen T-1

maximum and the strain development is unavoidably deviated due to the randomness of the initial imperfection, the deviation has no significant effects on the design of T-CDSCWs. Moreover, the finite element can well predict the failure mode, maximum load, and corresponding displacement of T-CDSCWs.

Based on the solid–shell finite element model, the cross-sectional capacity test can be further analysed, such as the effect of concrete on the stability of the T-CDSCW. Without infilled concrete, the T-CDSCW turns into a double steel corrugated-plate wall with T-section (T-DSCW). By removing infilled concrete from the finite element model of specimen T-1, the load-carrying capacity of the T-DSCW is analysed with the tie constraint, boundary conditions, load actions, and initial geometric imperfection remaining unchanged. The displacement and stress development (see Fig. 3.13) corresponds to points A, B, and C in Fig. 3.14. Although only the local initial geometric imperfection is introduced in the finite element of

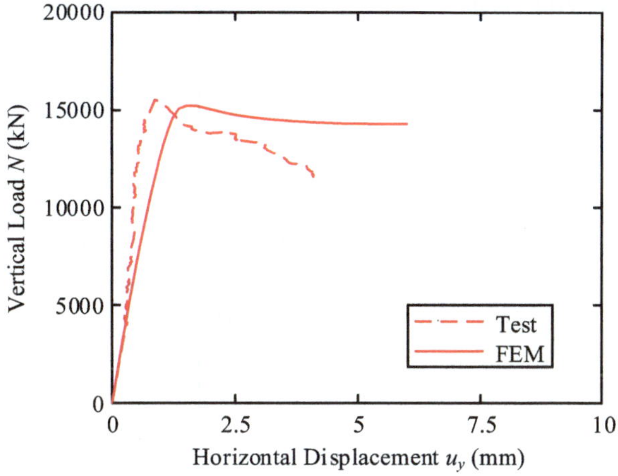

Fig. 3.10 Vertical load–horizontal displacement curve of specimen T-1

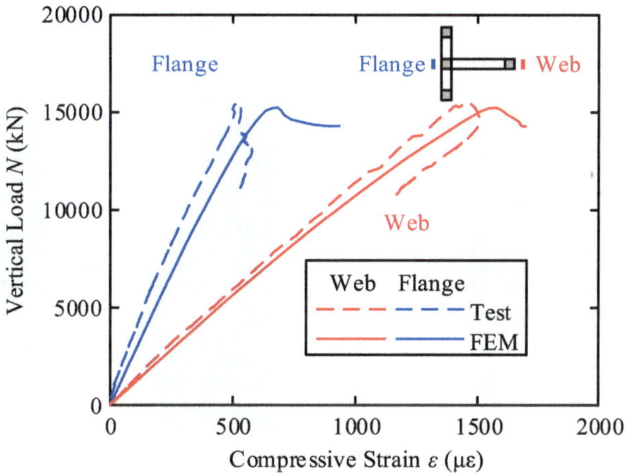

Fig. 3.11 Strain development of the steel parts at the mid-height section of specimen T-1

the T-DSCW and the global initial geometric imperfection is not introduced, the failure mode of the T-DSCW is not the local buckling of steel corrugated plates and rectangular tubes but the buckling of the web wall. It indicates that infilled concrete supports the outer steel corrugated plates and changes the failure mode.

The load-carrying capacity of specimen T-1 can be divided to the contributions of steel and concrete (see Fig. 3.14). The maximum of the vertical load shared by the steel in specimen T-1 is slightly less than the maximum of the vertical load of the T-DSCW. After the buckling occurs in the web wall of the T-DSCW, the vertical load of the T-DSCW degrades continuously, while after the local buckling after yielding occurs in the steel corrugated plates of specimen T-1, the vertical load of steel corrugated plates remains basically unchanged. It indicates that the lateral concrete expansion force may reduce the load-carrying capacity of steel corrugated plates, while the existence of infilled concrete improves the stability of the wall.

Fig. 3.12 Strain distribution of the steel parts at the mid-height section of the T-1 section

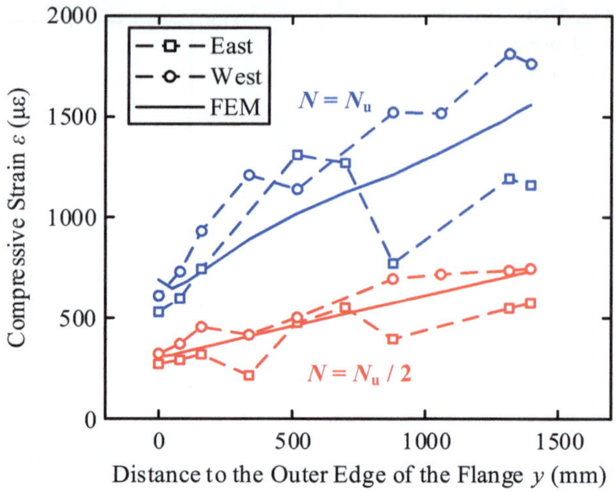

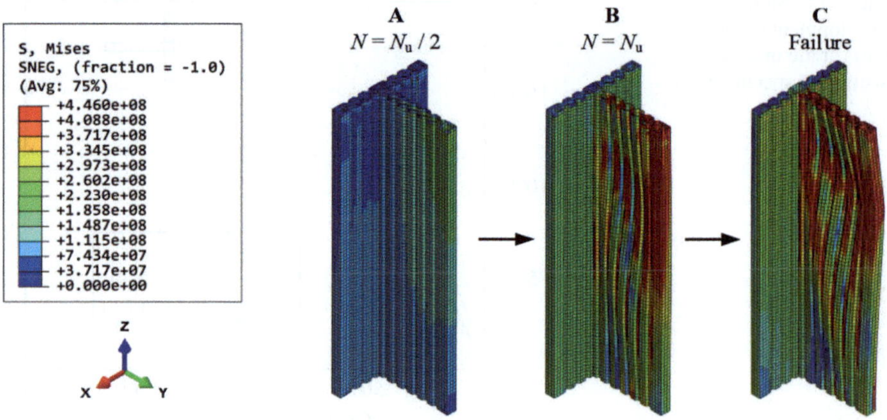

Fig. 3.13 Finite element simulation of the T-DSCW: strain distribution of steel (unit: Pa)

To further validate the solid–shell finite element model, the finite element simulation results are compared with the existing experimental results (Othuman Mydin and Wang 2011; Prabha et al. 2013; Zhu et al. 2019). The specimens have various sizes and detailing (see Table 3.1). All shear connectors in the specimens are aligned intermediate bolts. In the finite element modelling, the distribution of initial geometric imperfection is follows the lateral deformation caused by uniform pressure and the amplitude of the initial geometric imperfection is $d_v/750$.

The displacement and stress development (see Fig. 3.15) corresponds to points A, B, and C in Fig. 3.17, Fig. 3.18, and Fig. 3.19. The failure modes of steel corrugated plates are mainly the outward buckling at the boundaries and between two rows of intermediate bolts. The shape of the damage and the region where local buckling occurs are highly dependent on the distribution of initial geometrical

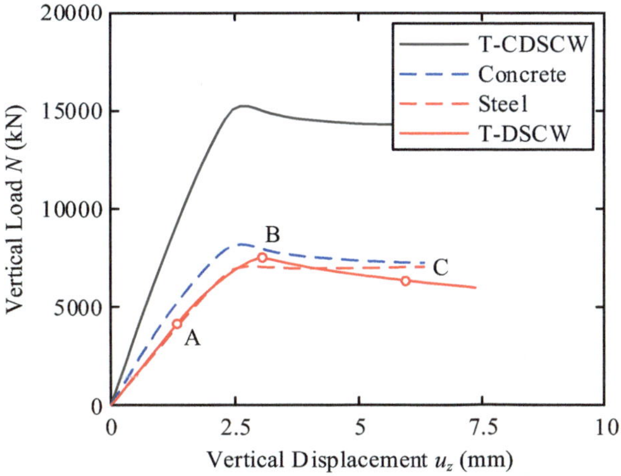

Fig. 3.14 Vertical load–vertical displacement curves of specimen T-1 and the T-DSCW

Table 3.1 Parameters and finite element simulation results of the axial compressive tests of CDSCWs

Authors	No.	H_w (mm)	b_w (mm)	d_w (mm)	d_v (mm)	q (mm)	q_1 (mm)	q_2 (mm)	q_3 (mm)	a (mm)	t_s (mm)	$N_{u,n}/N_{u,t}$
Othuman Mydin and Wang (2011)	1	400	400	100	160	120	50	10	50	10	0.4	1.12
	2	400	400	100	160	120	50	10	50	10	0.8	1.09
	3	400	400	100	160	120	50	10	50	10	0.4	1.05
	4	400	400	100	160	120	50	10	50	10	0.8	1.14
	5	400	400	100	160	120	50	10	50	10	0.4	0.97
	6	400	400	100	160	120	50	10	50	10	0.8	1.06
Prabha et al. (2013)	1	870	685	130	200	200	110	27.5	35	35	0.8	0.90
	2	870	685	130	200	200	110	27.5	35	35	0.8	1.04
	3	870	685	130	114	200	110	27.5	35	35	0.8	0.94
	4	870	685	130	400	200	110	27.5	35	35	0.8	1.15
	5	870	685	130	–*	200	110	27.5	35	35	0.8	0.99
Zhu et al. (2019)	1	1500	1070	168	750	200	100	25	50	25	4.0	0.96
Average ± Standard error												1.03 ± 0.08

$N_{u,n}$ denotes the maximum vertical load obtained from finite element simulation and $N_{u,t}$ denotes the maximum vertical load obtained from the experiments

imperfection, which is not presented in the literature and hard to simulate. However, the failure mode of steel corrugated plates is consistent with the experimental phenomena, which is the out-of-plane buckling in the middle of intermediate bolts.

The vertical load–displacement relationship obtained from finite element simulation agree well with the experimental results (see Table 3.1 and Fig. 3.16, Fig. 3.17, Fig. 3.18, Fig. 3.19, and Fig. 3.20). The maximum load, the ascending section, and the descending section obtained from finite element simulation are consistent with the experimental results. The average ratio of the maximum vertical load obtained

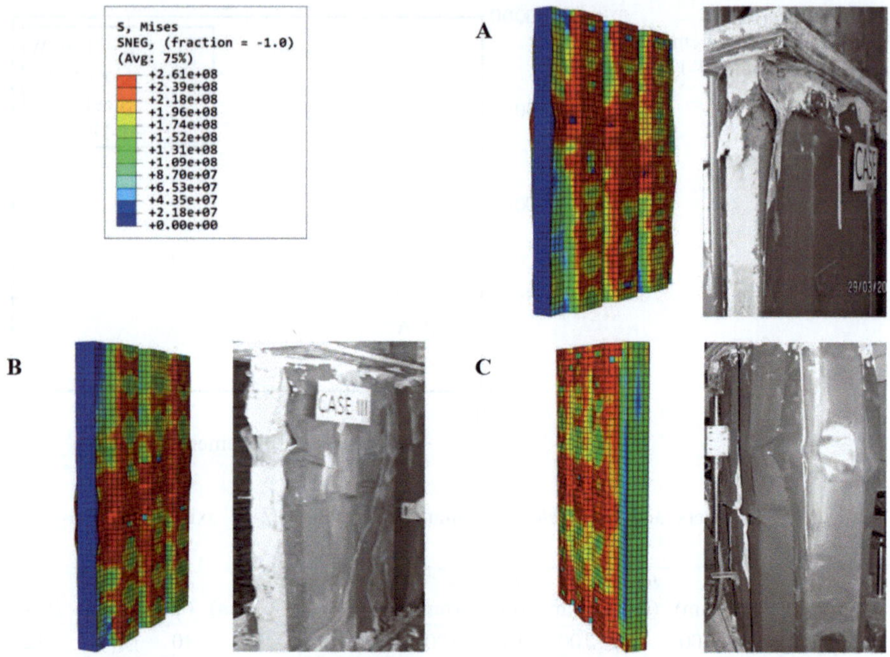

Fig. 3.15 Finite element simulation of the CDSCWs: strain distribution of steel (unit: Pa)

Fig. 3.16 Vertical load–vertical displacement curve of specimen 1 (Prabha et al. 2013)

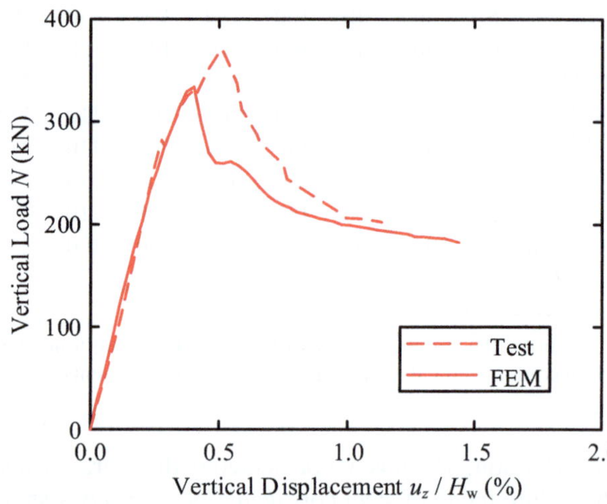

from finite element simulation to the experimental results is 1.03, and the standard deviation is 0.08. It is validated that the solid–shell finite element model can reflect the load-carrying performance of CDSCWs.

As can be seen from the comparison between the finite element simulation results and the experimental results, it is reasonable to introduce the initial geometric

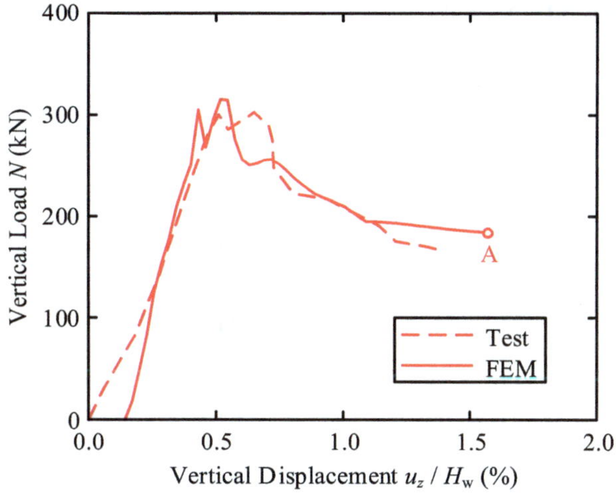

Fig. 3.17 Vertical load–vertical displacement curve of specimen 2 (Prabha et al. 2013)

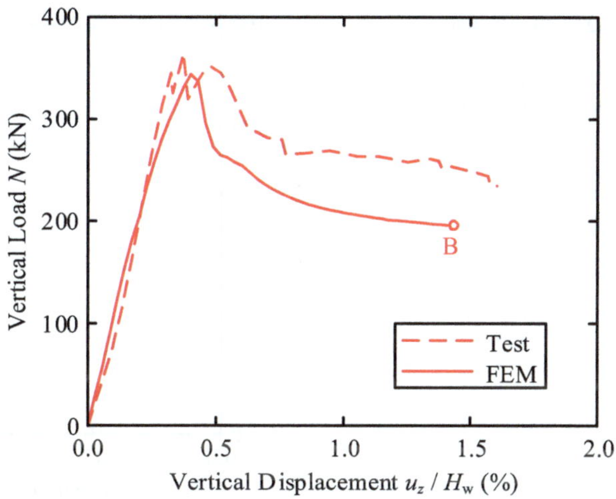

Fig. 3.18 Vertical load–vertical displacement curve of specimen 3 (Prabha et al. 2013)

imperfection of steel corrugated plates with the amplitude of $d_v/750$ in the analysis of the load-carrying performance of T-CDSCWs. The solid–shell finite element model can well simulate the load-carrying performance of T-CDSCWs.

3.2.2 Finite Element Simulations on the Global Stability

The finite element simulation results and the experimental results of the global stability test of specimen T-2 are compared. The material properties of steel and concrete are set according to the material property test results. In the finite element model (see Fig. 3.21), the bottom section of specimen T-2 is coupled with reference

Fig. 3.19 Vertical load–vertical displacement curve of specimen 4 (Prabha et al. 2013)

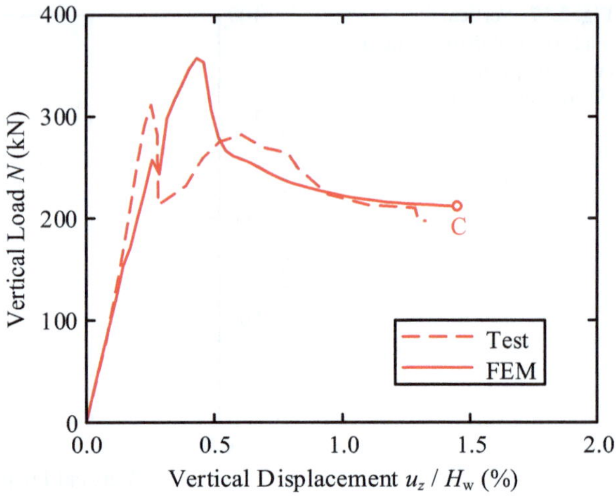

Fig. 3.20 Vertical load–vertical displacement curve of specimen 5 (Prabha et al. 2013)

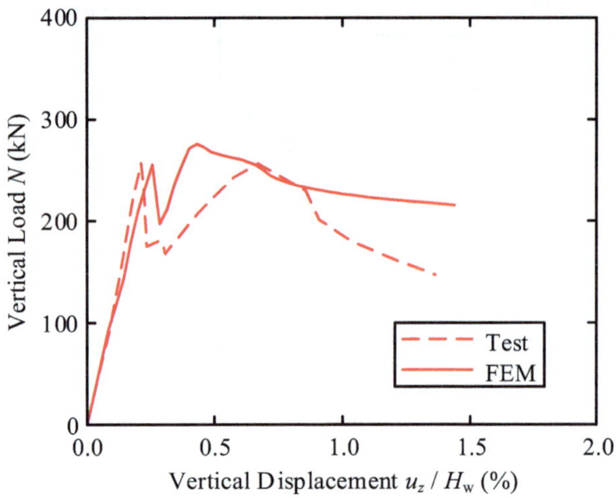

point A, which is located at the centroid of the bottom section. To achieve the fixed boundary condition, all degrees of freedom of reference point A are restrained. The top section of specimen T-2 is coupled with reference point B, which is right above the centroid of the top section. Reference point B acts as the role of the spherical joint, and the vertical distance between reference point B and the top section of the specimen is 1150 mm. The translational degree of freedom of reference point B is restrained in the direction of *x*- and *y*-axis. The vertical load is applied at reference point B.

Since only the global buckling of specimen T-2 is observed in the global stability test, the global initial geometric imperfection of specimen T-2 is introduced in the finite element model. The initial geometric imperfection is assumed in accordance with the consistent first-order buckling mode with the flange on the convex side.

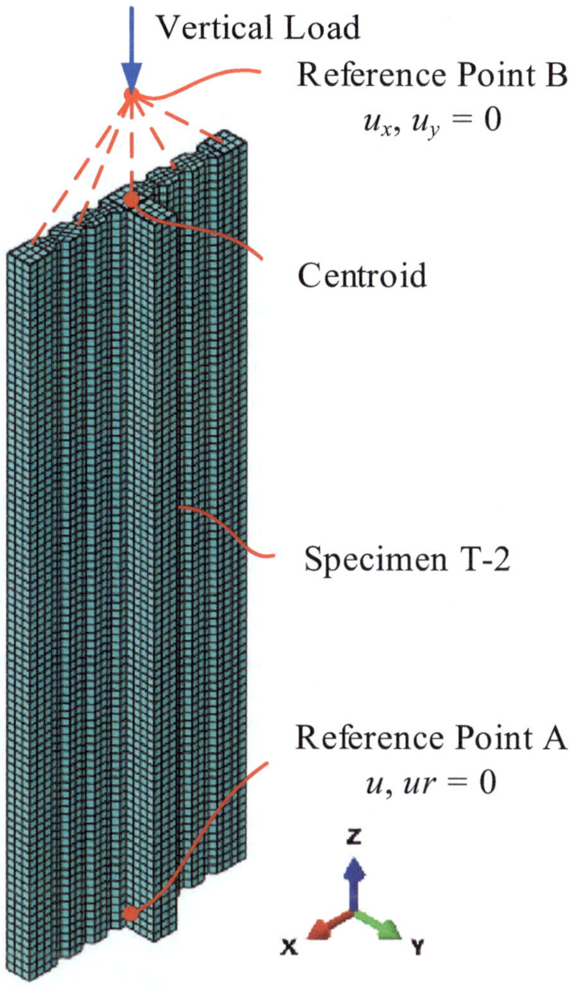

Fig. 3.21 Finite element model of the global stability test

The amplitude of the initial geometric imperfection is enlarged from the measured result to 1/500 of the wall height H_w. Such enlargement of the imperfection takes into consideration the eccentricity caused by the loading beam and the residual stress caused by welding operation between steel corrugated plates and rectangular tubes (Guo et al. 2018).

After introducing the initial geometric imperfection, the load-carrying capacity of specimen T-2 under axial compression is analysed by the static analysis method. The displacement and stress development (see Fig. 3.22) corresponds to points A, B, and C in Fig. 3.24. In the initial stage, the stress is uniformly distributed and no obvious horizontal displacement occurs. As the loading continues, the steel yields and horizontal displacement occurs in the negative direction of y-axis (see Fig. 3.23). As the loading continues, the plastic zone concentrates on the concave side of specimen T-2, and the deformation of the T-CDSCW develops in the form of flexural buckling about the asymmetry axis, which is consistent with the experimental phenomenon.

The load–displacement responses obtained from finite element simulation agree well with the experimental results (see Figs. 3.24 and 3.25). The maximum vertical load obtained from finite element simulation is 16,401 kN, 2.7% less than the experimental result. The corresponding vertical displacement obtained from finite element simulation is significantly less than the experimental result. In the initial stage, the stiffness of the finite element simulation results is close to that of the experimental results until the stiffness of the experiment begins to degrade.

The strain development and distribution obtained from finite element simulation agree well with the experimental results (see Figs. 3.26 and 3.27). The strain development on the flange and web sides is close to the experimental results. In the initial stage, the strain obtained from finite element simulation is distributed uniformly and close to the experimental results. When the vertical load reaches the maximum, the strain distribution obtained from the finite element simulation basically lies in the middle of the experimental results. Although the strain obtained from finite element simulation does not fluctuate as the experimental results, its relative relations are the same as the strain at the outer and inner edges of the flange. The greater fluctuation of the experimental results may be caused by the difference of the global imperfection of specimen T-2 and the local imperfection of steel corrugated plates between the assumption in the finite element model and the actual distribution.

Although the load–displacement curves obtained from the finite element simulation are different from the experimental results after the vertical load reaches the maximum and the strain development is unavoidably deviated due to the randomness of the initial imperfection, the deviation has no significant effect on the design of T-CDSCWs. Moreover, the finite element can well predict the failure mode and the maximum load of T-CDSCWs.

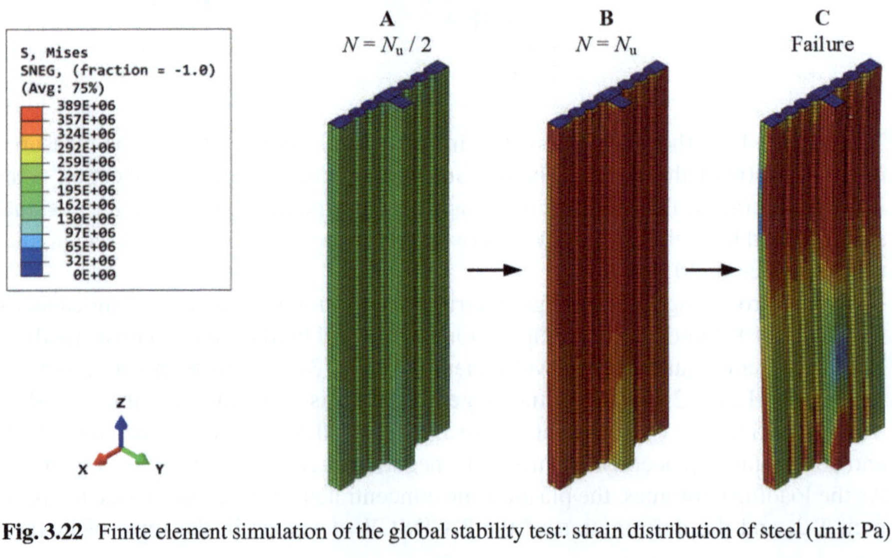

Fig. 3.22 Finite element simulation of the global stability test: strain distribution of steel (unit: Pa)

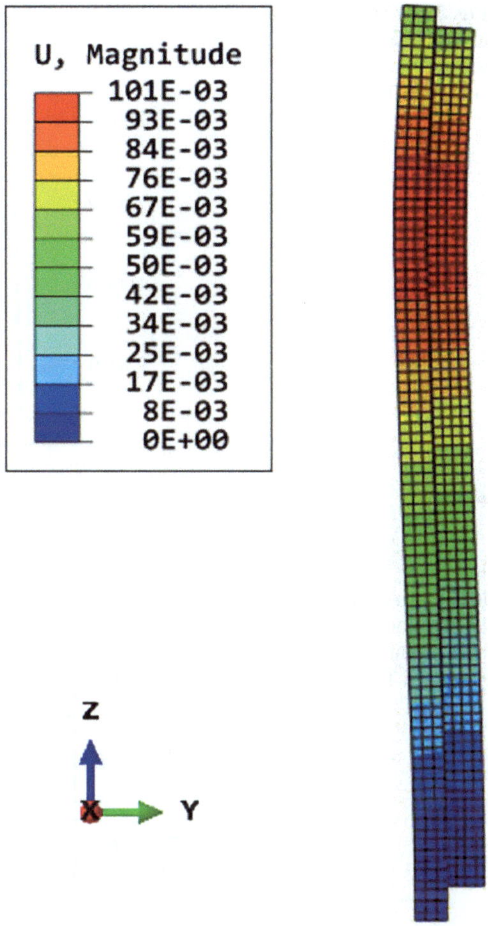

Fig. 3.23 Finite element simulation of the global stability test: side view (unit: m)

As can be seen from the comparison between the finite element simulation results and the experimental results, it is reasonable to introduce the initial geometric imperfection of T-CDSCWs with the amplitude of $H_w/500$ in the analysis of the global stability of T-CDSCWs. The solid–shell finite element model can well simulate the global stability of T-CDSCWs.

3.2.3 Finite Element Simulations on the Seismic Performance

The finite element simulation results of the seismic performance test of specimen T-3 are compared with the experimental results. The material properties of steel and concrete are set according to the material property test results. In the finite element model (see Fig. 3.28), the bottom section of specimen T-3 is coupled with reference

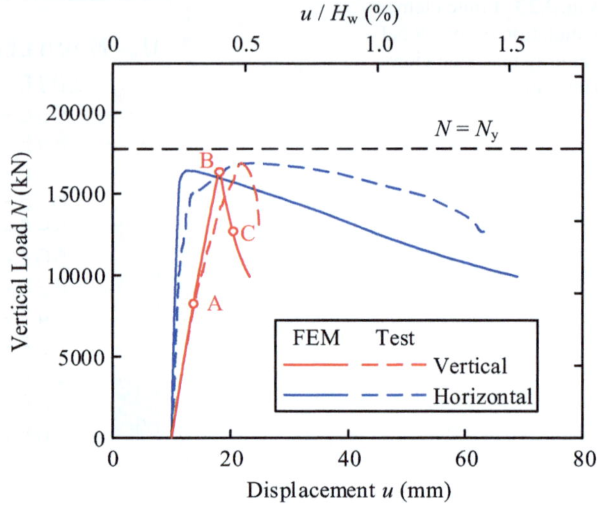

Fig. 3.24 Vertical load–displacement at the top section curve of specimen T-2

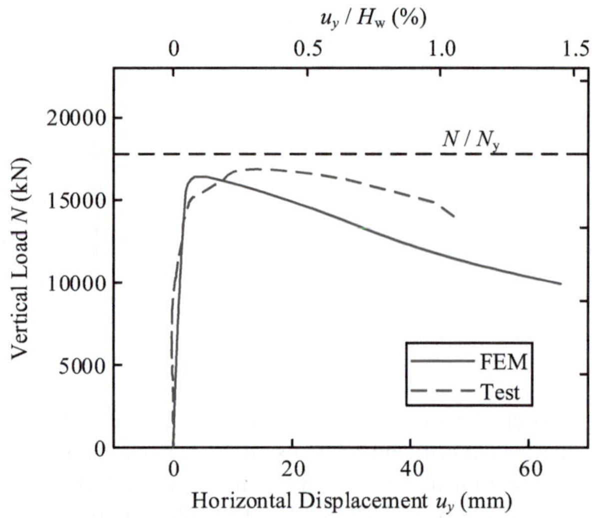

Fig. 3.25 Vertical load–horizontal displacement at the mid-height section curve of specimen T-2

point A, which is located at the centroid of the bottom section. To achieve the fixed boundary condition, all degrees of freedom of reference point A are restrained. The top section of specimen T-3 is coupled with reference point B, which is right above the centroid of the top section. Reference point B acts as the roles of the spherical joint and the horizontal actuator, and the vertical distance between reference point B and the top section of the specimen is 400 mm. The translational degree of freedom of reference point B is restrained in the direction of x-axis. The vertical and horizontal loads are applied at reference point B.

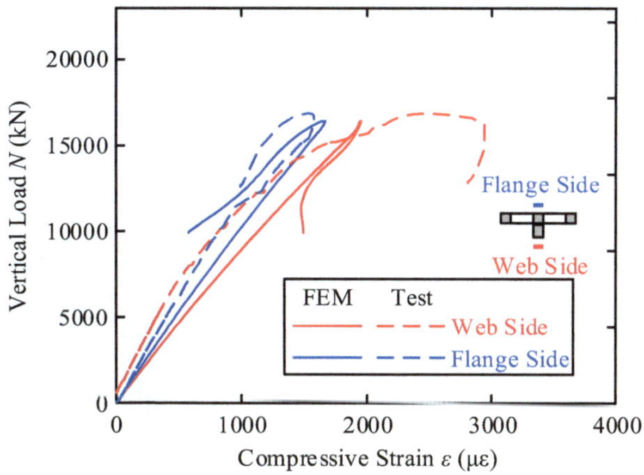

Fig. 3.26 Strain development of the steel parts at the mid-height section of specimen T-2

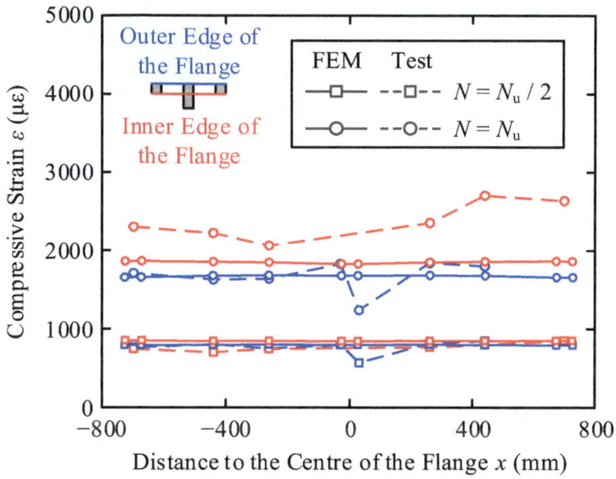

Fig. 3.27 Strain distribution of the steel parts at the mid-height section of specimen T-2

The load–displacement responses obtained from finite element simulation agree well with that of the experimental results (see Fig. 3.29). The positive direction of the horizontal load is the positive direction of y-axis and the flange is under compression. The negative direction of the horizontal load is the negative direction of y-axis and the web is under compression. Under the same loading amplitude, the load-carrying capacity of specimen T-3 with the flange under compression does not degrade very much, while the load-carrying capacity with the web under compression degrades significantly. The maximum horizontal load in the positive direction obtained from finite element simulation is 2229 kN, 1.9% less than the experimental result. The maximum horizontal load in the negative direction obtained from finite element simulation is 1422 kN, 5.3% less than the experimental result.

To further validate the shell–beam finite element model, the finite element simulation results are compared with the existing experimental results (Zhu et al. 2020;

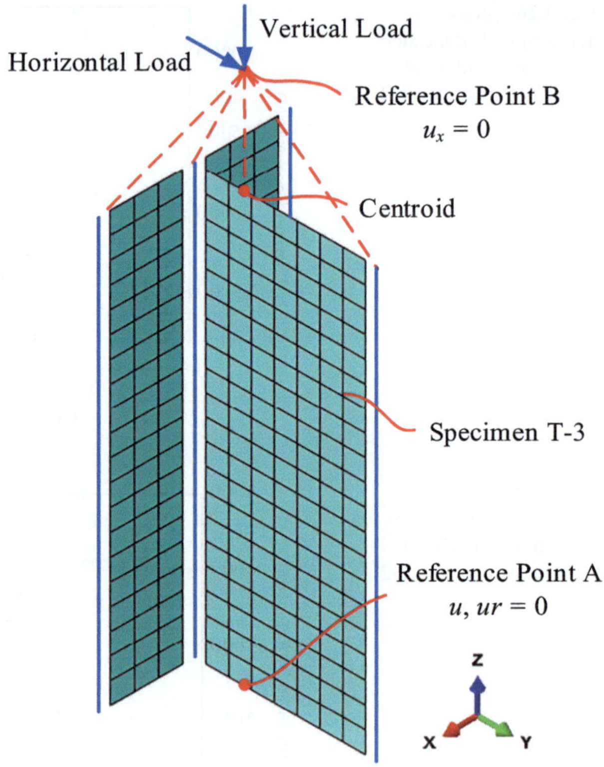

Fig. 3.28 Finite element model of the seismic performance test

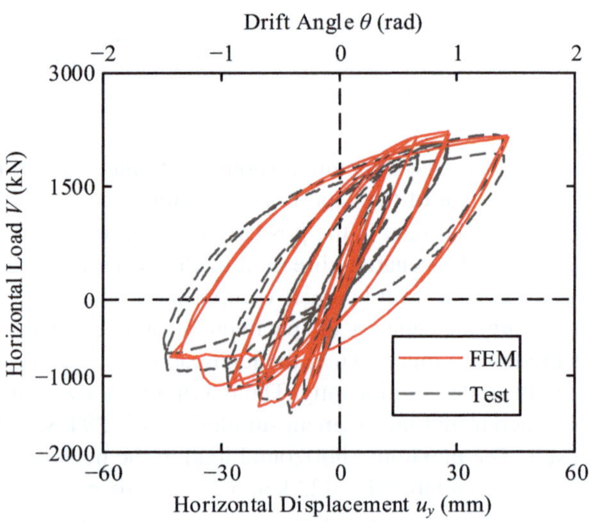

Fig. 3.29 Horizontal load–horizontal displacement hysteretic curve of specimen T-3

Table 3.2 Parameters and finite element simulation results of the seismic performance tests of CDSCWs

No.	H_w (mm)	b_w (mm)	d_w (mm)	d_v (mm)	n	$V_{u,n}/V_{u,t}$	
						+	−
W-1.5-1	1500	1077	150	280	0.28	0.98	1.05
W-1.5-2	1500	1077	150	280	0.42	1.07	0.95
W-2.0-1	2000	1077	150	280	0.42	1.01	0.92
W-2.0-2	2000	1077	150	280	0.42	0.97	1.00
W-2.0-3	2000	1077	150	500	0.30	0.92	0.95
W-2.0-4	2000	1077	150	500	0.42	0.98	0.91
W-2.5-1	2500	1077	150	280	0.50	1.04	0.95
W-2.5-2	2500	1077	150	280	0.58	0.97	0.94
Average ± Standard error						0.98 ± 0.05	

$V_{u,n}$ denotes the maximum horizontal load obtained from finite element simulation and $V_{u,t}$ denotes the maximum horizontal load obtained from the experiments

Zhu 2020). The specimens have various sizes (see Table 3.2). In the experiments, the wave width of steel corrugated plates q is 180 mm, the crest width q_1 is 85 mm, the transition width q_2 is 27.5 mm, the trough width q_3 is 40 mm, the wave height a is 35 mm, and the thickness t_s is 4 mm.

It is observed from the finite element simulation results that the finite element model cannot simulate the pinch effect of W-1.5-1 and W-2.0-3 (see Figs. 3.30 and 3.34), of which the failure modes are the fracture of tensile steel rectangular tubes. The fracture is not considered in the finite element model and should be avoided by construction measures. However, the horizontal load–displacement relationship obtained from finite element simulation agree well with other experimental results (see Figs. 3.31, 3.32, 3.33, 3.35, 3.36, and 3.37), including the backbone curves, unloading, and reloading. Not only the maximum horizontal load and corresponding horizontal displacement are in good agreement, but also the failure displacement. The average ratio of the maximum horizontal load obtained from finite element simulation to the experimental results is 0.98, and the standard deviation is 0.05. Zhu (2020) simulate the seismic performance of CDSCWs by the finite element model similar as the solid–shell finite element model. Since simpler elements are adopted in the shell–beam finite element model, compared with the finite element model used by Zhu (2020), the calculation efficiency of the shell–beam finite element model is more than 10 times faster.

As can be seen from the comparison between the finite element simulation results and the experimental results, the shell–beam finite element model can well simulate the seismic performance of T-CDSCWs.

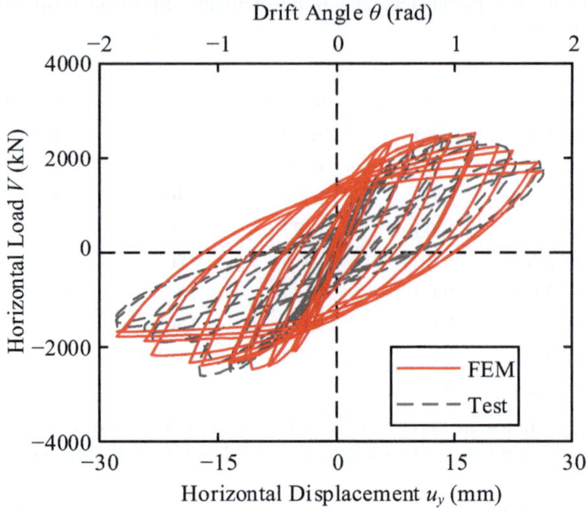

Fig. 3.30 Horizontal load–horizontal displacement hysteretic curve of specimen W-1.5-1 (Zhu et al. 2020)

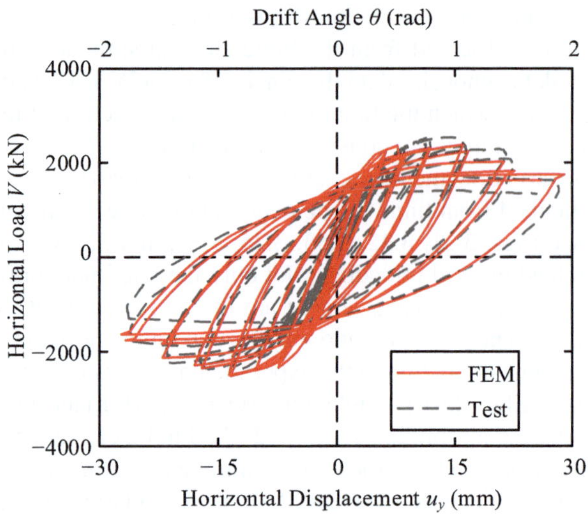

Fig. 3.31 Horizontal load–horizontal displacement hysteretic curve of specimen W-1.5-2 (Zhu et al. 2020)

3.3 Chapter Summary

Two kinds of finite element models of T-CDSCWs are developed, namely the solid–shell finite element model and the shell–beam finite element model. The finite element simulation results are compared with the results of T-CDSCW experiments and other existing experiments. Based on the comparison, the main conclusions are as follows.

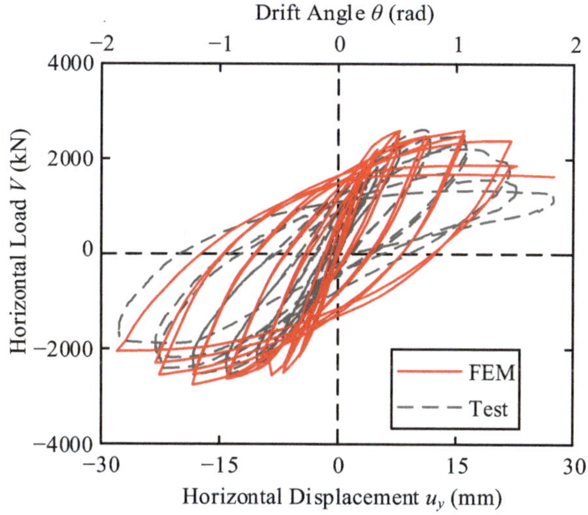

Fig. 3.32 Horizontal load–horizontal displacement hysteretic curve of specimen W-2.0-1 (Zhu et al. 2020)

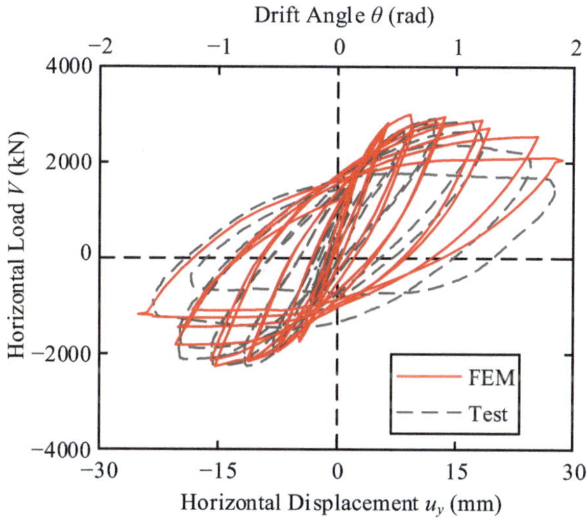

Fig. 3.33 Horizontal load–horizontal displacement hysteretic curve of specimen W-2.0-2 (Zhu et al. 2020)

1. The solid–shell finite element model can effectively simulate the load-carrying performance of T-CDSCWs and the local buckling performance of steel corrugated plates, especially accurately simulating the cross-sectional capacity of T-CDSCWs and the load-carrying capacity of steel corrugated plates.
2. The solid–shell finite element model can also effectively simulate the global stability of T-CDSCWs, especially the flexural buckling about the asymmetry axis with T-CDSCWs under axial compression. The load-carrying capacity of T-CDSCWs can be accurately simulated.

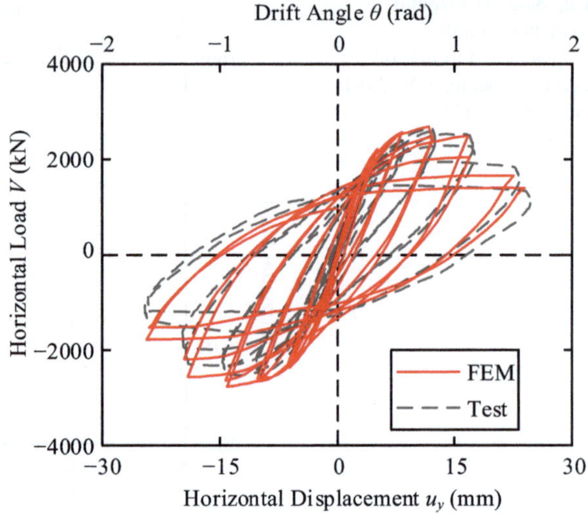

Fig. 3.35 Horizontal load–horizontal displacement hysteretic curve of specimen W-2.0-4 (Zhu et al. 2020)

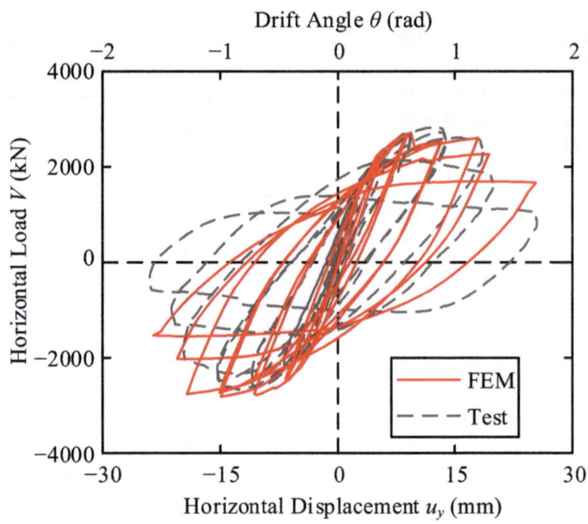

Fig. 3.36 Horizontal load–horizontal displacement hysteretic curve of specimen W-2.5-1 (Zhu et al. 2020)

3. The shell–beam finite element model can effectively simulate the seismic performance of T-CDSCWs. For CDSCWs with rectangular and T-sections, the shell–beam finite element model can accurately simulate the maximum horizontal load, failure displacement, unloading, and reloading.

Therefore, in the following chapters, the finite element models will be used to analyse the cross-sectional capacity, global stability, and seismic performance of T-CDSCWs. In the analysis of the cross-sectional capacity, only the initial geometric imperfection of steel corrugated plates with the amplitude of $d_v/750$ is

Fig. 3.37 Horizontal load–horizontal displacement hysteretic curve of specimen W-2.5-2 (Zhu et al. 2020)

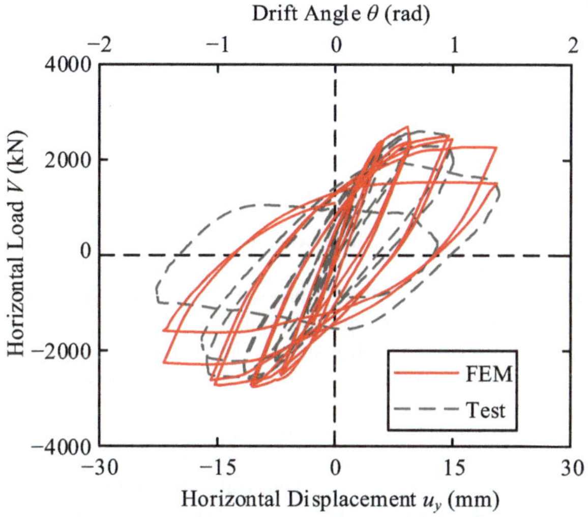

Fig. 3.34 Horizontal load–horizontal displacement hysteretic curve of specimen W-2.0-3 (Zhu et al. 2020)

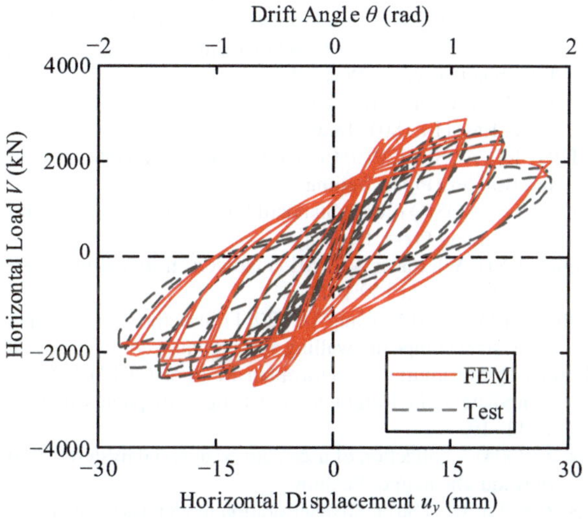

introduced and the distribution follows the out-of-plane deformation caused by lateral pressure. In the analysis of the global stability, only the initial geometric imperfection of T-CDSCWs with the amplitude of $H_w/500$ is introduced and the distribution is in accordance with the consistent first-order buckling mode. Based on experimental investigations and finite element simulations, the design methods of the cross-sectional capacity, global stability, and seismic performance of T-CDSCWs will be proposed in the following chapters.

References

Chaboche JL (1986) Time-independent constitutive theories for cyclic plasticity. Int J Plast 2:149–188

Chen Z, Zhu J, Wu P (1992) High strength concrete and its application (in Chinese). Tsinghua University Press, Beijing

Dassault Systèmes (2014a) Abaqus analysis user's guide. Dassault Systèmes Simulia Corp., Johnston, RI

Dassault Systèmes (2014b) Abaqus benchmarks guide. Dassault Systèmes Simulia Corp., Johnston, RI

Dassault Systèmes (2014c) Abaqus user subroutines reference guide, Version 6.14. Dassault Systèmes Simulia Corp., Johnston, RI

Dassault Systèmes (2014d) Abaqus (Version 6.14). Dassault Systèmes Simulia Corp, Johnston, RI. https://www.3ds.com/products-services/simulia/products/abaqus/abaqusstandard/

Guo Y, Zhu J, Wang M et al (2017) Research on corrugated-steel-concrete composite wall structure system in high-rise building (in Chinese). Tsinghua University, Beijing

Guo Y, Zhu J, Wang M et al (2018) Overall instability performance of concrete-infilled double steel corrugated plate wall. Thin Wall Struct 130:372–394

Guo Z (2014) Principle of reinforced concrete. Tsinghua University Press, Beijing

Han L, Zhao X, Tao Z (2001) Tests and mechanics model for concrete-filled SHS stub columns, columns and beam-columns. Steel Compos Struct 1:51–74

Hossain KMA, Rafiei S, Lachemi M et al (2016) Finite element modeling of impact shear resistance of double skin composite wall. Thin Wall Struct 107:101–118

Jiang J, Lu X, Ye L (2013) Finite element analysis of concrete structures (in Chinese), 2nd edn. Tsinghua University Press, Beijing

Mander JB, Priestley MJN, Park R (1988) Theoretical stress-strain model for confined concrete. J Struct Eng 114:1804–1826

MOHURD, AQSIQ (2010) Code for design of concrete structures (in Chinese). China Architecture and Building Press, Beijing

MOHURD, AQSIQ (2017) Standard for design of steel structures (in Chinese). China Architecture and Building Press, Beijing

Nguyen NH, Whittaker AS (2017) Numerical modelling of steel-plate concrete composite shear walls. Eng Struct 150:1–11

Othuman Mydin MA, Wang YC (2011) Structural performance of lightweight steel-foamed concrete–steel composite walling system under compression. Thin Wall Struct 49:66–76

Prabha P, Marimuthu V, Saravanan M et al (2013) Effect of confinement on steel-concrete composite light-weight load-bearing wall panels under compression. J Constr Steel Res 81:11–19

Qu Z (2004) A collection of user-defined uniaxial hysteretic models for ABAQUS/standard. http://www.qu-zhe.net/pqfiber.htm

Qu Z, Ye L (2011) Strength deterioration model based on effective hysteretic energy dissipation for RC members under cyclic loading (in Chinese). Eng Mech 28:45–51

Shi Y, Wang M, Wang Y (2012) Experimental study of structural steel constitutive relationship under cyclic loading (in Chinese). J Build Mater 15:293–300

Tao M, Nie J (2015) Element mesh, section discretization and material hysteretic laws for fiber beam–column elements of composite structural members. Mater Struct 48:2521–2544

Zhou Y, Zhu J, Guo Y et al (2021) Numerical and experimental studies on sectional load capacity of concrete-infilled double steel corrugated-plate walls under combined compression and in-plane bending. Thin Wall Struct 159:107250

Zhu J (2020) Theoretical and experimental study on design methods for I-section concrete-infilled double steel corrugated-plate walls (in Chinese). Ph.D. thesis, Tsinghua University, Beijing

Zhu J, Guo Y, Wang M et al (2019) Strength design of concrete-infilled double steel corrugated-plate walls under uniform compressions. Thin Wall Struct 141:153–174

Zhu J, Guo Y, Wang M et al (2020) Seismic performance of concrete-infilled double steel corrugated-plate walls: experimental research. Eng Struct 215:110601

Zhuang Z, You X, Liao J et al (2009) Finite element analyses and applications based on Abaqus (in Chinese). Tsinghua University Press, Beijing

Chapter 4
Cross-Sectional Capacity of T-CDSCWs

Based on the solid–shell finite element model, this chapter studies the load-carrying capacities of steel corrugated plates and T-CDSCWs, including the load-carrying capacity of steel corrugated plates under compression and the cross-sectional capacity of T-CDSCWs under axial compression and bending moment in the web plane, bending moment in the flange plane, and biaxial bending moment.

The analysis of the cross-sectional capacity of T-CDSCWs does not consider the global instability, while considers the effect of the local buckling of steel corrugated plates. The buckling behaviour of steel corrugated plates is analysed first, including the bilateral buckling behaviour of steel corrugated plates with only intermediate bolt constraints and the contact buckling behaviour of steel corrugated plates with intermediate bolt and infilled concrete constraints. The buckling loads of steel corrugated plates are derived based on the orthotropic plate theory and the energy method (Timoshenko and Gere 2009; Timoshenko and Woinowsky-Krieger 1959). To evaluate the load-carrying capacity of steel corrugated plates with intermediate bolt and infilled concrete constraints, the curve of the strength reduction factor is developed based on the normalised slenderness ratio of contact buckling. According to the requirement of compact section, the upper limit of the normalised slenderness ratio is recommended.

Considering the local buckling of steel corrugated plates, the cross-sectional capacity of T-CDSCWs is then analysed, including the stress distribution and the bending moment–curvature relationship. Based on the full section plasticity distribution, the simplified interaction curves are proposed on the cross-sectional capacity of T-CDSCWs under axial compression and bending moment in the web plane, bending moment in the flange plane, and biaxial bending moment.

M.-Z. Wang, *Design Method of Concrete-Infilled Double Steel Corrugated-Plate Walls with T-section*, Springer Theses,
https://doi.org/10.1007/978-981-95-7789-7_4

4.1 Local Buckling of Steel Corrugated Plates

To consider the effect of the local buckling of steel corrugated plates on the cross-sectional capacity of T-CDSCWs, this section studies the buckling behaviour and the load-carrying capacity of steel corrugated plates. The buckling modes of steel corrugated plates are analysed, and the buckling loads are derived. The stress distribution and the load–displacement relationship of steel corrugated plates are analysed, and the curve of the strength reduction factor is developed based on the normalised slenderness ratio of contact buckling. The upper limit of the normalised slenderness ratio is recommended, below which the effect of the local buckling of steel corrugated plates can be neglected.

4.1.1 Bilateral Buckling

The bilateral buckling of steel corrugated plates with only intermediate bolt constraints is analysed first. Only steel corrugated plates are developed in the finite element model (see Fig. 4.1). The material properties of steel corrugated plates are as follows: the elastic modulus $E_s = 2.06 \times 10^5$ MPa, and the Poisson's ratio $\upsilon_s = 0.3$ (MOHURD and AQSIQ 2017). The boundary conditions of the steel corrugated plate are simple supported. The top and bottom edges are coupled with the reference points located at the centroids, and the reference points are only allowed to move along the z-axis. Vertical loads with the same value but opposite directions are applied at the top and bottom reference points simultaneously. The lateral

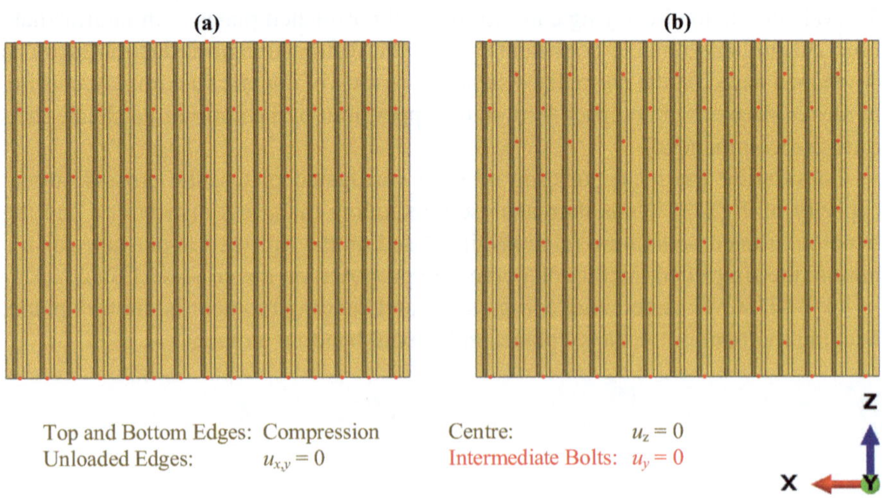

Fig. 4.1 Bilateral buckling finite element models of steel corrugated plates: (**a**) with aligned intermediate bolts; (**b**) with staggered intermediate bolts

Table 4.1 Numerical examples of bilateral buckling analysis of steel corrugated plates

NO.	Wave type	Connector type	Wave number n_w	Wall height h (mm)	Vertical bolt spacing d_v (mm)	Steel plate thickness t_s (mm)
A-1	A	Aligned	5–15	$5d_v$	250–1500	3–7
A-2	A	Staggered	5–15	$5d_v$	250–1500	3–7
A-3	B	Aligned	5–15	$5d_v$	250–1500	3–7
A-4	B	Staggered	5–15	$5d_v$	250–1500	3–7

displacement along the x- and y-axis of unloaded edges is constrained. At the same time, the vertical displacement along the z-axis of the central node of the steel corrugated plate is restrained to avoid rigid body displacement. To simulate the constraining effect of intermediate bolts on the steel corrugated plate, the lateral displacement along the y-axis of the nodes is restrained where intermediate bolts are connected to the steel corrugated plate.

The finite element simulations analyse the effects of geometric parameters, intermediate bolt arrangement, the number of waves n_w, the vertical bolt spacing d_v, and the steel plate thickness t_s on the bilateral buckling behaviour (see Table 4.1). The parameters of numerical examples in Table 4.1 are within the range commonly used in engineering practice. The geometric parameters of waves include types A and B. The parameters of wave type A are: q = 200 mm, q_1 = 100 mm, q_2 = 25 mm, q_3 = 50 mm, a = 25 mm. The parameters of wave type B are: q = 180 mm, q_1 = 85 mm, q_2 = 27.5 mm, q_3 = 40 mm, a = 35 mm. Two kinds of intermediate bolt arrangement are taken into consideration: the aligned and staggered ways. Intermediate bolts are arranged in the middle of the wave trough, and the vertical bolt spacing d_v refers to the distance between two adjacent intermediate bolts in one column. The height of steel corrugated plates is equal to $5d_v$, which is enough to show the constraining effect of intermediate bolts.

The bilateral buckling mode and load of simply supported steel corrugated plates with intermediate bolts can be obtained by the finite element buckling analysis (see Fig. 4.2). For aligned intermediate bolts, steel corrugated plates exhibit column behaviour and the buckling occurs between two rows of intermediate bolts. The wider the steel corrugated plates are, the closer the buckling mode is to that of the axially compressed column between two rows of intermediate bolts. For staggered intermediate bolts, steel corrugated plates exhibit plate behaviour or column behaviour. With relatively small vertical bolt spacing d_v, steel corrugated plates exhibit plate behaviour and the one-wave buckling occurs among four adjacent intermediate bolts. With relatively large vertical bolt spacing d_v, steel corrugated plates exhibit column behaviour and the buckling behaviour is the same as the steel corrugated plates with aligned intermediate bolts of half vertical spacing. That is, staggered intermediate bolts with half number achieves the same constraining effect as aligned intermediate bolts.

According to the bilateral buckling mode, the lateral displacement ω of the bilateral buckling mode can be assumed as Eq. (4.1).

$$\omega = \omega_0 \sin\frac{\pi x}{a_x}\sin\frac{\pi z}{a_z} \tag{4.1}$$

Where ω_0 denotes the amplitude of the lateral displacement. For aligned intermediate bolts, $a_x = n_w d_h$, $a_z = d_v$; for staggered intermediate bolts with small vertical spacing, $a_x = d_h$, $a_z = d_v$, and with large vertical spacing, $a_x = n_w d_h$, $a_z = d_v/2$; d_h denotes the horizontal bolt spacing, which is equal to q in this study.

The assumed bilateral buckling mode shows good agreement with numerical results (see Fig. 4.3). The contour plot illustrates the numerical results of the bilateral buckling mode of steel corrugated pates. The top illustrates the comparison

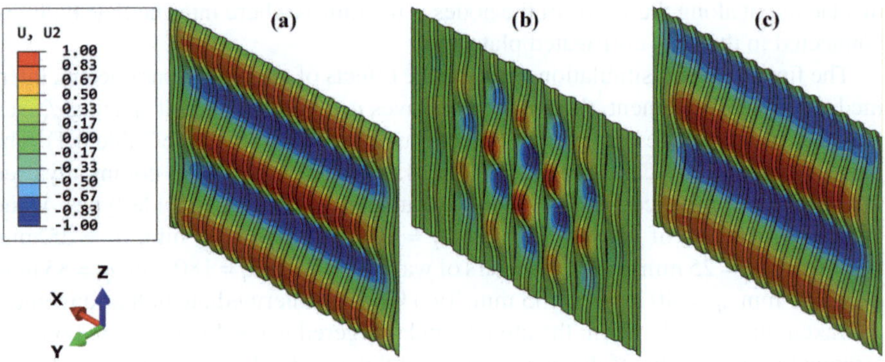

Fig. 4.2 Bilateral buckling modes of steel corrugated plates: (**a**) with aligned intermediate bolts (d_v = 500 mm); (**b**) with staggered intermediate bolts (d_v = 500 mm); (**c**) with staggered intermediate bolts (d_v = 1250 mm)

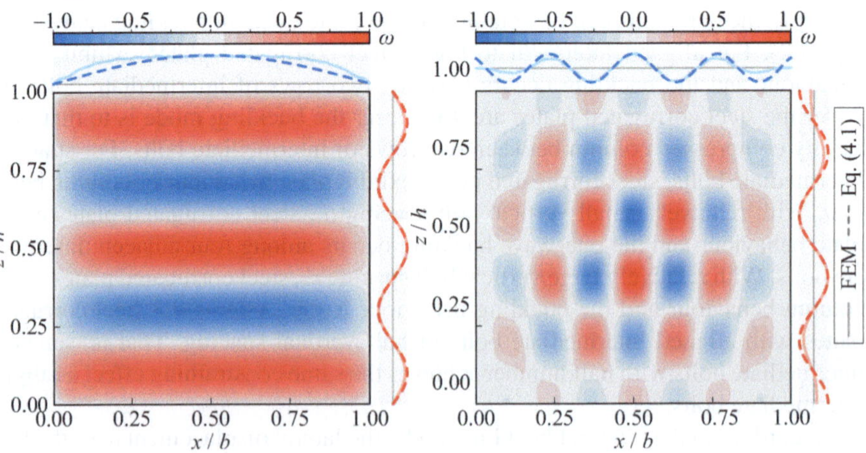

Fig. 4.3 Bilateral buckling mode shapes of steel corrugated plates: (**a**) with aligned intermediate bolts (d_v = 500 mm); (**b**) with staggered intermediate bolts (d_v = 500 mm)

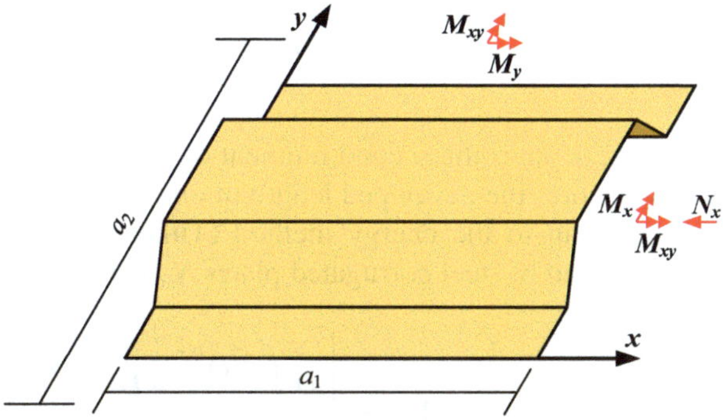

Fig. 4.4 Local coordinate of steel corrugated plates

between Eq. (4.1) and the numerical results of the xy cross-section at $z = h/2$, and the right illustrates the comparison of the yz cross-section at $x = b/2$, respectively. Although there exists some difference between the assumption and numerical results near the boundaries due to boundary conditions, the assumed bilateral buckling mode accords with the main characteristics of the numerical results, especially at the central region of steel corrugated plates.

Based on Eq. (4.1), the bilateral buckling load of steel corrugated plates can be derived according to the elastic buckling theory of orthotropic plates (Timoshenko and Gere 2009; Timoshenko and Woinowsky-Krieger 1959). In conformity to the local coordinate of steel corrugated plates shown in Fig. 4.4, the equilibrium differential equation of steel corrugated plates is expressed as Eq. (4.2).

$$D_x \frac{\partial^4 \omega}{\partial x^4} + 2H \frac{\partial^4 \omega}{\partial x^2 \partial y^2} + D_y \frac{\partial^4 \omega}{\partial y^4} = N_x \frac{\partial^2 \omega}{\partial x^2} \tag{4.2}$$

Where Nx denotes the distributed line load. D_x, D_y, and H denote the stiffness parameters of steel corrugated plates, which can be determined by Eqs. (4.3)–(4.7).

$$D_x = \frac{E_s I_x}{q} \tag{4.3}$$

$$D_y = \frac{q}{q_0} \frac{E_s t_s^3}{12\left(1 - v_s^2\right)} \tag{4.4}$$

$$H = D_1 + 2D_{xy} \tag{4.5}$$

$$D_1 = v_s \sqrt{D_x D_y} \tag{4.6}$$

$$D_{xy}=\frac{q_0}{q}\frac{E_s t_s^3}{12(1+\nu_s)} \tag{4.7}$$

Where Ix denotes the second moment of inertia of one wave about the major axis, and q_0 denotes the developed length of one wave.

According to the energy method (Timoshenko and Gere 2009), the bilateral buckling load of steel corrugated plates N_{xb} can be derived as Eq. (4.8).

$$N_{xb}=\frac{\frac{1}{2}\int_0^{a_1}\int_0^{a_2}\left[\begin{array}{l}D_x\left(\frac{\partial^2\omega}{\partial x^2}\right)^2+D_y\left(\frac{\partial^2\omega}{\partial y^2}\right)^2\\+2D_1\frac{\partial^2\omega}{\partial x^2}\frac{\partial^2\omega}{\partial y^2}+4D_{xy}\left(\frac{\partial^2\omega}{\partial x\partial y}\right)^2\end{array}\right]dxdy}{\frac{1}{2}\int_0^{a_1}\int_0^{a_2}\left(\frac{\partial^2\omega}{\partial x^2}\right)^2 dxdy}=\frac{\pi^2}{a_2^2}\left(D_x\frac{a_2^2}{a_1^2}+2H+D_x\frac{a_1^2}{a_2^2}\right) \tag{4.8}$$

By replacing a_1 and a_2 in the local coordinate with a_z and a_x in the global coordinate, the bilateral buckling stress of steel corrugated plates $\sigma_{cr,b}$ can be expressed as Eq. (4.9).

$$\sigma_{cr,b}=\frac{q}{q_0 t_s}N_{xb}=\frac{\pi^2 q}{a_x^2 q_0 t_s}\left(D_x\frac{a_x^2}{a_z^2}+2H+D_y\frac{a_z^2}{a_x^2}\right) \tag{4.9}$$

To evaluate the bilateral buckling behaviour of steel corrugated plate, a dimensionless variable, the normalised slenderness ratio of bilateral buckling λ_{pb} as defined in Eq. (4.10), is introduced.

$$\lambda_{pb}=\sqrt{\frac{f_y}{\sigma_{cr,b}}} \tag{4.10}$$

Where f_y denotes the yield strength of the steel. Although bilateral buckling analysis is carried out here without considering material nonlinearity, the yield strength f_y takes 345 MPa, the value of Q345 steel (MOHURD and AQSIQ 2003), to compare the values of the normalised slenderness ratio equally. The greater the normalised slenderness ratio λ_{pb}, the less the bilateral buckling stress $\sigma_{cr,b}$.

The normalised slenderness ratio of bilateral buckling λ_{pb} obtained from Eq. (4.10) shows good agreement with the numerical results of $\lambda_{pb,n}$ (see Fig. 4.5). For the numerical examples with aligned intermediate bolts, the deviation of theoretical results from numerical results is small and ranges within ±10%. However, for the

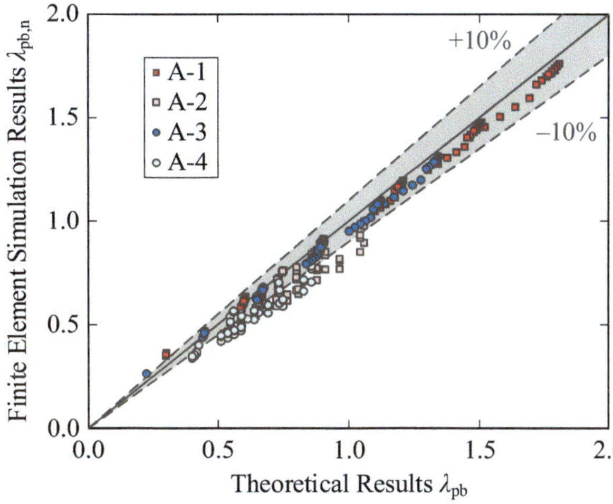

Fig. 4.5 Numerical and theoretical results of the normalised slenderness ratio of bilateral buckling λ_{pb}

numerical examples with staggered intermediate bolts, Eq. (4.9) underestimates the bilateral buckling load of steel corrugated plates. The reason behind the deviation is that steel corrugated plates are simplified to orthotropic plates, and the stiffness parameters are taken as the average values of one wave for the derivation of bilateral buckling load. Such simplification introduces unavoidable deviation since the horizontal bolt spacing d_h is equal to only the breadth of one wave. Generally, the bilateral buckling load of steel corrugated plates with staggered intermediate bolts is large and the normalised slenderness ratio λ_{pb} is less than 1.0. As a result, the instability of steel corrugated plates with staggered intermediate bolts is not as significant as that of steel corrugated plates with aligned intermediate bolts. Furthermore, the overestimation of normalised slenderness ratio λ_{pb} is conservative consideration. Therefore, Eq. (4.9) is able to evaluate the bilateral buckling stress σ_{cr1} of steel corrugated plates with intermediate bolts under compression.

4.1.2 *Contact Buckling*

The eigenvalue buckling analysis is not able to solve the geometric nonlinear problem of the contact between steel corrugated plates and infilled concrete. To investigate the contact buckling behaviour of steel corrugated plates, the finite element model for contact buckling analysis is developed. Typically, two steel corrugated plates are connected by intermediate bolts in the composite structures and concrete is infilled. In the finite element modelling, only half the structure is developed according to the symmetricity about the *zx*-plane to facilitate the finite element analysis (see Fig. 4.6). The material properties of steel corrugated plates are the same as those in the finite element model for bilateral buckling analysis: the elastic modulus $E_s = 2.06 \times 10^5$ MPa, and the Poisson's ratio $\upsilon_s = 0.3$ (MOHURD and AQSIQ 2017). The material properties of

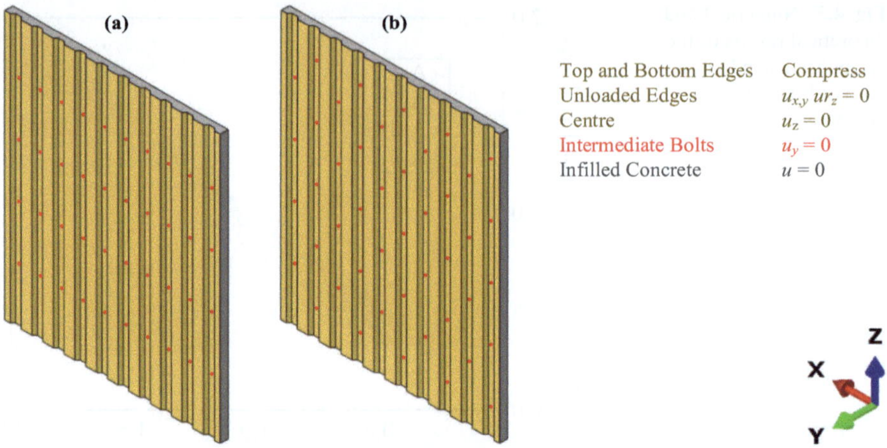

Fig. 4.6 Contact buckling finite element models of steel corrugated plates: (**a**) with aligned intermediate bolts; (**b**) with staggered intermediate bolts

concrete are as follows: the elastic modulus $E_c = 3.25 \times 10^4$ MPa, and the Poisson's ratio $\upsilon_c = 0.2$ (MOHURD and AQSIQ 2010). The contact behaviour between steel corrugated plates and infilled concrete is considered.

Corrugated composite structures generally have boundary elements, so the boundary conditions of the steel corrugated plate are clamped. The top and bottom surfaces of the steel corrugated plate and infilled concrete are coupled to the reference points located at the centre respectively, and the reference points are only allowed to move along the *z*-axis. Vertical loads with the same value but opposite directions are applied at the top and bottom reference points at the same time. For the unloaded surfaces of the steel corrugated plate and infilled concrete, both lateral displacement and rotation about the *z*-axis are restrained. To satisfy the symmetric boundary conditions, the back surface of infilled concrete is restrained to be symmetric about the *zx*-plane, which implies that $u_y = 0$, $ur_x = 0$, and $ur_z = 0$. The vertical displacement along the *z*-axis of the central node of the steel corrugated plate and infilled concrete is restrained to avoid rigid body displacement. To simulate the constraining effect of intermediate bolts on the steel corrugated plate, the lateral displacement along the *y*-axis of the nodes is restrained where intermediate bolts are connected to the steel corrugated plate.

The finite element simulations analyse the effects of geometric parameters, intermediate bolt arrangement, the number of waves n_w, the vertical bolt spacing d_v, and the steel plate thickness t_s on the contact buckling behaviour (see Table 4.2). The finite element analysis is not able to obtain the contact buckling mode if steel corrugated plates are ideally perfect. To obtain the buckling deflection of steel corrugated plates, initial geometric imperfection with small amplitude is introduced. In the construction of composite structures, the placement of infilled concrete causes lateral pressure on steel corrugated plates and results to lateral deflection. Therefore, the initial geometric imperfection of steel corrugated plates is in the form of the deflection under uniform lateral pressure, and the amplitude is equal to $d_v/10000$.

Table 4.2 Numerical examples of contact buckling analysis of steel corrugated plates

NO.	Wave type	Connector type	Wave number n_w	Wall height h (mm)	Vertical bolt spacing d_v (mm)	Steel plate thickness t_s (mm)
B-1	A	Aligned	5–15	$5d_v$	250–1500	3–7
B-2	A	Staggered	5–15	$5d_v$	250–1500	3–7
B-3	B	Aligned	5–15	$5d_v$	250–1500	3–7
B-4	B	Staggered	5–15	$5d_v$	250–1500	3–7

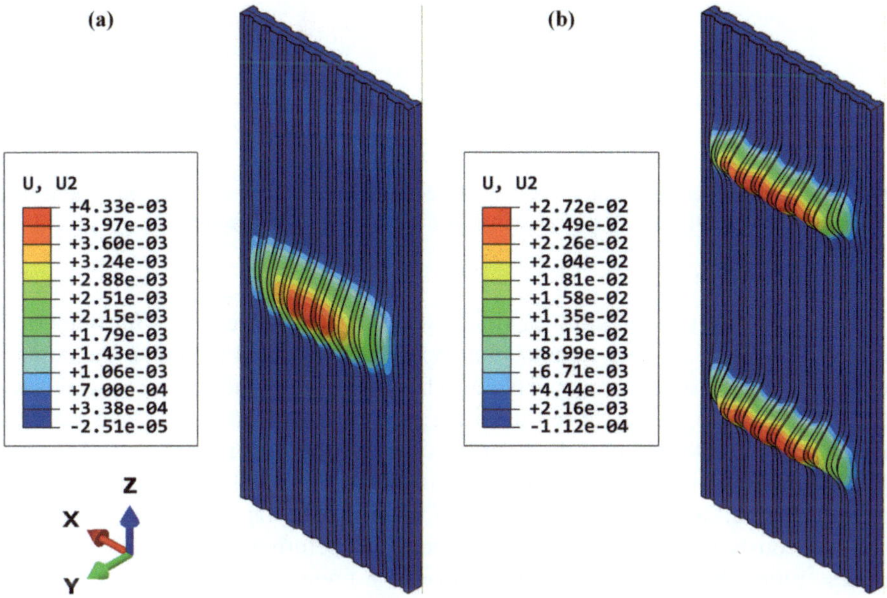

Fig. 4.7 Contact buckling modes of steel corrugated plates (d_v = 1000 mm): (**a**) with aligned intermediate bolts; (**b**) with staggered intermediate bolts

The contact buckling mode and load of steel corrugated plates can be obtained by the finite element static analysis (Fig. 4.7). For most numerical examples, steel corrugated plates with aligned and staggered intermediate bolts have the same buckling mode, the buckling of steel corrugated plates between two rows of intermediate bolts. This further proves the higher restraining efficiency of staggered intermediate bolts. Moreover, the buckling is localised between two rows of intermediate bolts, because the load will degrade and the buckling does not occur in another region.

According to the contact buckling mode, the lateral displacement ω of the bilateral buckling mode can be assumed as Eq. (4.11).

$$\omega = \omega_0 \left(1 - \cos\frac{2\pi x}{a_x}\right)\left(1 - \cos\frac{2\pi z}{a_z}\right) \tag{4.11}$$

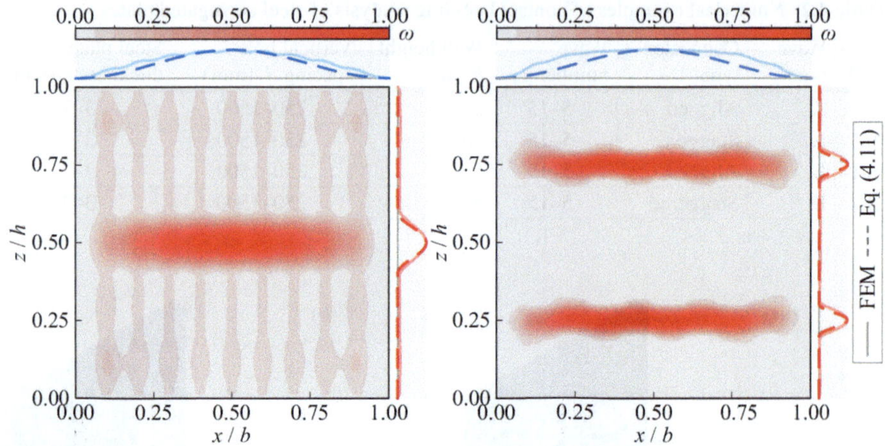

Fig. 4.8 Contact buckling mode shapes of steel corrugated plates: (**a**) with aligned intermediate bolts (d_v = 1000 mm); (**b**) with staggered intermediate bolts (d_v = 1000 mm)

The assumed contact buckling mode show acceptable agreement with numerical results (see Fig. 4.8). The contour plot shows the numerical results of the contact buckling mode of steel corrugated plates. The top illustrates the comparison between Eq. (4.11) and the numerical results of the xy cross-section at $z = h/2$, and the right illustrates the comparison of the yz cross-section at $x = b/2$, respectively. The constraining effects of intermediate bolts are assumed in Eq. (4.11) as clamped boundary conditions, which is not fully satisfied since the buckling does not occur simultaneously in adjacent region. Therefore, the deformation of steel corrugated plates around the boundaries obtained from the finite element analysis are larger than assumed deflection in Eq. (4.11). However, the assumed contact buckling mode accords with the main characteristics of the numerical results, especially along the height.

Based on Eq. (4.11), the contact buckling load of steel corrugated plates N_{xc} can be derived as Eq. (4.12)

$$N_{xc} = \frac{4\pi^2}{a_2^2}\left(D_x \frac{a_2^2}{a_1^2} + \frac{2}{3} H + D_y \frac{a_1^2}{a_2^2} \right) \tag{4.12}$$

By replacing a_1 and a_2 in the local coordinate with a_z and a_x in the global coordinate, the contact buckling stress of steel corrugated plates $\sigma_{cr,c}$ can be expressed as Eq. (4.13).

$$\sigma_{cr,c} = \frac{q}{q_0 t_s} N_{xc} = \frac{4\pi^2 q}{a_x^2 q_0 t_s}\left(D_x \frac{a_x^2}{a_z^2} + \frac{2}{3} H + D_y \frac{a_z^2}{a_x^2} \right) \tag{4.13}$$

To evaluate the contact buckling behaviour of steel corrugated plate, a dimensionless variable, the normalised slenderness ratio of bilateral buckling λ_{pc} as defined in Eq. (4.14), is introduced.

$$\lambda_{pc} = \sqrt{\frac{f_y}{\sigma_{cr,c}}} \tag{4.14}$$

Where f_y denotes the yield strength of the steel. Although contact buckling analysis is carried out here without considering material nonlinearity, the yield strength f_y takes 345 MPa, the value of Q345 steel (MOHURD and AQSIQ 2003), to compare the values of the normalised slenderness ratio equally. The greater the normalised slenderness ratio λ_{pc}, the less the contact buckling stress $\sigma_{cr,c}$.

Compared with the normalised slenderness ratio proposed by Zhu (2020) and Zhu et al. (2019), Eq. (4.14) does not need to introduce correction coefficients, which lack physical significance and are merely applicable to a certain wave type or intermediate bolt arrangement. This makes Eq. (4.14) more applicable. Within the common range, the stiffness parameters H and D_y are much less than D_x. Therefore, with the same buckling length a_z, the contact buckling stress of steel corrugated plates $\sigma_{cr,c}$ is approximately four times as the bilateral buckling stress $\sigma_{cr,b}$, which implies that the normalised slenderness ratio of contact buckling λ_{pc} is approximately half of the normalised slenderness ratio of bilateral buckling λ_{pb}. As mentioned above, the constraining effects of intermediate bolts and infilled concrete on steel corrugated plates are not as strong as the ideal clamped boundary conditions, so Eq. (4.13) overestimates the contact stress $\sigma_{cr,c}$, thereby underestimating the normalised slenderness ratio of contact buckling λ_{pc} (see Fig. 4.9). Generally, the normalised slenderness ratio of contact buckling λ_{pc} is less than 1.0. From the perspective of convenience, the boundary conditions of steel corrugated

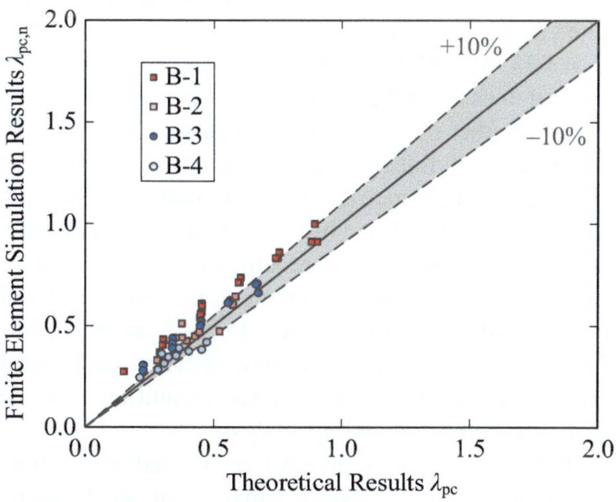

Fig. 4.9 Numerical and theoretical results of the normalised slenderness ratio of contact buckling λ_{pc}

Table 4.3 Numerical examples of load-carrying analysis of steel corrugated plates

NO.	Wave type	Connector type	Wave number n_w	Wall height h (mm)	Vertical bolt spacing d_v (mm)	Steel plate thickness t_s (mm)
C-1	A	Aligned	5–15	$5d_v$	250–1500	3–7
C-2	A	Staggered	10	$5d_v$	250–1500	3–7
C-3	B	Aligned	5–15	$5d_v$	250–1500	3–7
C-4	B	Staggered	10	$5d_v$	250–1500	3–7

plates between two rows of intermediate bolts are still treated as clamped, and the deviation will be considered in the following formulas of the strength reduction factor.

4.1.3 *Stress Distribution, Load–Displacement Curve, and Design Method of Load-Carrying Capacity*

In the analysis of bilateral and contact buckling, the material is linearly elastic, while in the structures, the buckling load of steel corrugated plates is so large that the stress may reach the yield strength under compression. Meanwhile, there exists initial imperfection on steel corrugated plates, including the initial imperfection caused by manufacture and the lateral deflection caused by the placement of infilled concrete. Therefore, the evaluation of load-carrying capacity should take into consideration the effects of material and geometric nonlinearity on steel corrugated plates.

Based on the finite element model for contact buckling analysis, the finite element model for load-carrying analysis is developed. The plasticity is introduced according to the material properties in Chap. 3. The material properties of steel are as follows: the yield strength f_y = 235 MPa or 345 MPa, the elastic modulus $E_s = 2.06 \times 10^5$ MPa, and the Poisson's ratio $\upsilon_s = 0.3$ (MOHURD and AQSIQ 2017). The material properties of concrete are as follows: the elastic modulus $E_c = 3.25 \times 10^4$ MPa, and the Poisson's ratio $\upsilon_c = 0.2$ (MOHURD and AQSIQ 2010). Since it does not influence on the load–displacement relationship of steel corrugated plates whether infilled concrete is elastic or inelastic (Song et al. 2019), the infilled concrete is still linearly elastic to improve the efficiency of the finite element analysis. The initial geometric imperfection of steel corrugated plates is in the form of the lateral deflection caused by uniform pressure, and the amplitude of the imperfection is $d_v/750$. The finite element simulations analyse the effects of geometric parameters, intermediate bolt arrangement, the number of waves n_w, the vertical bolt spacing d_v, and the steel plate thickness t_s on the contact buckling behaviour (see Table 4.3). The contact behaviour between steel corrugated plates and infilled concrete is considered. The boundary conditions are the same as the finite element model for contact buckling analysis.

The failure modes of steel corrugated plates are related to intermediate bolt arrangement. The stress distribution of steel corrugated plates (see Fig. 4.10)

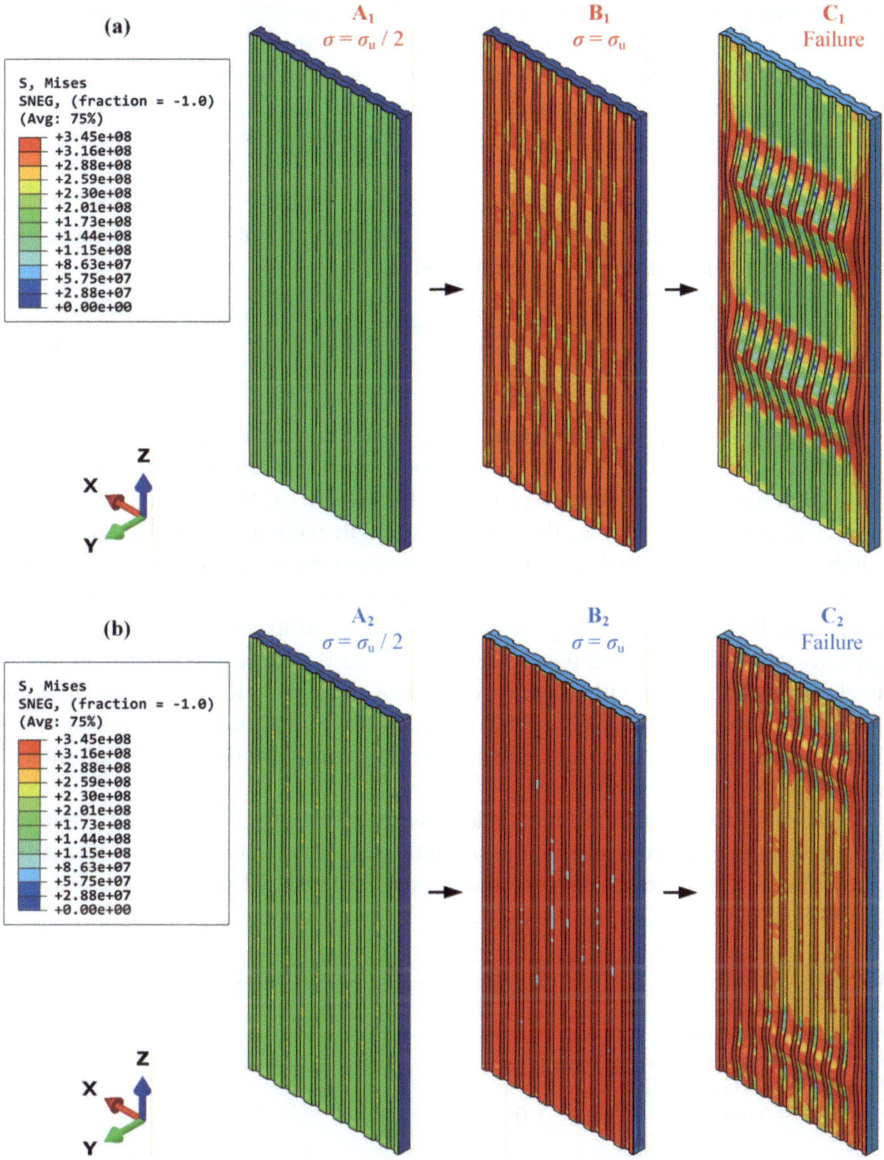

Fig. 4.10 Stress distribution (unit of stress: Pa) and failure modes (lateral deformation magnified by five times) of typical numerical examples (d_v = 1000 mm): (**a**) with aligned intermediate bolts; (**b**) with staggered intermediate bolts

corresponds to the points on the vertical load–displacement curves (see Figs. 4.11 and 4.13). For aligned intermediate bolts, steel corrugated plates are restrained from inward buckling and deflect outwards, while for most numerical examples with staggered intermediate bolts, the failure modes of steel corrugated plates are all close to full-section plasticity and the peak compressive stress σ_u is close to the yield strength f_y.

The load-carrying capacity of steel corrugated plates can be evaluated by the strength reduction factor φ_s, which is defined as the ratio of the peak compressive stress σ_u to the yield strength f_y as expressed in Eq. (4.15).

$$\varphi_s = \frac{\sigma_u}{f_y} \tag{4.15}$$

The effects of geometric parameters on the load-carrying capacity of steel corrugated plates are different. It is the vertical bolt spacing d_v and the steel plate thickness t_s that have large influence (see Figs. 4.11, 4.12, 4.13, 4.14, 4.15, and 4.16): the greater d_v and the smaller t_s, the less the strength reduction factor φ_s. The wave number n_w only has a small effect on the steel corrugated plates with aligned intermediate bolts: the larger n_w, the closer the steel corrugated plates behave like columns and the less φ_s. For the geometric parameters of the wave, the load-carrying capacity of steel corrugated plates with wave type B is always greater than that of steel corrugated plates with wave type A when other parameters are remained the same. This is because wave type B has a greater wave amplitude a and thus a greater out-of-plane stiffness; therefore, the stability of wave type B is better than that of wave type A.

According to the normalised slenderness ratio of contact buckling λ_{pc} obtained from the finite element analysis, the strength reduction factor curve of steel corrugated plates is proposed as expressed in Eqs. (4.16) and (4.17).

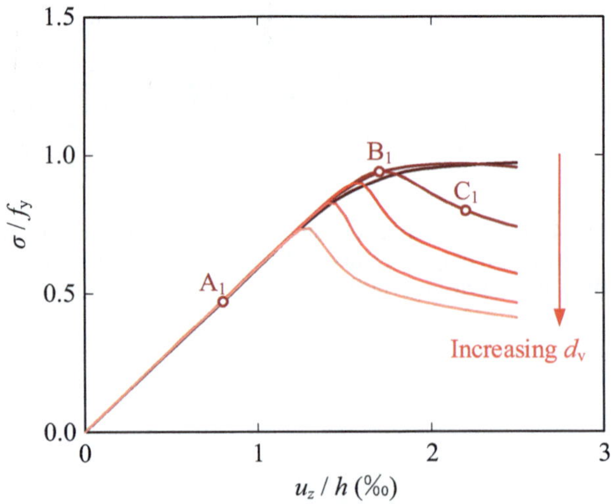

Fig. 4.11 Vertical load–displacement relationship of typical numerical examples (f_y = 345 MPa): Group C-1, n_w = 10, t_s = 5 mm

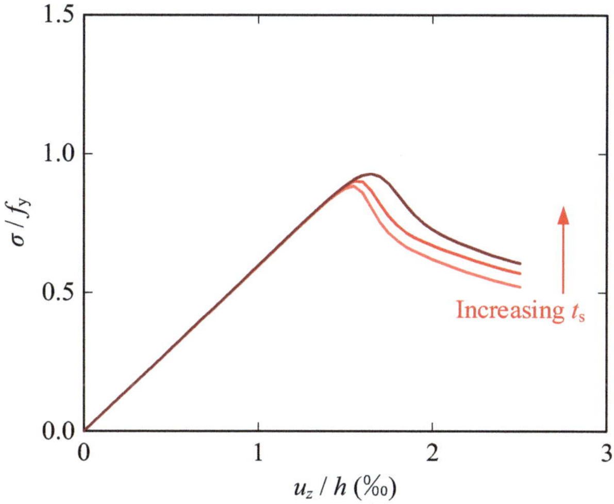

Fig. 4.12 Vertical load–displacement relationship of typical numerical examples (f_y = 345 MPa): Group C-1, n_w = 10, d_v = 1000 mm

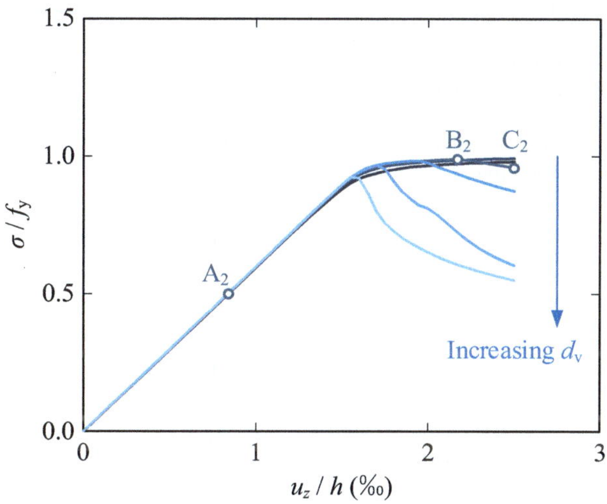

Fig. 4.13 Vertical load–displacement relationship of typical numerical examples (f_y = 345 MPa): Group C-2, n_w = 10, t_s = 5 mm

$$\varphi_s = \frac{1}{\Phi + \sqrt{\Phi^2 - \lambda_{pc}^2}} \tag{4.16}$$

$$\Phi = 0.5\left(1 + \alpha\lambda_{pc} + \lambda_{pc}^2\right) \tag{4.17}$$

Where α is the coefficient that effects the curve shape: for the lower envelop curve, $\alpha = 0.207$; for the average curve, $\alpha = 0.109$; for the upper envelop curve, $\alpha = 0.0426$. Considering the range of the parameters, Eq. (4.16) is applicable with the normalised slenderness ratio λ_{pc} no greater than 1.0.

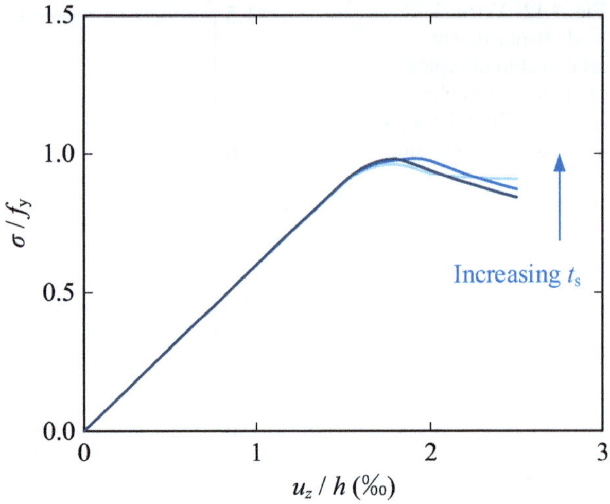

Fig. 4.14 Vertical load–displacement relationship of typical numerical examples (f_y = 345 MPa): Group C-2, n_w = 10, d_v = 1000 mm

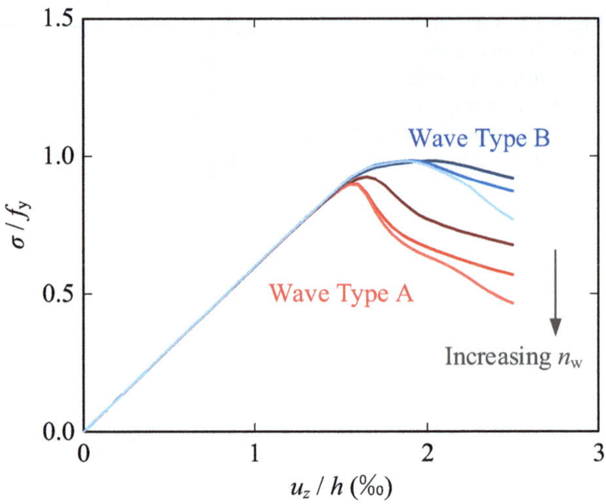

Fig. 4.15 Vertical load–displacement relationship of typical numerical examples (f_y = 345 MPa): aligned intermediate bolts, d_v = 1000 mm, t_s = 5 mm

According to the numerical results and the design curve (see Figs. 4.17 and 4.18), Eq. (4.16) can well evaluate the relationship of φ_s–λ_{pc}. In comparison with the design formulas proposed by Zhu (2020) and Zhu et al. (2019), Eq. (4.16) is based on the buckling theory of orthotropic plates and consider the effect of boundary conditions on the normalised slenderness ratio of contact buckling, which makes the design curve lie below the Euler curve and have clear physical significance.

With proper design parameters, the normalised slenderness ratio of steel corrugated plates λ_{pc} can be small enough that the strength reduction factor φ_s approaches to 1.0, and steel corrugated plates can be classified as compact section. According to the principle of φ_s no less than 0.95, steel corrugated plates can be treated as

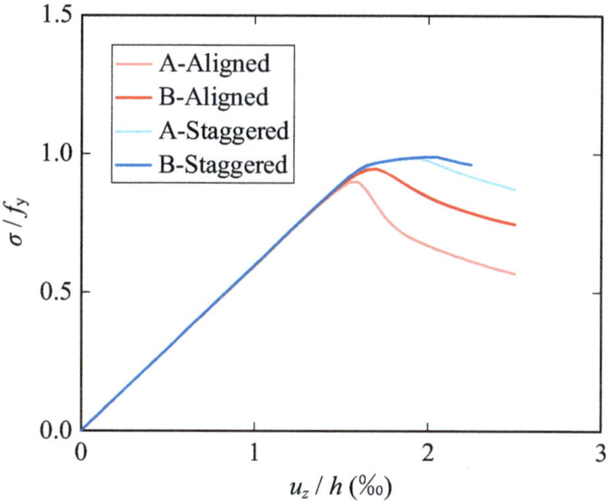

Fig. 4.16 Vertical load–displacement relationship of typical numerical examples (f_y = 345 MPa): n_w = 10, d_v = 1000 mm, t_s = 5 mm

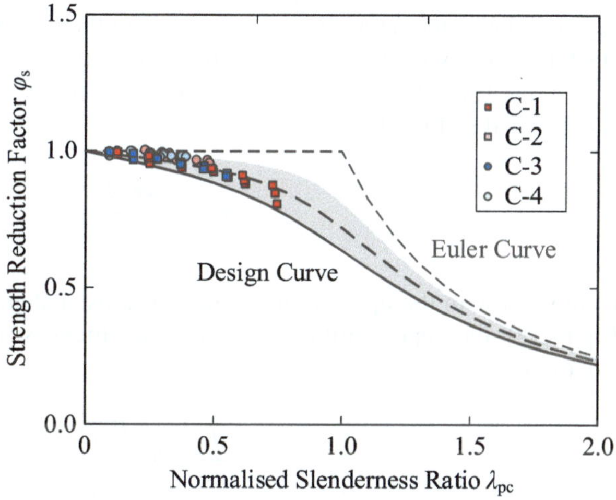

Fig. 4.17 Numerical results of the strength reduction factor and the normalised slenderness ratio of contact buckling φ_s–λ_{pc}: f_y = 235 MPa

full-section plasticity and the normalised slenderness ratio λ_{pc} should satisfy the requirement in Eq. (4.18) combined with the average design curve.

$$\lambda_{pc} \le 0.4 \tag{4.18}$$

From Eq. (4.18), the upper limit of the normalised slenderness ratio λ_{pc} of compact section can be derived according to the geometric parameters of the wave. Take the steel corrugated plate with wave type B, staggered intermediate bolts and t_s = 4 mm as an example, the limit value of the spacing-to-thickness ratio for steel corrugated plate is $181\varepsilon_k$, which is much greater than the limit value for steel flat plate.

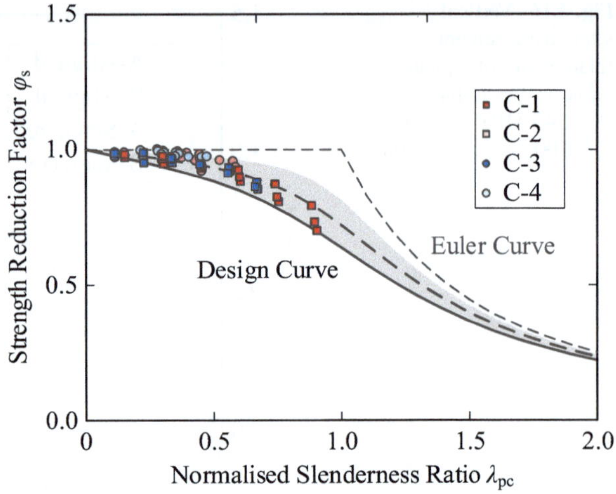

Fig. 4.18 Numerical results of the strength reduction factor and the normalised slenderness ratio of contact buckling φ_s–λ_{pc}: f_y = 345 MPa

Although Eq. (4.18) is derived according to the buckling behaviour of steel corrugated plates, the upper limit also applies to the design of steel flat plates. For the steel flat plates with aligned intermediate bolts, the contact buckling load is approximately the Euler load of an axially loaded column, and the contact buckling stress is expressed as Eq. (4.19).

$$\sigma_{cr,c} = \frac{4\pi^2 E_s t_s^2}{12 d_v^2} \tag{4.19}$$

Combined with the upper limit of the normalised slenderness ratio λ_{pc} proposed in Eq. (4.18), the upper limit of the spacing-to-thickness ratio d_v/t_s can be obtained as Eq. (4.20).

$$\frac{d_v}{t_s} \leq 20\sqrt{\frac{235}{f_y}} \tag{4.20}$$

Eq. (4.20) is close to the upper limit of the spacing-to-thickness ratio obtained from the existing experimental results (Yang et al. 2016; Huang and Liew 2016).

Above all, Eq. (4.16) can well evaluate the load-carrying capacity of the steel corrugated plates constrained by infilled concrete and intermediate bolts under compression. The upper limit of the normalised slenderness ratio λ_{pc} in Eq. (4.20) forms an accurate basis for the design of the spacing-to-thickness ratio to ensure the composite effect of CDSCWs. The design of compressive strength provides fundamental for the design of steel corrugated plates under compression and bending moment. In the following analysis of load-carrying capacities, when the normalised slenderness ratio λ_{pc} is greater than the upper limit 0.4, the effect of the local buckling of steel corrugated plates is considered according to Eq. (4.16).

4.2 Cross-Sectional Capacity under Axial Compression and Bending Moment in the Web Plane

T-CDSCWs are mainly subjected to bending moment in the web plane. Since the T-section is not symmetric about the *y*-axis, T-CDSCWs exhibit different moment capacities with the flange wall under compression and tension. The load-carrying performance of T-CDSCWs under bending moment in the web plane will be investigated, including the stress distribution and the bending moment–curvature relationship. Based on the numerical results and theoretical analyses, the interaction equations will be proposed corresponding to different bending directions.

4.2.1 *Stress Distribution and Bending Moment–Curvature Curve*

Based on the solid–shell finite element model, the cross-sectional capacity of T-CDSCWs under axial compression and bending moment in the web plane is analysed. The material properties of steel are as follows: the yield strength $f_y = 345$ MPa, the elastic modulus $E_s = 2.06 \times 10^5$ MPa, and the Poisson's ratio $\upsilon_s = 0.3$ (MOHURD and AQSIQ 2017). The material properties of concrete are as follows: the axial compressive strength $f_c = 26.8$ MPa, the elastic modulus $E_c = 3.25 \times 10^4$ MPa, and the Poisson's ratio $\upsilon_c = 0.2$ (MOHURD and AQSIQ 2010). The other material properties are the same as Chap. 3. The initial geometric imperfection of steel corrugated plates is in the form of the lateral deflection caused by uniform pressure, and the amplitude of the imperfection is $d_v/750$.

The boundary condition of the T-CDSCW is pin ended. The top and bottom sections are coupled with the reference points at the centroid, which is weighted with the elastic modulus of the materials. The T-CDSCW is loaded at the reference points. The displacement and the rotation about the *z*-axis of the bottom reference point are restrained. The displacement along *x*- and *y*-axis of the top reference point are restrained. The axial compression N is first applied at the top reference point, and then the bending moment M_x in the web plane is applied at both reference points. The stress distribution and the bending moment–curvature relationship can be obtained by the finite element static analysis.

The axial compression N is applied to the T-CDSCW according to the axial compression ratio n as defined in Eq. (4.21).

$$n = \frac{N}{N_u} \tag{4.21}$$

Where N_u denotes the ultimate compression capacity of the T-CDSCW, which is evaluated by Eq. (4.23).

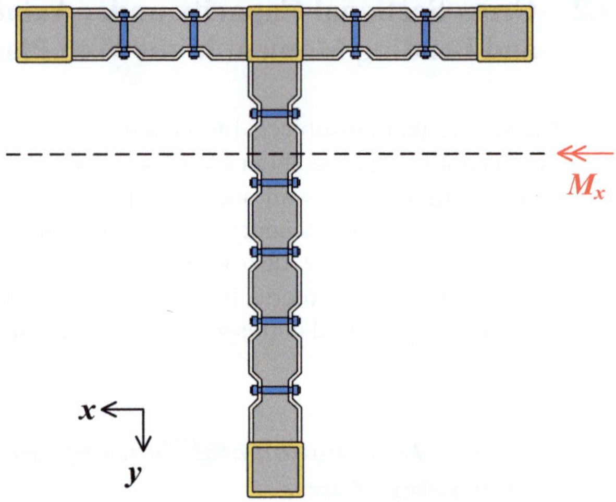

Fig. 4.19 T-CDSCW under bending moment in the web plane

Table 4.4 Numerical examples of load-carrying analysis of T-CDSCWs

NO.	Flange breadth b_f (mm)	Web breadth b_w (mm)	Vertical bolt spacing d_v (mm)	Steel plate thickness t_s (mm)	Axial compression ratio n
D-1	400–1200	1200	250–1500	5	0–0.75
D-2	1200	400–1200	250–1500	5	0–0.75
D-3	1200	1200	250	3–7	0–0.75
D-4	4800	4800	250	5	0–0.75

When the bending moment makes the flange under compression (see Fig. 4.19), the bending direction is along the positive x-axis, and bending moment in this direction is defined as positive. Otherwise, when the bending moment makes the flange under tension, the bending direction is along the negative x-axis, and bending moment in this direction is defined as negative.

The finite element simulations analyse the effects of the flange breadth b_f, the web breadth b_w, the vertical bolt spacing d_v and the steel plate thickness t_s on the cross-sectional capacity of T-CDSCWs (see Table 4.4). The wall height H_w is the maximum of 3 m and $3d_v$. The wall thickness d_w is 150 mm. The parameters of steel corrugated plates are: q = 200 mm, q_1 = 100 mm, q_2 = 25 mm, q_3 = 50 mm, a = 25 mm. The dimensions of the CFST $b_c \times d_c$ are 160 × 160 mm, and the steel tube thickness t_c is 4 mm greater than t_s. Intermediate bolts are arranged in aligned and staggered ways. The minimum vertical bolt spacing d_v is 250 mm, which is so small that steel corrugated plates do not buckle before yielding, and the local buckling of steel corrugated plates does not influence the cross-sectional capacity of T-CDSCWs. The maximum vertical bolt spacing d_v is 1500 mm, which is a very large value in practice and sufficiently large to consider the local buckling. The numerical examples cover a broad range and consider the local buckling and yielding of steel corrugated plates.

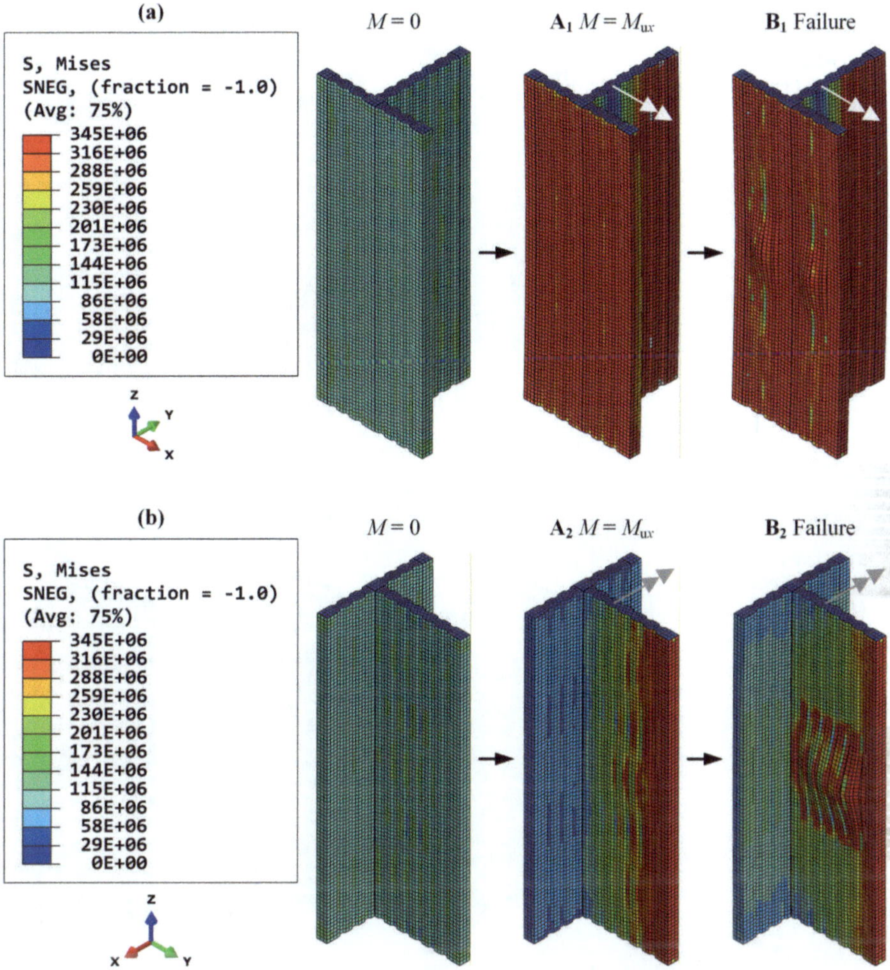

Fig. 4.20 Stress distribution (unit of stress: Pa) and failure modes of typical numerical examples (with aligned intermediate bolts, d_v = 1500 mm, n = 0.5): (**a**) positive bending moment; (**b**) negative bending moment

Under bending moment in different directions, T-CDSCWs exhibit different stress distribution. The stress distribution (see Figs. 4.20 and 4.21) corresponds to the points on the bending moment–curvature curves (see Figs. 4.26 and 4.27). Under positive bending moment, the steel located at the edge of the flange and web walls has already yielded when the ultimate bending moment is reached, and the stress distribution is close to the full-section plastic stress distribution. Under negative bending moment, the steel at the edge of the web wall has yielded when the ultimate bending moment is reached, but the steel at the flange wall has not yet yielded and the normal stress is even compressive when the axial compression ratio n is great. This is due to the large area of the flange wall, which makes the centroid of the cross-section located close to the flange wall. Under positive bending moment, the area of concrete under compression increases as the axial compression ratio n

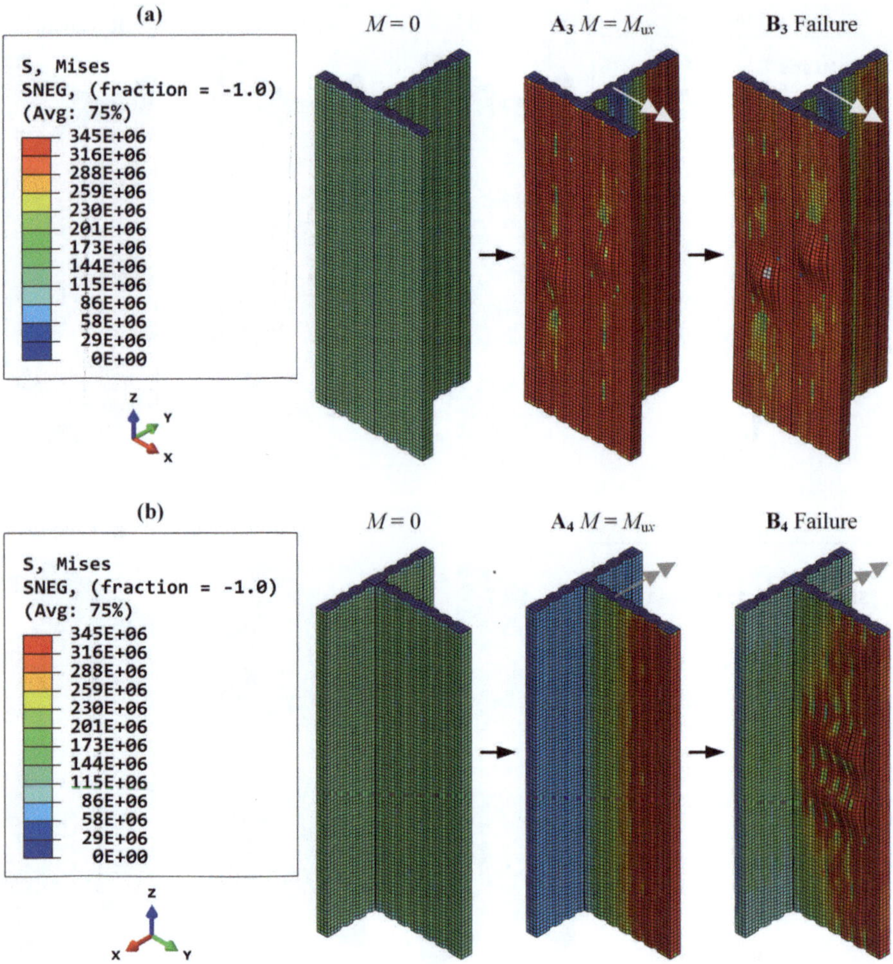

Fig. 4.21 Stress distribution (unit of stress: Pa) and failure modes of typical numerical examples (with staggered intermediate bolts, $d_v = 1500$ mm, $n = 0.5$): (**a**) positive bending moment; (**b**) negative bending moment

increases, and the position of the neutral axis gradually moves from the side of the flange wall to the side of the web edge, and the steel on both sides yields when the ultimate bending moment is reached. Under negative bending moment, the increase of the axial compression ratio n makes the neutral axis constantly close to the flange wall, and the distance between the steel corrugated plate in the flange wall and the neutral axis becomes smaller and smaller, thus the steel in the flange wall is under compression even at the ultimate state.

The bolt arrangement has a significant effect on the buckling performance of steel corrugated plates (see Figs. 4.20 and 4.21), and the staggered arrangement of bolts is more conducive to enhance the load-carrying capacity of steel corrugated

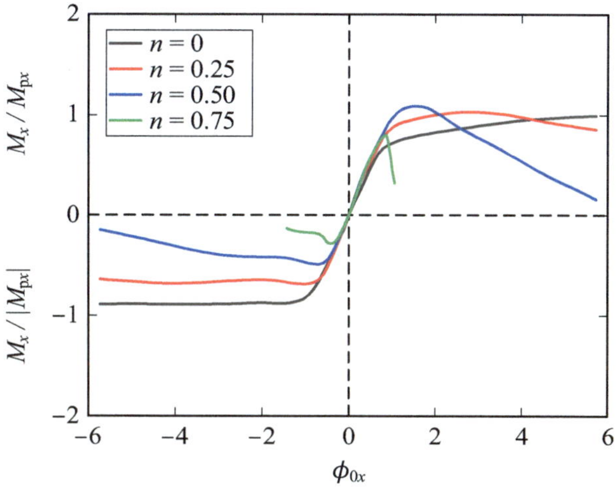

Fig. 4.22 Bending moment–curvature relationship of typical numerical examples: with aligned intermediate bolts, $b_f = 1200$ mm, $b_w = 400$ mm, $d_v = 1500$ mm

plates. Under bending moment in different directions, the compression zone of the cross-section is different, and, obviously, local buckling only occurs to the steel corrugated plates in the compression zone. Under positive bending moment, the flange wall is compressed, so the steel corrugated plates at the flange wall buckle locally; while, under the negative bending, the web wall is compressed, so the steel corrugated plates at the web wall buckle locally. Whether local buckling occurs at the flange or web walls, steel corrugated plates always deflect outward between two adjacent rows of bolts. This means that the constraining effect of staggered bolts is equal to that of aligned bolts with half the vertical bolt spacing.

T-CDSCWs exhibit different bending moment–curvature curves in different bending directions (see Figs. 4.22, 4.23, 4.24, 4.25, 4.26, and 4.27). The ordinate shows the normalised bending moment, the bending moment M_x divided by the plastic moment capacity M_{px} in corresponding bending direction, which will be given by Eq. (4.23). The abscissa shows the normalised curvature ϕ_{0x}, which is given by Eq. (4.22).

$$\phi_{0x} = \frac{\phi_x}{2\varepsilon_y / B_w} \tag{4.22}$$

Where ϕ_x denotes the curvature of the T-CDSCW about the x-axis, ε_y denotes the yield strain of steel, and B_w denotes the overall breadth of the web wall.

T-CDSCWs exhibit different trends of the cross-sectional capacities in different bending directions. With the increase of the axial compression ratio n, the positive moment capacity first increases and then decreases, while the negative moment capacity continues to decrease. With the bending direction remaining the same, the bending moment of a cross-section about the centroidal axis under axial compression reaches the maximum when the neutral axis passes through the centroid. For positive bending, with the initial increase of the axial compression ratio n, the

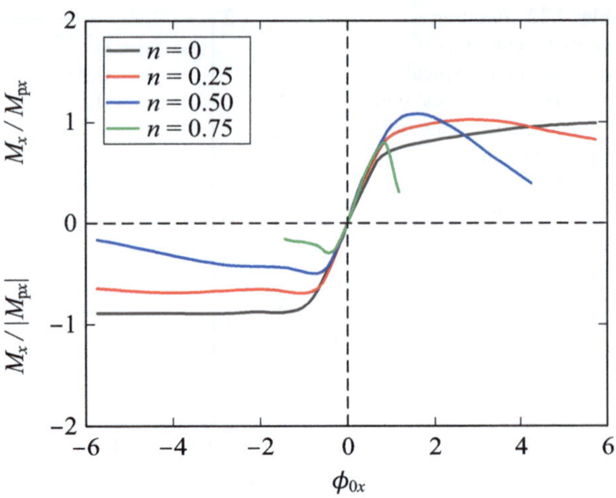

Fig. 4.23 Bending moment–curvature relationship of typical numerical examples: with staggered intermediate bolts, $b_f = 1200$ mm, $b_w = 400$ mm, $d_v = 1500$ mm

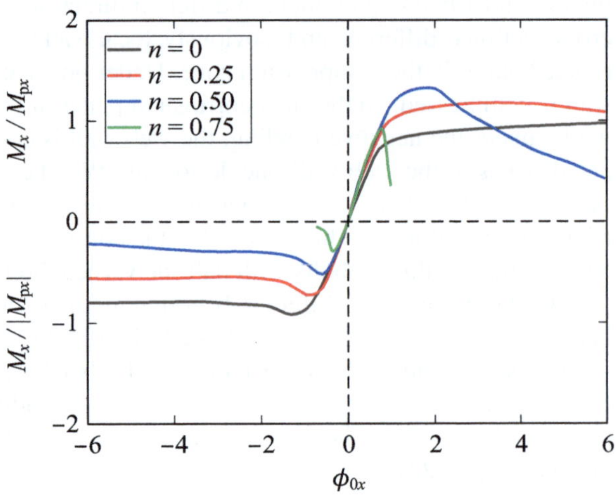

Fig. 4.24 Bending moment–curvature relationship of typical numerical examples: with aligned intermediate bolts, $b_f = 1200$ mm, $b_w = 800$ mm, $d_v = 1500$ mm

neutral axis moves from the side of the flange wall towards the centroid. When the neutral axis arrives at the centroid, the positive moment capacity reaches the maximum. With the further movement of the neutral axis towards the side of the web, the positive moment capacity begins to decrease. Until the neutral axis moves to infinitely far, the moment capacity vanishes under pure compression. For pure negative bending, the neutral axis is located near the flange wall. The increase of the axial compression ratio n causes the neutral axis to constantly move away from the centroid, and the moment capacity constantly decreases.

The curvature corresponding to the maximum moment capacity decreases continuously with the increase of the axial compression ratio n. This is because, under axial compression and bending moment, the compressed extreme fibre has the maximum compressive strain. As the axial compression ratio n increases, the distance between the compressed extreme fibre and the neutral axis increases, and, as a result, the curvature decreases.

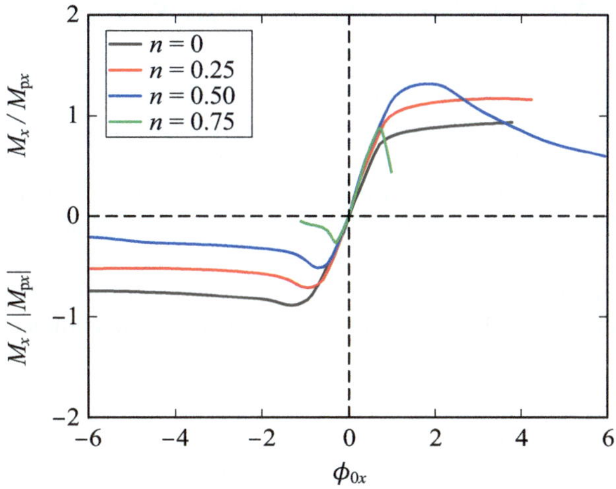

Fig. 4.25 Bending moment–curvature relationship of typical numerical examples: with staggered intermediate bolts, $b_f = 1200$ mm, $b_w = 800$ mm, $d_v = 1500$ mm

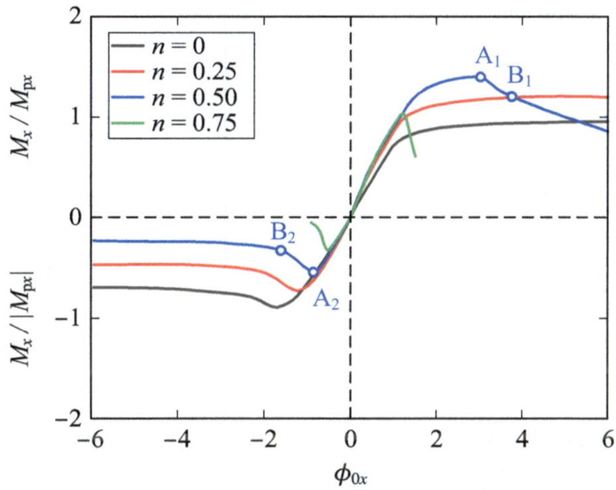

Fig. 4.26 Bending moment–curvature relationship of typical numerical examples: with aligned intermediate bolts, $b_f = 1200$ mm, $b_w = 1200$ mm, $d_v = 1500$ mm

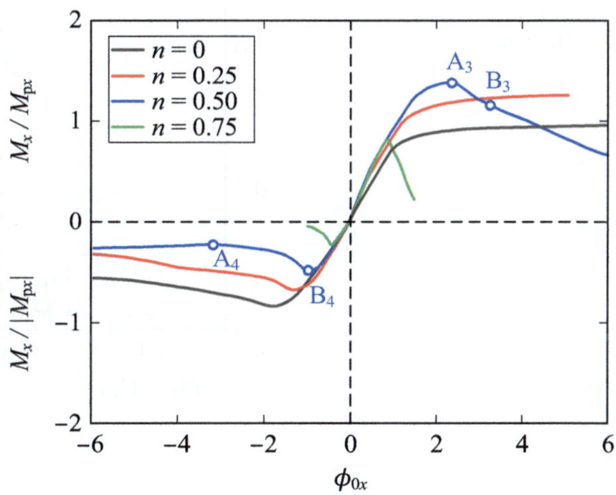

Fig. 4.27 Bending moment–curvature relationship of typical numerical examples: with staggered intermediate bolts, $b_f = 1200$ mm, $b_w = 1200$ mm, $d_v = 1500$ mm

4.2.2 Design Method of Cross-Sectional Capacity

The cross-sectional capacity of T-CDSCWs obtained by the finite element analysis can be used as the basis for the design formula. The interaction curves of T-CDSCWs under bending moment in the web plane reflect the effects of geometric parameters, such as web breadth b_w, steel plate thickness t_s, and vertical bolt spacing d_v (see Figs. 4.28, 4.29, 4.30, 4.31, 4.32, 4.33, 4.34, and 4.35). Overall, the interaction curves of N–positive M_x bulge, and the degree of the bulge is influenced by the geometric parameters; the interaction curves of N–negative M_x are linearly distributed and basically not influenced by the geometric parameters.

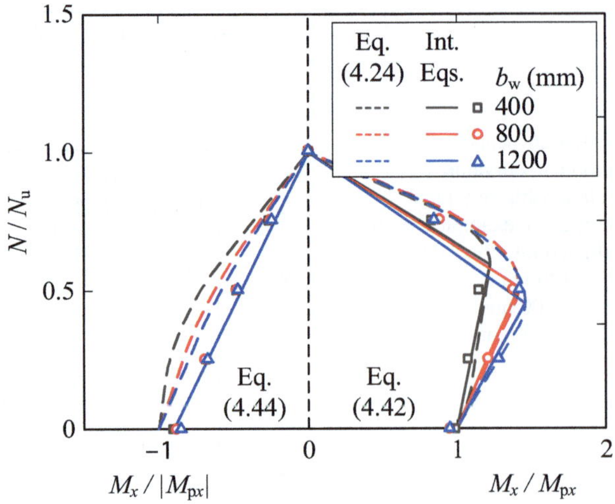

Fig. 4.28 Interaction curves of typical numerical examples: with aligned intermediate bolts, $b_f = 1200$ mm, $d_v = 250$ mm

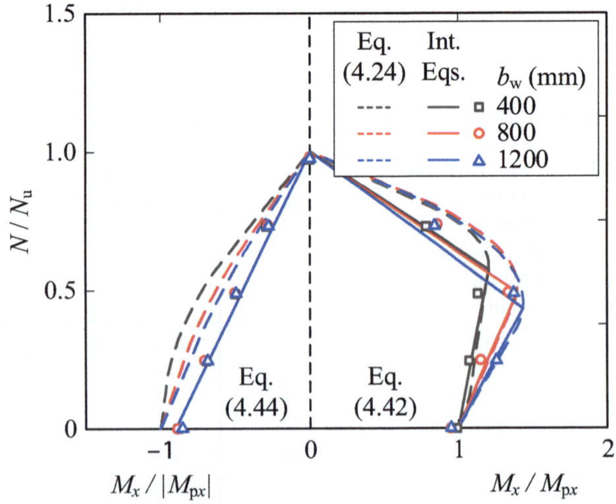

Fig. 4.29 Interaction curves of typical numerical examples: with aligned intermediate bolts, $b_f = 1200$ mm, $d_v = 1000$ mm

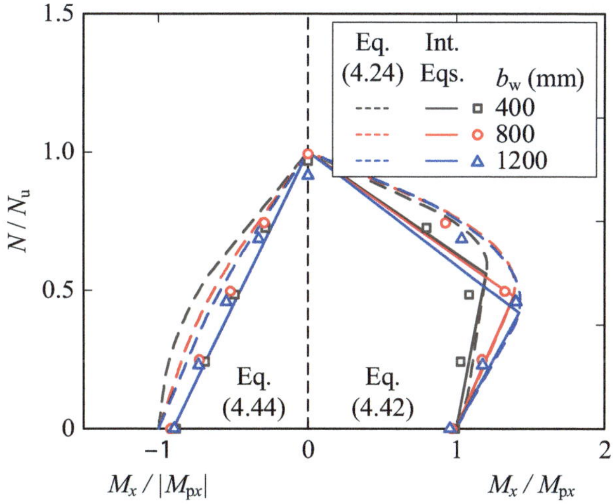

Fig. 4.30 Interaction curves of typical numerical examples: with aligned intermediate bolts, $b_f = 1200$ mm, $d_v = 1500$ mm

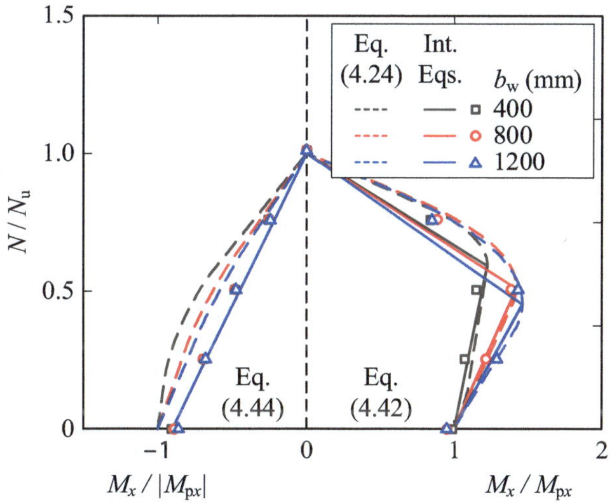

Fig. 4.31 Interaction curves of typical numerical examples: with staggered intermediate bolts, $b_f = 1200$ mm, $d_v = 250$ mm

The web breadth b_w has a significant effect on the shape of the interaction curves. The interaction curves with the same web breadth b_w are essentially the same and are not significantly influenced by the vertical bolt spacing d_v and arrangement of intermediate bolts, although the variation of the intermediate bolts can influence the local buckling performance of steel corrugated plates. A greater web breadth b_w contributes to a more significant bulge of the positive interaction curves, and a smaller axial compression corresponding to the maximum moment capacity. This is because the increase of the web breadth b_w makes the compression zone expand. When the neutral axis moves from the original position under pure bending moment to the centroid, the movement causes a larger relative change of the bending moment M_x while a smaller relative change of the axial compression N.

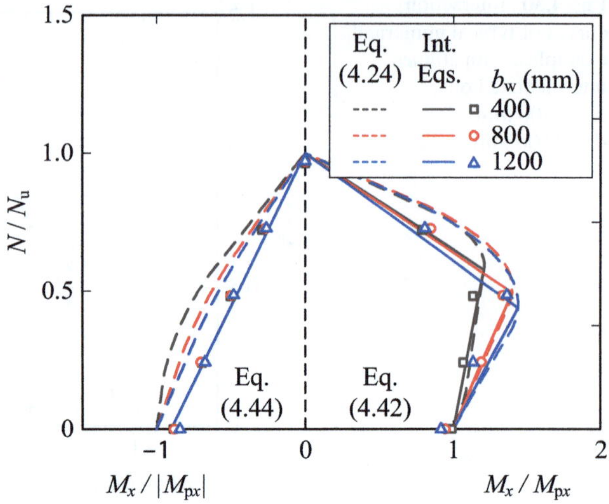

Fig. 4.32 Interaction curves of typical numerical examples: with staggered intermediate bolts, $b_f = 1200$ mm, $d_v = 1000$ mm

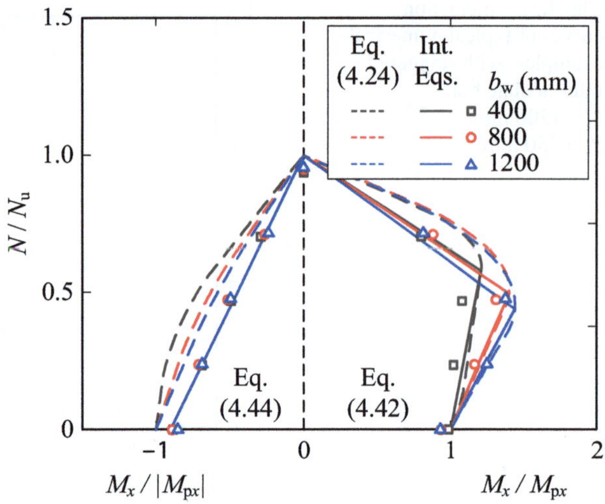

Fig. 4.33 Interaction curves of typical numerical examples: with staggered intermediate bolts, $b_f = 1200$ mm, $d_v = 1500$ mm

For the numerical examples with the same web breadth b_w, the steel plate thickness t_s has a significant effect on the positive interaction curves. The thicker steel corrugated plates are, the less the positive interaction curves bulge, while the axial compression corresponding to the maximum moment capacity is also reduced. The former results come from the reduced contribution of concrete in the moment capacity, and the latter results come from the shorter distance between the centroid and the neutral axis under pure bending moment.

To evaluate the moment capacity of T-CDSCWs, the following assumptions are made based on the numerical results.

1. When the vertical bolt spacing d_v is sufficiently small so that the normalised slenderness ratio of contact buckling λ_{pc} of steel corrugated plates is not greater

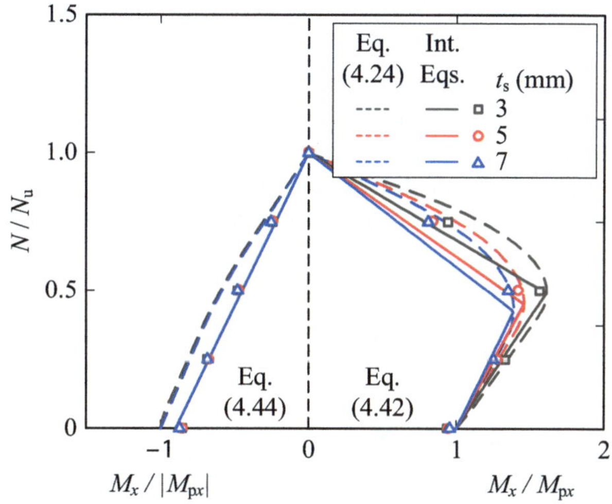

Fig. 4.34 Interaction curves of typical numerical examples: with aligned intermediate bolts, $b_f = 1200$ mm, $b_w = 1200$ mm

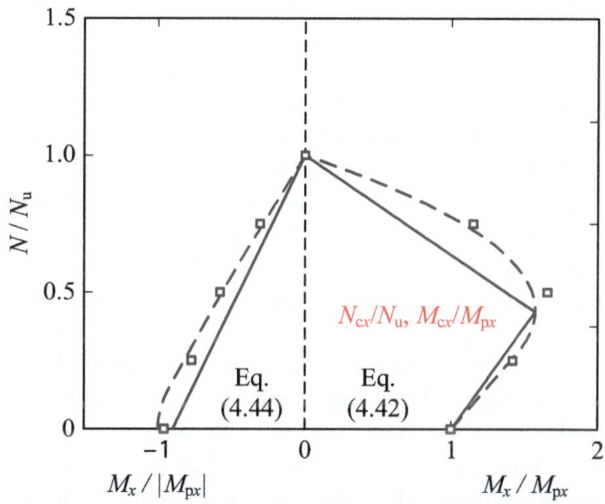

Fig. 4.35 Interaction curves of typical numerical examples: with aligned intermediate bolts, $b_f = 4800$ mm, $b_w = 4800$ mm

than the upper limit 0.4 given by Eq. (4.18), T-CDSCWs reach the full-section plastic stress distribution. The steel yields in the compression and tension zones. The concrete in the compression zone reaches the axial compressive strength, while the contribution of the concrete in the tension zone to the cross-sectional capacity is neglected.

2. When the vertical bolt spacing d_v is large so that the normalised slenderness ratio of contact buckling λ_{pc} of steel corrugated plates is greater than the upper limit 0.4 given by Eq. (4.18), for the sake of simplicity, only the stress of steel in compression zone is corrected to the buckling stress. The effect of the local buckling of steel corrugated plates is neglected on the steel in the tension zone and the concrete. The steel yields in the tension zone. The concrete in the compression

zone reaches the axial compressive strength, while the contribution of the concrete in the tension zone to the cross-sectional capacity is neglected.

Based on the above assumptions, the equilibrium equations of axial force and bending moment are derived as Eqs. (4.23) and (4.24). Since the cross-section is symmetric about the y-axis, the neutral axis is obviously parallel to the x-axis. The position of the neutral axis is determined according to the equilibrium of axial force N, and then the equilibrium of bending moment yields the plastic moment capacity M_{px}.

$$N = \sum_{\mathrm{C}} \left(\varphi_{\mathrm{s}} \sum_{\mathrm{WE}} \left(f_{\mathrm{y}} A_{\mathrm{s}} \right) + \sum_{\mathrm{BE}} \left(f_{\mathrm{y}} A_{\mathrm{s}} \right) + f_{\mathrm{c}} A_{\mathrm{c}} \right) - \sum_{\mathrm{T}} \left(f_{\mathrm{y}} A_{\mathrm{s}} \right) \tag{4.23}$$

$$M_{\mathrm{px}} = \sum_{\mathrm{C}} \left(\sum_{\mathrm{WE}} \left(\varphi_{\mathrm{s}} f_{\mathrm{y}} A_{\mathrm{s}} y_{\mathrm{sn}} \right) + \sum_{\mathrm{BE}} \left(f_{\mathrm{y}} A_{\mathrm{s}} y_{\mathrm{sn}} \right) + f_{\mathrm{c}} A_{\mathrm{c}} y_{\mathrm{cn}} \right) + \sum_{\mathrm{T}} \left(f_{\mathrm{y}} A_{\mathrm{s}} y_{\mathrm{sn}} \right) \tag{4.24}$$

In the subscript of the summation symbols, C and T represent the summation with respect to the compression and tension zone; y_{sn} and y_{cn} denote the distance of steel and concrete to the centroidal axis parallel to the x-axis; φ_s denotes the strength reduction factor of steel corrugated plates, which is determined by Eq. (4.16).

Based on the assumed stress distribution, the full-section plastic moment capacity is derived as Eq. (4.25). In the following formulas, the material properties of steel and concrete in different parts are assumed to be the same. If the material properties of certain parts are different in the application, such as the strength of boundary elements is higher than the strength of steel corrugated plates, the moment capacity can be obtained just by modifying the strength of the corresponding part. The following formulas assumes that the neutral axis is located in the web wall or the concrete part of the flange wall. Otherwise, if the neutral axis is in the steel corrugated plates at the flange wall, the calculation of the moment capacity will be too complicated, and thus the moment capacity should be obtained by the interpolation. At this time, the axial force must lie between the axial force corresponding to the neutral axes at the wave trough and crest. Therefore, the moment capacity can be interpolated from the corresponding moment capacities according to the axial force.

The plastic moment capacity M_{p} can be expressed as the sum of the moment capacities of the flange wall M_{f}, the web wall M_{w}, and the boundary elements M_{c}, as shown in Eq. (4.25).

$$M_{\mathrm{p}} = M_{\mathrm{f}} + M_{\mathrm{w}} + M_{\mathrm{c}} \tag{4.25}$$

For positive bending, when the neutral axis is located in the concrete part of the flange wall, the height of the compression zone at the flange wall is denoted as y_{f} (see Fig. 4.36). The axial force and the moment capacities are derived as Eqs. (4.26)–(4.29).

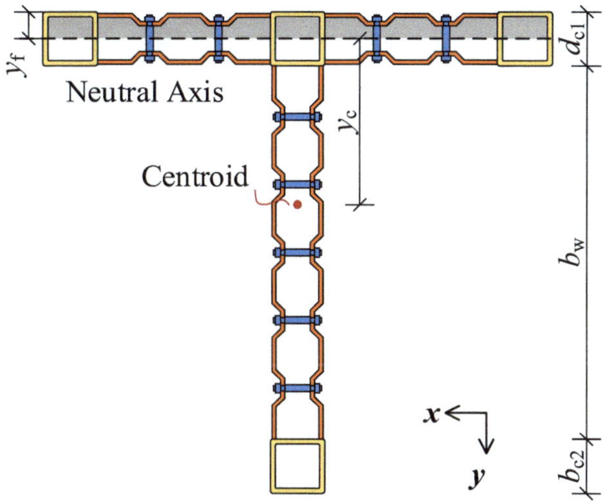

Fig. 4.36 Stress distribution under positive bending moment in the web plane: neutral axis in the concrete part of the flange wall

$$N = f_y\left[\frac{1}{2}(\varphi_{sf}-1)A_{sf}+12t_{c1}y_f-6d_{c1}t_{c1}-A_{sw}-A_{sc2}\right] + f_c\left(A_{cf}\frac{y_f-d_0}{d_f}+3A_{cc1}\frac{y_f}{d_{c1}}\right) \tag{4.26}$$

$$M_f = \frac{1}{2}f_yA_{sf}\left[\varphi_{sf}\left(y_f+y_c-d_0-a\right)+\left(d_0+d_f-a-y_f-y_c\right)\right] + f_cA_{cf}\frac{y_f-d_0}{d_f}\left(\frac{y_f-d_0}{2}+y_c\right) \tag{4.27}$$

$$M_w = f_yA_{sw}\left(d_{c1}+\frac{b_w}{2}-y_f-y_c\right) \tag{4.28}$$

$$M_c = f_y\left\{\begin{array}{l} 3b_{c1}d_{c1}t_{c1}+6d_{c1}t_{c1}\left[\dfrac{y_f^2+\left(d_{c1}-y_f\right)^2}{2d_{c1}}+\left(\dfrac{2y_f}{d_{c1}}-1\right)y_c\right] \\ +A_{sc2}\left(d_{c1}+b_w+\dfrac{b_{c2}}{2}-y_f-y_c\right) \end{array}\right\} + 3f_cA_{cc1}\frac{y_f}{d_{c1}}\left(\frac{y_f}{2}+y_c\right) \tag{4.29}$$

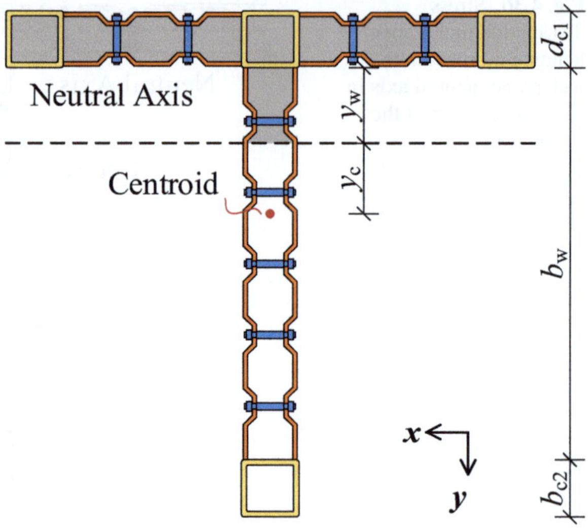

Fig. 4.37 Stress distribution under positive bending moment in the web plane: neutral axis in the web wall

For positive bending, when the neutral axis is located in the web wall, the height of the compression zone at the web wall is denoted as y_w (see Fig. 4.37). The axial force and the moment capacities are derived as Eqs. (4.30)–(4.33).

$$\begin{aligned} N = f_y \left\{ \varphi_{sf} A_{sf} + 3A_{sc1} + A_{sw} \left[\frac{(\varphi_{sw} + 1) y_w}{b_w} - 1 \right] - A_{sc2} \right\} \\ + f_c \left(A_{cf} + 3A_{cc1} + A_{cw} \frac{y_w}{b_w} \right) \end{aligned} \tag{4.30}$$

$$M_f = \varphi_{sf} f_y A_{sf} \left(\frac{d_{c1}}{2} + y_w + y_c \right) + f_c A_{cf} \left(\frac{d_{c1}}{2} + y_w + y_c \right) \tag{4.31}$$

$$\begin{aligned} M_w = f_y A_{sw} \left\{ \frac{\varphi_{sw} y_w^2 + (b_w - y_w)^2}{2b_w} + \left[\frac{(\varphi_{sw} + 1) y_w}{b_w} - 1 \right] y_c \right\} \\ + f_c A_{cw} \frac{y_w}{b_w} \left(\frac{y_w}{2} + y_c \right) \end{aligned} \tag{4.32}$$

$$\begin{aligned} M_c = f_y \left[3A_{sc1} \left(\frac{d_{c1}}{2} + y_w + y_c \right) + A_{sc2} \left(b_w + \frac{b_{c2}}{2} - y_w - y_c \right) \right] \\ + 3 f_c A_{cc1} \left(\frac{d_{c1}}{2} + y_w + y_c \right) \end{aligned} \tag{4.33}$$

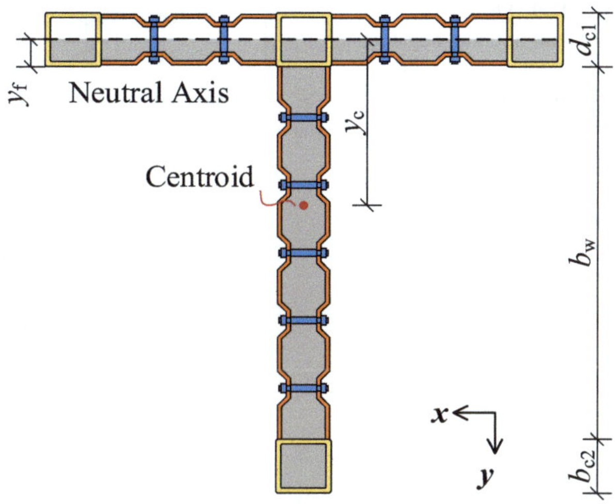

Fig. 4.38 Stress distribution under negative bending moment in the web plane: neutral axis in the concrete part of the flange wall

For negative bending, when the neutral axis is located in the concrete part of the flange wall, the height of the compression zone at the flange wall is denoted as y_f (see Fig. 4.38). The axial force and the moment capacities are derived as Eqs. (4.34)–(4.37).

$$N = f_y\left[\frac{1}{2}(\varphi_{sf}-1)A_{sf} + 12t_{c1}y_f - 6d_{c1}t_{c1} + \varphi_{sw}A_{sw} + A_{sc2}\right] + f_c\left(A_{cf}\frac{y_f - d_0}{d_f} + 3A_{cc1}\frac{y_f}{d_{c1}} + A_{cw} + A_{cc2}\right) \tag{4.34}$$

$$M_f = \frac{1}{2}f_yA_{sf}\left[\varphi_{sf}(y_f - y_c - d_0 - a) + (d_0 + d_f - a - y_f + y_c)\right] + f_cA_{cf}\frac{y_f - d_0}{d_f}\left(\frac{y_f - d_0}{2} - y_c\right) \tag{4.35}$$

$$M_w = \varphi_{sw}f_yA_{sw}\left(\frac{b_w}{2} + y_f - y_c\right) + f_cA_{cw}\left(\frac{b_w}{2} + y_f - y_c\right) \tag{4.36}$$

$$M_c = f_y\left\{\begin{aligned}&3b_{c1}d_{c1}t_{c1} + 6d_{c1}t_{c1}\left[\frac{y_f^2 + (d_{c1} - y_f)^2}{2d_{c1}} - \left(\frac{2y_f}{d_{c1}} - 1\right)y_c\right]\\ &+A_{sc2}\left(b_w + \frac{b_{c2}}{2} + y_f - y_c\right)\end{aligned}\right\} + f_c\left[3A_{cc1}\frac{y_f}{d_{c1}}\left(\frac{y_f}{2} - y_c\right) + A_{cc2}\left(b_w + \frac{b_{c2}}{2} + y_f - y_c\right)\right] \tag{4.37}$$

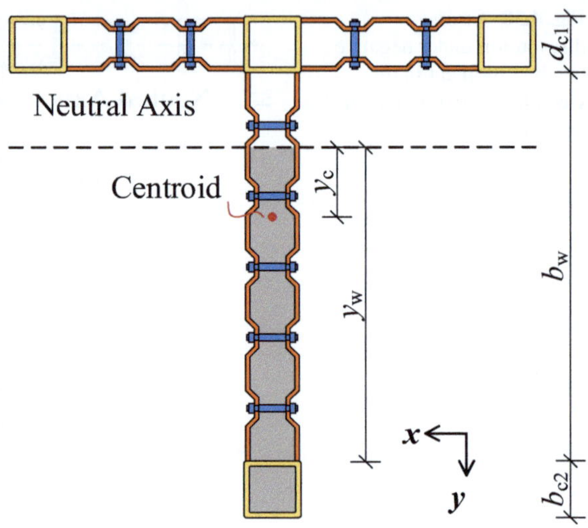

Fig. 4.39 Stress distribution under negative bending moment in the web plane: neutral axis in the web wall

For negative bending, when the neutral axis is located in the web wall, the height of the compression zone at the web wall is denoted as y_w (see Fig. 4.39). The axial force and the moment capacities are derived as Eqs. (4.38)–(4.41).

$$N = f_y \left\{ -A_{sf} - 3A_{sc1} + A_{sw} \left[\frac{(\varphi_{sw}+1)y_w}{b_w} - 1 \right] + A_{sc2} \right\} + f_c \left(A_{cw} \frac{y_w}{b_w} + A_{cc2} \right) \tag{4.38}$$

$$M_f = f_y A_{sf} \left(\frac{d_{c1}}{2} + b_w - y_w + y_c \right) \tag{4.39}$$

$$\begin{aligned} M_w = f_y A_{sw} &\left\{ \frac{\varphi_{sw} y_w^2 + (b_w - y_w)^2}{2b_w} - \left[\frac{(\varphi_{sw}+1)y_w}{b_w} - 1 \right] y_c \right\} \\ &+ f_c A_{cw} \frac{y_w}{b_w} \left(\frac{y_w}{2} - y_c \right) \end{aligned} \tag{4.40}$$

$$\begin{aligned} M_c = f_y &\left[3A_{sc1} \left(\frac{d_{c1}}{2} + b_w - y_w + y_c \right) + A_{sc2} \left(\frac{b_{c2}}{2} + y_w - y_c \right) \right] \\ &+ f_c A_{cc2} \left(\frac{b_{c2}}{2} + y_w - y_c \right) \end{aligned} \tag{4.41}$$

Where A_{sf}, A_{sw}, A_{sc1}, and A_{sc2} denote the cross-sectional area of the steel part in the flange wall, the web wall, the boundary element in the flange wall, and the boundary elements in the web wall; A_{cf}, A_{cw}, A_{cc1}, and A_{cc2} denote the cross-sectional area of the concrete part in the flange wall, the web wall, the boundary element in the flange wall, and the boundary elements in the web wall; y_c denotes the distance between the neutral axis and the centroid; φ_{sf} and φ_{sw} denote the strength reduction factor of the steel corrugated plates in the flange and web walls.

However, the theoretical results of the plastic moment capacity obtained from Eq. (4.24) deviate from the numerical results (see Figs. 4.28, 4.29, 4.30, 4.31, 4.32, 4.33, 4.34, and 4.35), which mainly results from the violation of the assumption (1). Under positive bending moment and large axial compression ($n > 0.5$), or negative bending moment, the theoretical results are slightly greater than the numerical results because the cross-section does not yield around the neutral axis. Especially under negative bending moment, according to the theoretical results, the bulge of the interaction curves varies with the web breadth b_w, which is inconsistent with the numerical results. Under positive bending moment and small axial compression ($n \leq 0.5$), the theoretical results are close to the numerical results. As the flange breadth b_f and the web breadth b_w increase (see Fig. 4.35), the stress distribution is closer to the assumption, and then the numerical results are even greater than the theoretical results due to the confinement strength of concrete.

Despite the deviation, the assumption can reflect the mechanism of the cross-sectional capacity. The theoretical results are consistent with the numerical results in the degree of the bulge when T-CDSCWs are under positive bending moment. To make the design formulas more accurate, simplified formulas are proposed based on the plastic stress distribution method. The interaction equations regarding positive and negative bending are in the bilinear and linear form, respectively, as shown in Eqs. (4.42)–(4.45).

Positive bending:

$$\begin{cases} \dfrac{N}{N_{cx}} + \dfrac{M_{cx} - M_x}{M_{cx} - M_{ux}} = 1 & 0 \leq N \leq N_{cx} \\ \dfrac{N - N_{cx}}{N_u - N_{cx}} + \dfrac{M_x}{M_{cx}} = 1 & N_{cx} < N \leq N_u \end{cases} \tag{4.42}$$

$$M_{ux} = M_{px} \tag{4.43}$$

Negative bending:

$$\frac{N}{N_u} + \frac{M_x}{M_{ux}} = 1 \quad 0 < N \leq N_u \tag{4.44}$$

$$M_{ux} = 0.9M_{px} \tag{4.45}$$

For positive bending, the plastic moment capacity M_{px} obtained from Eq. (4.24) is still used for the pure moment capacity M_{ux} as shown in Eq. (4.43). The apex of the positive interaction curve is determined by the axial compression N_{cx} and bending moment M_{cx} when the neutral axis is located at the centroid. When the axial compression is less than N_{cx}, Eq. (4.42) is close to the plastic moment capacity M_{px} obtained from Eq. (4.24). When the axial compression is greater than N_{cx}, Eq. (4.24) has a parabolic distribution, which is higher than the line obtained from Eq. (4.42). For negative bending, the reduced plastic moment capacity is used for the pure moment capacity M_{ux}, and the reduction factor is 0.9. The linear interaction equation Eq. (4.44) agree with the numerical results well.

When the web breadth b_w is small, the simplified interaction equation Eq. (4.42) may overestimate the moment capacity. When the web breadth b_w is large, the simplified interaction equation Eq. (4.42) may underestimate the moment capacity. The application range of Eqs. (4.42)–(4.45) is limited with reference to JGJ 3 (MOHURD 2010). When the ratio of the overall web breadth B_w to the wall thickness d_w is not less than 4, Eqs. (4.42)–(4.45) can be used to design the cross-sectional capacity of T-CDSCWs under axial compression and bending moment in the web plane.

The design results of Eqs. (4.42)–(4.45) are close to the experimental results. According to the principle of remaining the ratio of the bending moment M_x to the axial compression N the same, the design results can be obtained from Eqs. (4.42) and (4.44). In the cross-sectional capacity test, the axial compression N of the T-1 specimen is 15,518 kN, and the corresponding moment capacity is −4417 kN m. The compression capacity N_u of the T-1 specimen is 30,733 kN, and the pure moment capacity M_{ux} is −8479 kN m. For the design results, the axial compression N of the T-1 specimen is 15,127 kN and the corresponding moment capacity is −4306 kN m, which is 2.5% lower than the experimental value. In the seismic performance test, the axial compression N of the T-3 specimen is 8443 kN, and the corresponding positive moment capacity is 7761 kN m and the corresponding negative moment capacity is −5201 kN m. The compression capacity N_u of the T-3 specimen is 23,270 kN, the positive pure moment capacity M_{ux} is 4972 kN m, and the negative pure moment capacity is −6513 kN m. For the design results: under positive bending moment, the axial compression N of the T-3 specimen is 7122 kN and the corresponding moment capacity is 6546 kN m, which is 15.6% smaller than the experimental value; under negative bending moment, the axial compression N of the T-3 specimen is 7270 kN, and the corresponding moment capacity is −4478 kN m, which is 13.9% smaller than the experimental value. This indicates that Eqs. (4.42)–(4.45) can ensure the safety of the T-CDSCW design and meet the requirements of engineering application.

4.3 Cross-Sectional Capacity under Axial Compression and Bending Moment in the Flange Plane

Besides bending moment in the web plane, T-CDSCWs are also subjected to bending moment in the flange plane. Since the T-section is mono-symmetric, there is no difference between two bending directions when T-CDSCWs are under bending moment in the flange plane. Thus, only one bending direction is investigated. The load-carrying performance of T-CDSCWs under bending moment in the flange plane will be investigated, including the stress distribution and the bending moment–curvature relationship. Based on the numerical results and theoretical analyses, the interaction equations will be proposed.

4.3.1 Stress Distribution and Bending Moment–Curvature Curve

Based on the solid–shell finite element model, the cross-sectional capacity of T-CDSCWs under axial compression and bending moment in the flange plane is analysed. The material properties of steel are as follows: the yield strength f_y = 345 MPa, the elastic modulus $E_s = 2.06 \times 10^5$ MPa, and the Poisson's ratio $\upsilon_s = 0.3$ (MOHURD and AQSIQ 2017). The material properties of concrete are as follows: the axial compressive strength f_c = 26.8 MPa, the elastic modulus $E_c = 3.25 \times 10^4$ MPa, and the Poisson's ratio $\upsilon_c = 0.2$ (MOHURD and AQSIQ 2010). The other material properties are the same as Chap. 3. The initial geometric imperfection of steel corrugated plates is in the form of the lateral deflection caused by uniform pressure, and the amplitude of the imperfection is $d_v/750$.

The boundary condition of the T-CDSCW is pin ended. The top and bottom sections are coupled with the reference points at the centroid, which is weighted with the elastic modulus of the materials. The T-CDSCW is loaded at the reference points. The displacement and the rotation about the z-axis of the bottom reference point are restrained. The displacement along x- and y-axis of the top reference point are restrained. The axial compression N is first applied at the top reference point, and then the bending moment M_y in the flange plane is applied at both reference points. The stress distribution and the bending moment–curvature relationship can be obtained by the finite element static analysis.

The axial compression N is applied to the T-CDSCW according to the axial compression ratio n as defined in Eq. (4.21).

Due to the symmetry, the finite element simulations only study the load-carrying behaviour of T-CDSCWs under the bending moment along the positive y-axis (see Fig. 4.40). The numerical examples are the same as Sect. 4.2. The finite element simulations analyse the effects of the flange breadth b_f, the web breadth b_w, the vertical bolt spacing d_v and the steel plate thickness t_s on the cross-sectional capacity of T-CDSCWs (see Table 4.4).The wall height H_w is the maximum of 3 m and $3d_v$. The wall thickness d_w is 150 mm. The parameters of steel corrugated plates are: q = 200 mm, q_1 = 100 mm, q_2 = 25 mm, q_3 = 50 mm, a = 25 mm. The dimensions of the CFST $b_c \times d_c$ are 160 × 160 mm, and the steel tube thickness t_c is 4 mm greater than t_s. Intermediate bolts are arranged in aligned and staggered ways. The minimum vertical bolt spacing d_v is 250 mm, which is so small that steel corrugated plates do not buckle before yielding, and the local buckling of steel corrugated plates does not influence the cross-sectional capacity of T-CDSCWs. The maximum vertical bolt spacing d_v is 1500 mm, which is a very large value in practice and sufficiently large to consider the local buckling. The numerical examples cover a broad range and consider the local buckling and yielding of steel corrugated plates.

Although the flange wall sustains most of the bending moment, the web wall changes stress development. The stress distribution (see Fig. 4.41) corresponds to the points on the bending moment–curvature curves (see Figs. 4.46 and 4.47). When the ultimate bending moment is reached, generally, only the steel in the compression zone of the flange wall yields, while the steel in the web wall and the tension

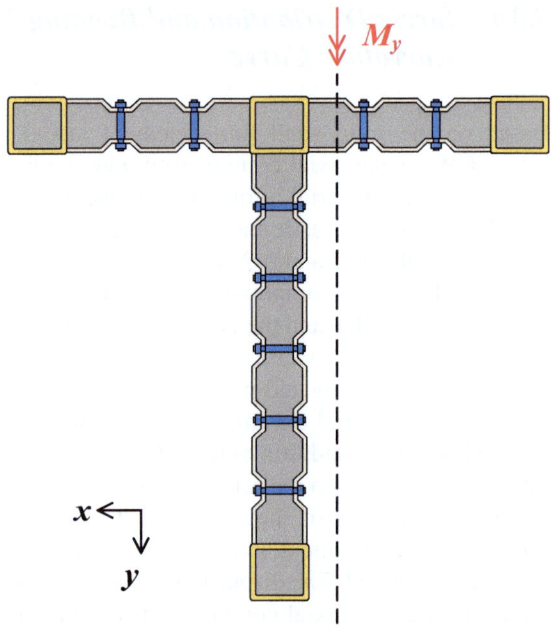

Fig. 4.40 T-CDSCW under bending moment in the flange plane

zone of the flange wall does not. As the loading continues, the plastic zone expands to the web wall, as well as the tension zone of the flange wall. The stress in the tension zone is related to the buckling behaviour of steel corrugated plates. Since both of the flange and web walls are partly under compression, the steel corrugated plates at both walls buckle. At first, it is the steel corrugated plates at the flange that buckle, and then the plates at the web wall. This makes the stress in the tension zone continue to develop after the ultimate bending moment is reached.

T-CDSCWs exhibit different bending moment–curvature curves from bending moment in the web plane (see Figs. 4.42, 4.43, 4.44, 4.45, 4.46, and 4.47). The ordinate shows the normalised bending moment, the bending moment M_y divided by the plastic moment capacity M_{py}, which will be given by Eq. (4.47). The abscissa shows the normalised curvature ϕ_{0y}, which is given by Eq. (4.46).

$$\phi_{0y} = \frac{\phi_y}{2\varepsilon_y / B_f} \tag{4.46}$$

Where ϕ_y denotes the curvature of the T-CDSCW about the y-axis, ε_y denotes the yield strain of steel, and B_f denotes the overall breadth of the flange wall.

As the axial compression ratio n increases, the moment capacity of T-CDSCWs decreases. If the neutral axis was always parallel to the y-axis, the increase of the axial compression ratio n would move the neutral axis from the compression zone of the flange wall towards the tension zone, and the moment capacity would increase first and then decrease as positive bending moment in the web plane. However, this

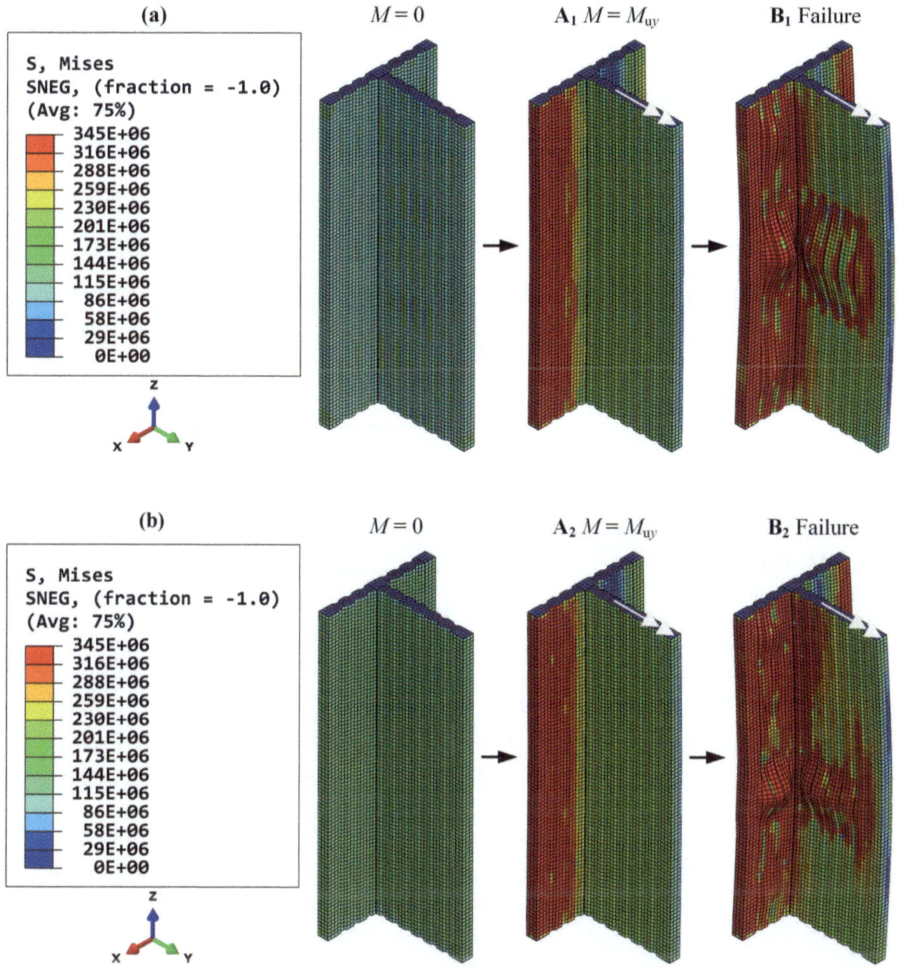

Fig. 4.41 Stress distribution (unit of stress: Pa) and failure modes of typical numerical examples (d_v = 1500 mm, n = 0.5): (**a**) with aligned intermediate bolts; (**b**) with staggered intermediate bolts

is inconsistent with the numerical results, so the neutral axis is not always parallel to the *y*-axis under bending moment in the flange plane.

Due to the successive buckling of the steel corrugated plates at the flange and web walls, the bending moment may rise again after reaching the extremum. Beyond the first extremum, the snap-through buckling occurs. Therefore, only the first extremum is meaningful in terms of the actual load carried by the structure. Only the first extremum of bending moment is considered as the moment capacity of T-CDSCWs, which will be derived then.

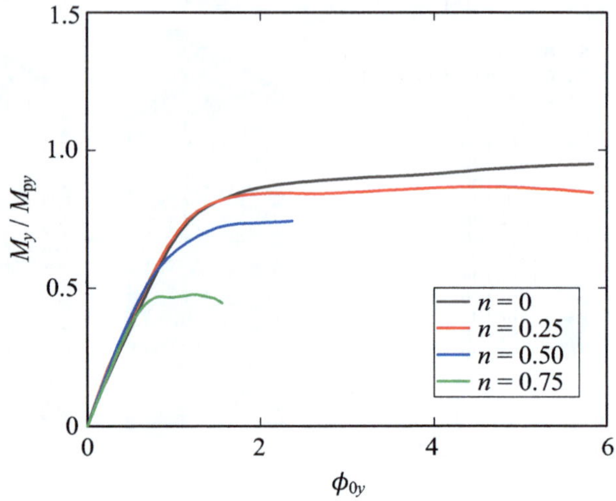

Fig. 4.42 Bending moment–curvature relationship of typical numerical examples: with aligned intermediate bolts, $b_f = 400$ mm, $b_w = 1200$ mm, $d_v = 1500$ mm

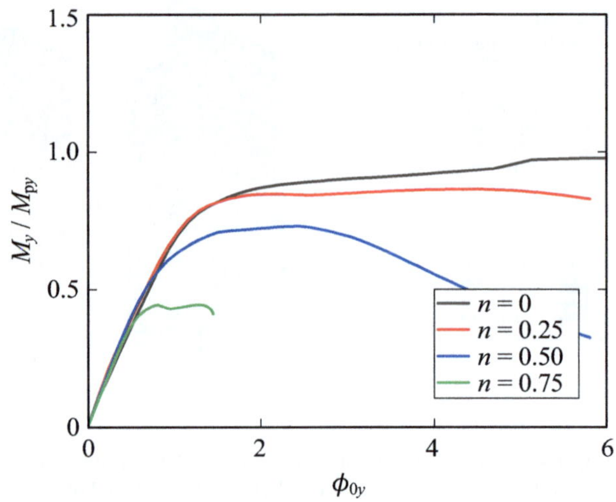

Fig. 4.43 Bending moment–curvature relationship of typical numerical examples: with staggered intermediate bolts, $b_f = 400$ mm, $b_w = 1200$ mm, $d_v = 1500$ mm

4.3.2 Design Method of Cross-Sectional Capacity

The cross-sectional capacity of T-CDSCWs obtained by the finite element analysis can be used as the basis for the design formula. The interaction curves of T-CDSCWs under bending moment in the flange plane reflect the effects of geometric parameters, such as flange breadth b_f, steel plate thickness t_s, and vertical bolt spacing d_v (see Figs. 4.48, 4.49, 4.50, 4.51, 4.52, 4.53, and 4.54). Overall, the interaction curves of N–M_y are not influenced by the geometric parameters significantly.

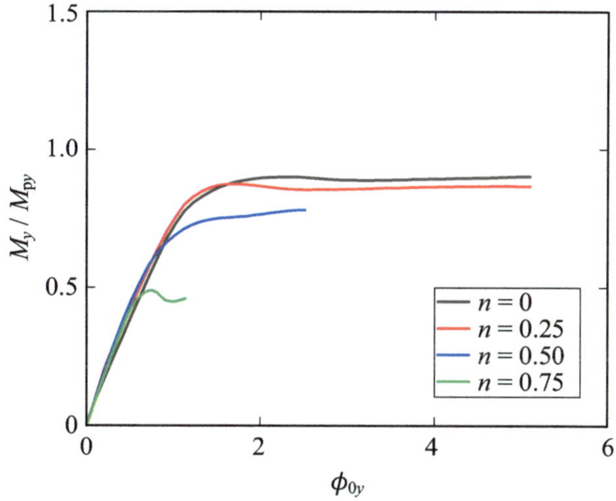

Fig. 4.44 Bending moment–curvature relationship of typical numerical examples: with aligned intermediate bolts, $b_f = 800$ mm, $b_w = 1200$ mm, $d_v = 1500$ mm

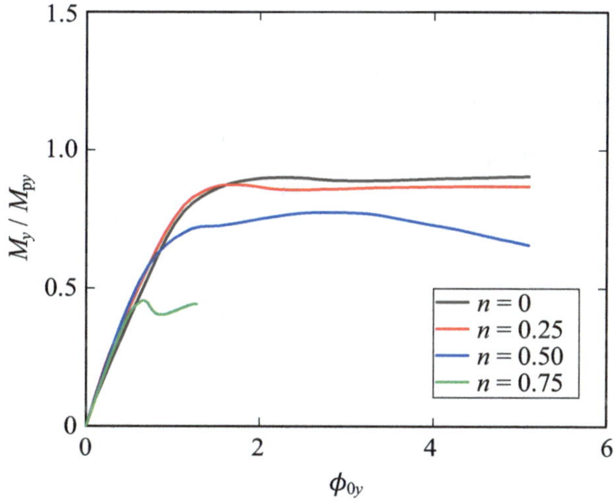

Fig. 4.45 Bending moment–curvature relationship of typical numerical examples: with staggered intermediate bolts, $b_f = 800$ mm, $b_w = 1200$ mm, $d_v = 1500$ mm

To evaluate the moment capacity of T-CDSCWs, the additional assumptions are made based on the assumptions (1) and (2).

3. When T-CDSCWs are under bending moment in the flange plane, the neutral axis can be approximated as parallel to the y-axis.

This is a very strong assumption since it is not valid according to the previous analysis of the relationship between the moment capacity and the axial compression ratio n. Moreover, if the full-section plastic stress distribution holds and the neutral axis is parallel to the y-axis, the integral of the stress concerning the x-axis must not be zero, that is, T-CDSCWs are also subjected to bending

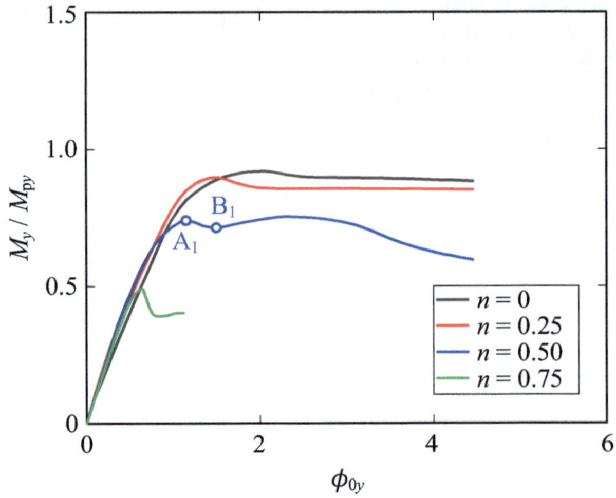

Fig. 4.46 Bending moment–curvature relationship of typical numerical examples: with aligned intermediate bolts, $b_f = 1200$ mm, $b_w = 1200$ mm, $d_v = 1500$ mm

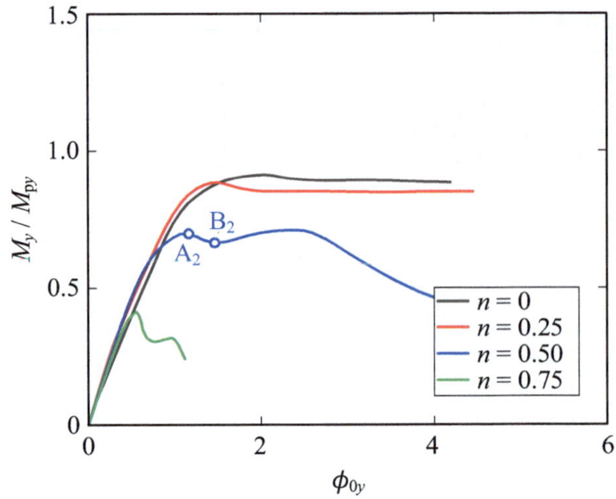

Fig. 4.47 Bending moment–curvature relationship of typical numerical examples: with staggered intermediate bolts, $b_f = 1200$ mm, $b_w = 1200$ mm, $d_v = 1500$ mm

moment in the web plane, which does not conform to the load conditions. However, due to the geometric characteristics of the T-section and steel corrugated plates, if the neutral axis is not parallel to the *y*-axis, then the formulas of the moment capacity according to the full-section plastic stress distribution will be more complicated and inconvenient to use. For the sake of simplicity, the moment capacity is derived according to assumption (3) first and checked on the reasonableness by the comparison with the numerical results.

Based on the above assumptions, the equilibrium equations of axial force and bending moment are derived. The equation of axial force N is in the same form as Eq. (4.23), except that the neutral axis is parallel to the *y*-axis here. The equation of bending moment M_{py} is rewritten as Eq. (4.47) to calculate the bending moment about the *y*-axis.

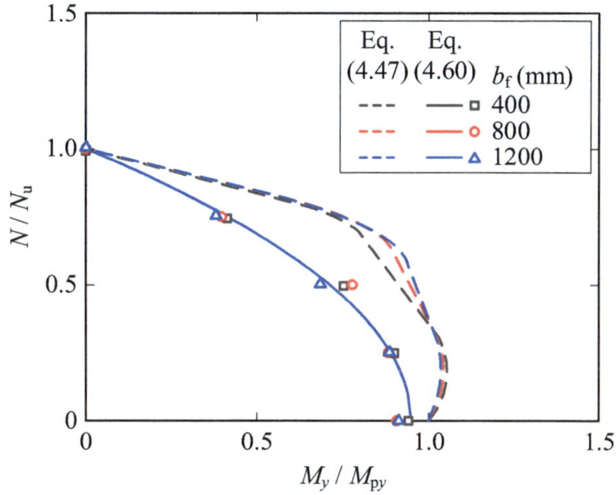

Fig. 4.48 Interaction curves of typical numerical examples: with aligned intermediate bolts, $b_w = 1200$ mm, $d_v = 250$ mm

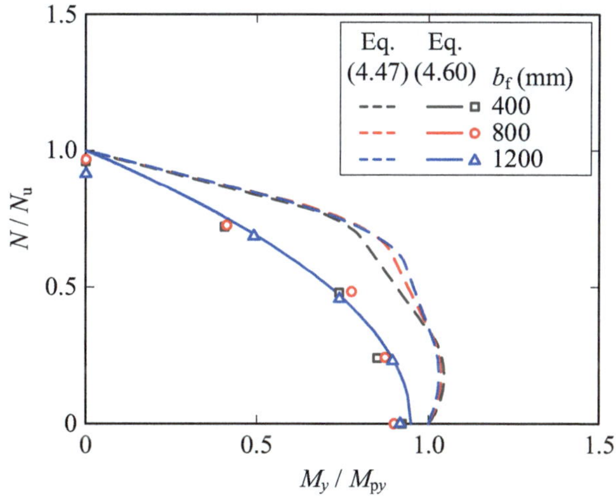

Fig. 4.49 Interaction curves of typical numerical examples: with aligned intermediate bolts, $b_w = 1200$ mm, $d_v = 1000$ mm

$$M_{py} = \sum_{C}\left(\sum_{WE}\left(\varphi_s f_y A_s x_{sn}\right) + \sum_{BE}\left(f_y A_s x_{sn}\right) + f_c A_c x_{cn}\right) + \sum_{T}\left(f_y A_s x_{cn}\right) \quad (4.47)$$

In the subscript of the summation symbols, C and T represent the summation with respect to the compression and tension zone; x_{sn} and x_{cn} denote the distance of steel and concrete to the centroidal axis parallel to the y-axis; φ_s denotes the strength reduction factor of steel corrugated plates, which is determined by Eq. (4.16).

Based on the assumed stress distribution, the full-section plastic moment capacity is derived as Eq. (4.25). In the following formulas, the material properties of steel and concrete in different parts are assumed to be the same. If the material properties of certain parts are different in the application, such as the strength of

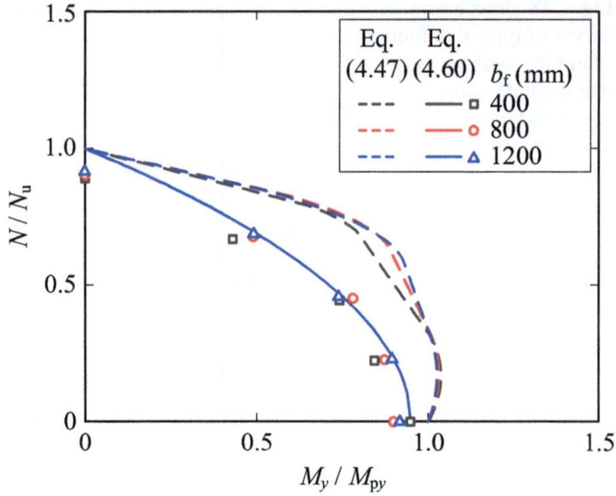

Fig. 4.50 Interaction curves of typical numerical examples: with aligned intermediate bolts, $b_w = 1200$ mm, $d_v = 1250$ mm

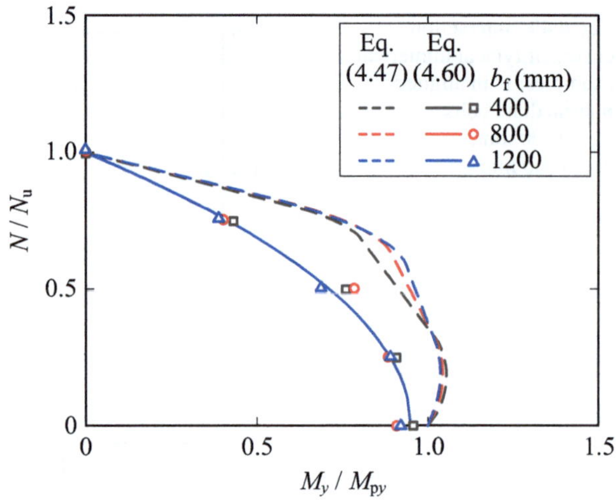

Fig. 4.51 Interaction curves of typical numerical examples: with staggered intermediate bolts, $b_w = 1200$ mm, $d_v = 250$ mm

boundary elements is higher than the strength of steel corrugated plates, the moment capacity can be obtained just by modifying the strength of the corresponding part. The following formulas assumes that the neutral axis is located in the flange wall or the concrete part of the web wall. Otherwise, if the neutral axis is in the steel corrugated plates at the web wall, the calculation of the moment capacity will be too complicated, and thus the moment capacity should be obtained by the interpolation. At this time, the axial force must lie between the axial force corresponding to the neutral axes at the wave trough and crest. Therefore, the moment capacity can be interpolated from the corresponding moment capacities according to the axial force.

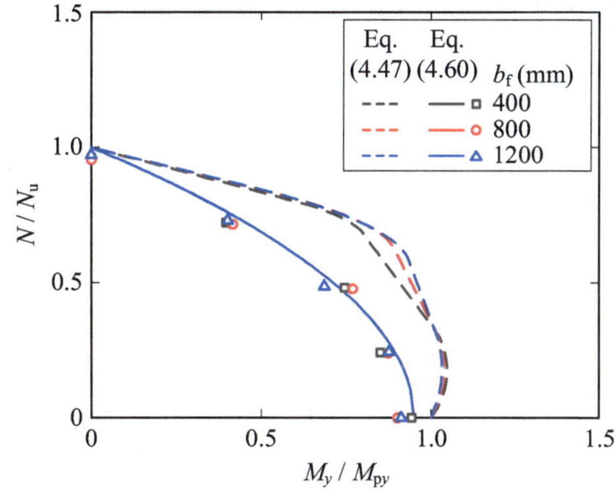

Fig. 4.52 Interaction curves of typical numerical examples: with staggered intermediate bolts, $b_w = 1200$ mm, $d_v = 1000$ mm

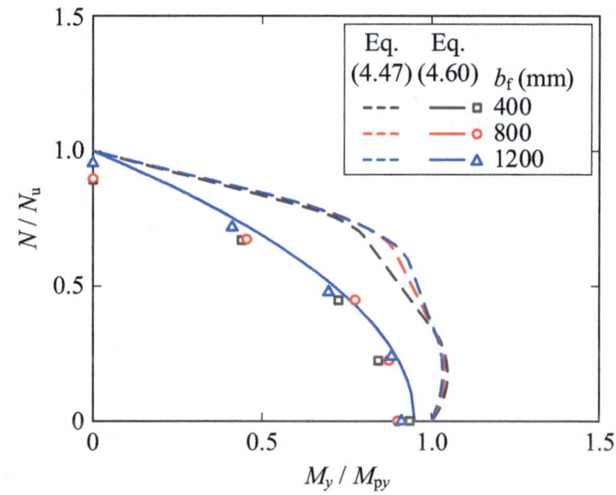

Fig. 4.53 Interaction curves of typical numerical examples: with staggered intermediate bolts, $b_w = 1200$ mm, $d_v = 1250$ mm

The plastic moment capacity M_p can be expressed as the sum of the moment capacities of the flange wall M_f, the web wall M_w, and the boundary elements M_c, as shown in Eq. (4.25).

When the neutral axis is located in the flange wall at the positive x-axis, the height of the compression zone at the flange wall is denoted as x_{f1} (see Fig. 4.55). The axial force and the moment capacities are derived as Eqs. (4.48)–(4.51).

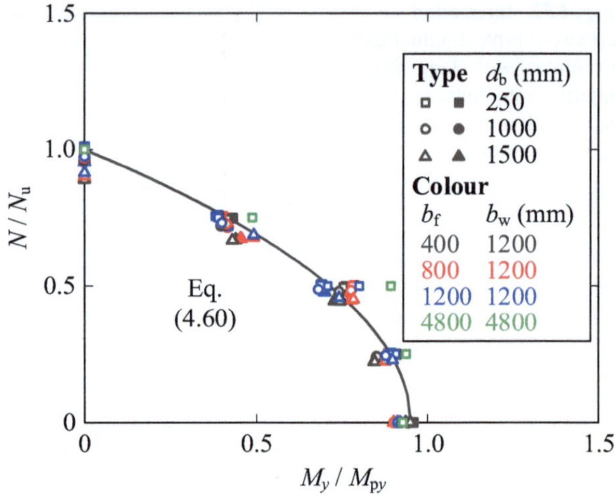

Fig. 4.54 Interaction curves of all numerical examples

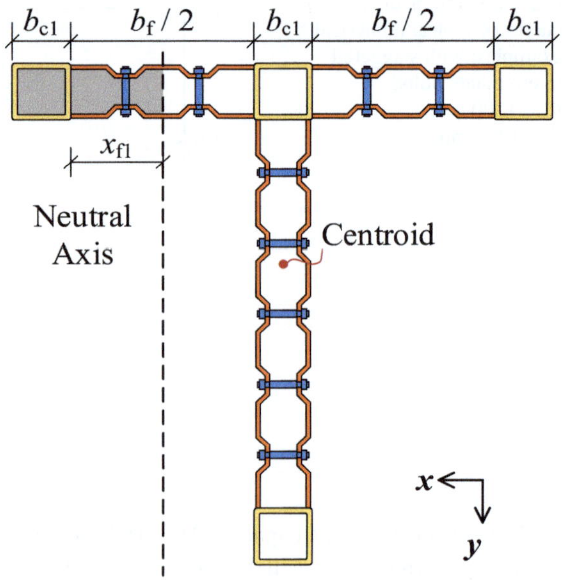

Fig. 4.55 Stress distribution under bending moment in the flange plane: neutral axis in the flange wall at the positive x-axis

$$N = f_{\mathrm{y}}\left\{\frac{1}{2}A_{\mathrm{sf}}\left[\frac{(\varphi_{\mathrm{sf}}+1)x_{\mathrm{f1}}}{b_{\mathrm{f}}/2}-2\right]-A_{\mathrm{sc1}}-A_{\mathrm{sw}}-A_{\mathrm{sc2}}\right\}+f_{\mathrm{c}}\left(\frac{1}{2}A_{\mathrm{cf}}\frac{x_{\mathrm{f1}}}{b_{\mathrm{f}}/2}+A_{\mathrm{cc1}}\right) \tag{4.48}$$

$$
\begin{aligned}
M_{\mathrm{f}} = {} & \frac{1}{2} f_{\mathrm{y}} A_{\mathrm{sf}} \left[\frac{\varphi_{\mathrm{sf}} x_{\mathrm{f1}}}{b_{\mathrm{f}}/2} \left(\frac{b_{\mathrm{f}}}{2} - \frac{x_{\mathrm{f1}}}{2} + \frac{b_{\mathrm{c1}}}{2} \right) + \left(1 - \frac{x_{\mathrm{f1}}}{b_{\mathrm{f}}/2} \right) \left(\frac{b_{\mathrm{f}}}{4} - \frac{x_{\mathrm{f1}}}{2} + \frac{b_{\mathrm{c1}}}{2} \right) \right] \\
& + \frac{1}{2} f_{\mathrm{c}} A_{\mathrm{cf}} \frac{x_{\mathrm{f1}}}{b_{\mathrm{f}}/2} \left(\frac{b_{\mathrm{f}}}{2} - \frac{x_{\mathrm{f1}}}{2} + \frac{b_{\mathrm{c1}}}{2} \right)
\end{aligned} \tag{4.49}
$$

$$
M_{\mathrm{w}} = 0 \tag{4.50}
$$

$$
M_{\mathrm{c}} = 2 f_{\mathrm{y}} A_{\mathrm{sc1}} \left(\frac{b_{\mathrm{f}}}{2} + b_{\mathrm{c1}} \right) + f_{\mathrm{c}} A_{\mathrm{cc1}} \left(\frac{b_{\mathrm{f}}}{2} + b_{\mathrm{c1}} \right) \tag{4.51}
$$

When the neutral axis is located in the web wall, the height of the compression zone at the flange wall is denoted as x_{w} (see Fig. 4.56). The axial force and the moment capacities are derived as Eqs. (4.52)–(4.55).

$$
\begin{aligned}
N = {} & f_{\mathrm{y}} \left[\frac{1}{2} (\varphi_{\mathrm{sf}} - 1) A_{\mathrm{sf}} + 2(2x_{\mathrm{w}} - b_{\mathrm{c1}}) t_{\mathrm{c1}} + 2(2x_{\mathrm{w}} - d_{\mathrm{c2}}) t_{\mathrm{c2}} + \frac{1}{2} (\varphi_{\mathrm{s}} - 1) A_{\mathrm{sw}} \right] \\
& + f_{\mathrm{c}} \left[\frac{1}{2} A_{\mathrm{cf}} + A_{\mathrm{cc1}} \left(1 + \frac{x_{\mathrm{w}}}{b_{\mathrm{c1}}} \right) + A_{\mathrm{cc2}} \frac{x_{\mathrm{w}}}{d_{\mathrm{c2}}} + A_{\mathrm{cw}} \frac{x_{\mathrm{w}} - d_0}{d_{\mathrm{w}}} \right]
\end{aligned} \tag{4.52}
$$

$$
M_{\mathrm{f}} = \frac{1}{2} f_{\mathrm{y}} A_{\mathrm{sf}} (\varphi_{\mathrm{sf}} + 1) \left(\frac{b_{\mathrm{f}}}{4} + \frac{b_{\mathrm{c1}}}{2} \right) + \frac{1}{2} f_{\mathrm{c}} A_{\mathrm{cf}} \left(\frac{b_{\mathrm{f}}}{4} + \frac{b_{\mathrm{c1}}}{2} \right) \tag{4.53}
$$

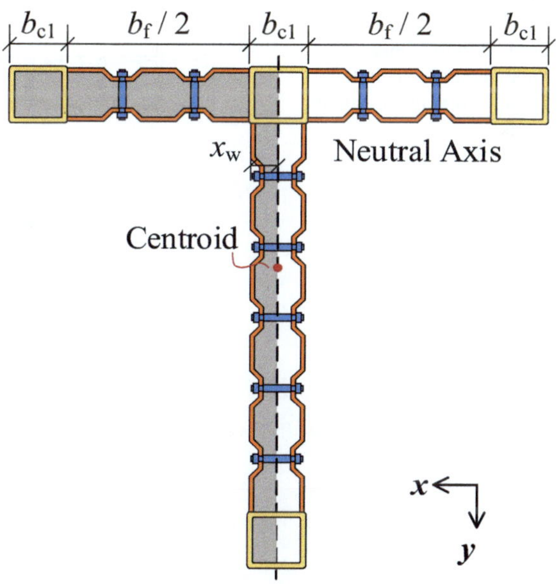

Fig. 4.56 Stress distribution under bending moment in the flange plane: neutral axis in the web wall

$$M_{\mathrm{w}} = \frac{1}{2} f_{\mathrm{y}} A_{\mathrm{sw}} \left(\varphi_{\mathrm{sw}} + 1\right) \left(\frac{d_{\mathrm{w}}}{2} - a\right) + f_{\mathrm{c}} A_{\mathrm{cw}} \frac{\left(x_{\mathrm{w}} - d_0\right)\left(b_{\mathrm{c1}} - x_{\mathrm{w}} + d_0\right)}{2d_{\mathrm{w}}} \tag{4.54}$$

$$\begin{aligned} M_{\mathrm{c}} = f_{\mathrm{y}} & \begin{bmatrix} 2A_{\mathrm{sc1}}\left(\dfrac{b_{\mathrm{f}}}{2} + b_{\mathrm{c1}}\right) + b_{\mathrm{c1}} d_{\mathrm{c1}} t_{\mathrm{c1}} + b_{\mathrm{c2}} d_{\mathrm{c2}} t_{\mathrm{c2}} \\ + 2\left(b_{\mathrm{c1}} - x_{\mathrm{w}}\right) x_{\mathrm{w}} t_{\mathrm{c1}} + 2\left(d_{\mathrm{c2}} - x_{\mathrm{w}}\right) x_{\mathrm{w}} t_{\mathrm{c2}} \end{bmatrix} \\ & + f_{\mathrm{c}} \left\{ A_{\mathrm{cc1}} \left[\frac{b_{\mathrm{f}}}{2} + b_{\mathrm{c1}} + \frac{x_{\mathrm{w}}\left(b_{\mathrm{c1}} - x_{\mathrm{w}}\right)}{2b_{\mathrm{c1}}}\right] + A_{\mathrm{cc2}} \frac{x_{\mathrm{w}}\left(d_{\mathrm{c2}} - x_{\mathrm{w}}\right)}{2d_{\mathrm{c2}}} \right\} \end{aligned} \tag{4.55}$$

When the neutral axis is located in the flange wall at the negative x-axis, the height of the compression zone at the flange wall is denoted as x_{f2} (see Fig. 4.57). The axial force and the moment capacities are derived as Eqs. (4.56)–(4.59).

$$\begin{aligned} N = f_{\mathrm{y}} & \left\{ \frac{1}{2} A_{\mathrm{sf}} \left[\varphi_{\mathrm{sf}} - 1 + \frac{\left(\varphi_{\mathrm{sf}} + 1\right) x_{\mathrm{f2}}}{b_{\mathrm{f}}/2}\right] + A_{\mathrm{sc1}} + \varphi_{\mathrm{sw}} A_{\mathrm{sw}} + A_{\mathrm{sc2}} \right\} \\ & + f_{\mathrm{c}} \left[\frac{1}{2} A_{\mathrm{cf}} \left(1 + \frac{x_{\mathrm{f2}}}{b_{\mathrm{f}}/2}\right) + 2A_{\mathrm{cc1}} + A_{\mathrm{cw}} + A_{\mathrm{cc2}}\right] \end{aligned} \tag{4.56}$$

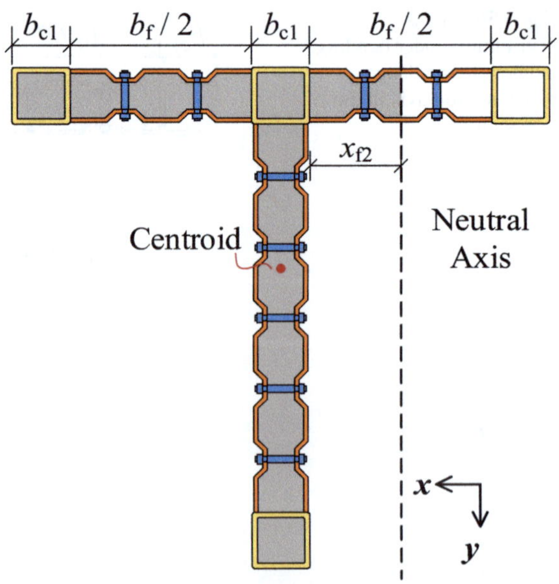

Fig. 4.57 Stress distribution under bending moment in the flange plane: neutral axis in the flange wall at the negative x-axis

$$M_{\mathrm{f}}=\frac{1}{2}f_{\mathrm{y}}A_{\mathrm{sf}}\left[\varphi_{\mathrm{sf}}\left(\frac{b_{\mathrm{f}}}{4}+\frac{b_{\mathrm{c1}}}{2}\right)+\frac{\varphi_{\mathrm{sf}}x_{\mathrm{f2}}}{b_{\mathrm{f}}/2}\left(\frac{b_{\mathrm{c1}}}{2}+\frac{x_{\mathrm{f2}}}{2}\right)+\left(1-\frac{x_{\mathrm{f2}}}{b_{\mathrm{f}}/2}\right)\left(\frac{b_{\mathrm{c1}}}{2}+\frac{b_{\mathrm{f}}}{4}-\frac{x_{\mathrm{f2}}}{2}\right)\right]$$
$$+\frac{1}{2}f_{\mathrm{c}}A_{\mathrm{cf}}\left[\frac{b_{\mathrm{f}}}{4}+\frac{b_{\mathrm{c1}}}{2}+\frac{x_{\mathrm{f2}}}{b_{\mathrm{f}}/2}\left(\frac{b_{\mathrm{c1}}}{2}+\frac{x_{\mathrm{f2}}}{2}\right)\right] \quad (4.57)$$

$$M_{\mathrm{w}}=0 \quad (4.58)$$

$$M_{\mathrm{c}}=2f_{\mathrm{y}}A_{\mathrm{sc1}}\left(\frac{b_{\mathrm{f}}}{2}+b_{\mathrm{c1}}\right)+f_{\mathrm{c}}A_{\mathrm{cc1}}\left(\frac{b_{\mathrm{f}}}{2}+b_{\mathrm{c1}}\right) \quad (4.59)$$

Where A_{sf}, A_{sw}, A_{sc1}, and A_{sc2} denote the cross-sectional area of the steel part in the flange wall, the web wall, the boundary element in the flange wall, and the boundary elements in the web wall; A_{cf}, A_{cw}, A_{cc1}, and A_{cc2} denote the cross-sectional area of the concrete part in the flange wall, the web wall, the boundary element in the flange wall, and the boundary elements in the web wall; x_{c} denotes the distance between the neutral axis and the centroid; φ_{sf} and φ_{sw} denote the strength reduction factor of the steel corrugated plates in the flange and web walls.

Compared with the numerical examples under bending moment in the web plane (see Figs. 4.28, 4.29, 4.30, 4.31, 4.32, 4.33, 4.34, and 4.35), the theoretical results obtained from Eq. (4.47) are far from the numerical results (see Figs. 4.48, 4.49, 4.50, 4.51, 4.52, 4.53, and 4.54). The reason is not only that the assumed full-section plastic stress distribution does not match the actual stress distribution, but more important that the neutral axis is not parallel to the y-axis. Fortunately, however, the deviation of the theoretical results from the numerical results is not significant for pure bending, suggesting that the assumption does not deviate much from the actual situation under pure bending.

To make the design formulas more accurate, simplified formulas are proposed based on the numerical results. The interaction equation in in the parabolic form, as shown in Eqs. (4.60) and (4.61).

$$\left(\frac{N}{N_{\mathrm{u}}}\right)^{2}+\frac{M_{y}}{M_{\mathrm{uy}}}\leq 1 \quad 0<N\leq N_{\mathrm{u}} \quad (4.60)$$

$$M_{\mathrm{uy}}=0.95M_{\mathrm{py}} \quad (4.61)$$

The reduced plastic moment capacity is used for the pure moment capacity M_{uy} to consider the deviation from assumption (3), and the reduction factor is 0.95. Since there is no obvious relationship between the geometry and the shape of the interaction curves, the simplified interaction equation is adopted in the parabolic form. The reason why the interaction equation is in parabolic form is that the bending moment is mainly sustained by the flange wall, which is similar to a rectangle. The interaction equation of a rectangular cross-section happens to be in parabolic form.

When the flange breadth b_f is small, the simplified interaction equation Eq. (4.60) may overestimate the moment capacity. When the flange breadth b_f is large, the simplified interaction equation Eq. (4.60) may underestimate the moment capacity. The application range of Eqs. (4.60) and (4.61) is limited with reference to JGJ 3 (MOHURD 2010). When the ratio of the overall flange breadth B_f to the wall thickness d_w is not less than 4, Eqs. (4.60) and (4.61) can be used to design the cross-sectional capacity of T-CDSCWs under axial compression and bending moment in the flange plane.

4.4 Cross-Sectional Capacity under Axial Compression and Biaxial Bending Moment

Following the investigation of the cross-sectional capacities of T-CDSCWs under axial compression and bending moment in the web plane and the flange plane, this section investigates the cross-sectional capacity of T-CDSCWs under axial compression and biaxial bending moment, including the stress distribution and the bending moment–curvature relationship. Based on the numerical results and theoretical analyses, the interaction equations will be proposed.

4.4.1 Stress Distribution and Bending Moment–Curvature Curve

Based on the solid–shell finite element model, the cross-sectional capacity of T-CDSCWs under axial compression and biaxial bending moment is analysed. The material properties of steel are as follows: the yield strength $f_y = 345$ MPa, the elastic modulus $E_s = 2.06 \times 10^5$ MPa, and the Poisson's ratio $\upsilon_s = 0.3$ (MOHURD and AQSIQ 2017). The material properties of concrete are as follows: the axial compressive strength $f_c = 26.8$ MPa, the elastic modulus $E_c = 3.25 \times 10^4$ MPa, and the Poisson's ratio $\upsilon_c = 0.2$ (MOHURD and AQSIQ 2010). The other material properties are the same as Chap. 3. The initial geometric imperfection of steel corrugated plates is in the form of the lateral deflection caused by uniform pressure, and the amplitude of the imperfection is $d_v/750$.

The boundary condition of the T-CDSCW is pin ended. The top and bottom sections are coupled with the reference points at the centroid, which is weighted with the elastic modulus of the materials. The T-CDSCW is loaded at the reference points. The displacement and the rotation about the z-axis of the bottom reference point are restrained. The displacement along x- and y-axis of the top reference point are restrained. The axial compression N is first applied at the top reference point, and then the bending moment M is applied at both reference points. The angle between the bending direction and the x-axis is denoted as α. When $\alpha = 0$, the

T-CDSCW is under positive bending moment in the web plane; when $\alpha = \pi/2$, the T-CDSCW is under bending moment in the flange plane; when $\alpha = \pi$, the T-CDSCW is under negative bending moment in the web plane. The stress distribution and the bending moment–curvature relationship can be obtained by the finite element static analysis.

The axial compression N is applied to the T-CDSCW according to the axial compression ratio n as defined in Eq. (4.21).

The numerical examples are the same as Sect. 4.2. The finite element simulations analyse the effects of the flange breadth b_f, the web breadth b_w, the vertical bolt spacing d_v and the steel plate thickness t_s on the cross-sectional capacity of T-CDSCWs (see Table 4.4).The wall height H_w is the maximum of 3 m and $3d_v$. The wall thickness d_w is 150 mm. The parameters of steel corrugated plates are: $q = 200$ mm, $q_1 = 100$ mm, $q_2 = 25$ mm, $q_3 = 50$ mm, $a = 25$ mm. The dimensions of the CFST $b_c \times d_c$ are 160 × 160 mm, and the steel tube thickness t_c is 4 mm greater than t_s. The minimum vertical bolt spacing d_v is 250 mm, which is so small that steel corrugated plates do not buckle before yielding, and the local buckling of steel corrugated plates does not influence the cross-sectional capacity of T-CDSCWs. The maximum vertical bolt spacing d_v is 1500 mm, which is a very large value in practice and sufficiently large to consider the local buckling. The numerical examples cover a broad range and consider the local buckling and yielding of steel corrugated plates.

From the previous numerical results, it is found that the staggered arrangement of bolts has a good constraining effect on the local buckling of steel corrugated plates and thus the local buckling is not significant. While for the aligned arrangement of bolts, the local buckling may be significant, but the shape of the interaction curves basically remains unchanged as the vertical bolt spacing d_v increases. As a result, only the numerical results with aligned bolts and $d_v = 250$ mm and 1500 mm are investigated.

The numerical results show that the behaviour of T-CDSCWs under biaxial bending moment is the superimposition of the behaviour under uniaxial bending moment. The asymmetry regarding the bending direction is still reflected in the performance of bending moment in the web plane (see Fig. 4.58). Under M_y and positive M_x, the flange wall is compressed to yield on only one side when the ultimate bending moment is reached. As the loading proceeds, the steel corrugated plates at both the flange and web walls buckle. Under M_y and negative M_x, when the ultimate bending moment is reached, the flange and web walls are compressed to yield. As the loading proceeds, only the steel corrugated plates at the web wall buckle. Different stress distribution effects the load–displacement and interaction curves.

Two normalised variables are used to equally compare the load–displacement responses of T-CDSCWs under bending moment in variable bending directions (see Fig. 4.59, 4.60, 4.61, 4.62, 4.63, and 4.64). The ordinate shows the normalised bending moment, the bending moment M divided by the resultant moment of the moment capacities M_{nx} and M_{ny} under axial compression, which is given by the interaction Eqs. (4.42), (4.44), and (4.60). The abscissa shows the normalised

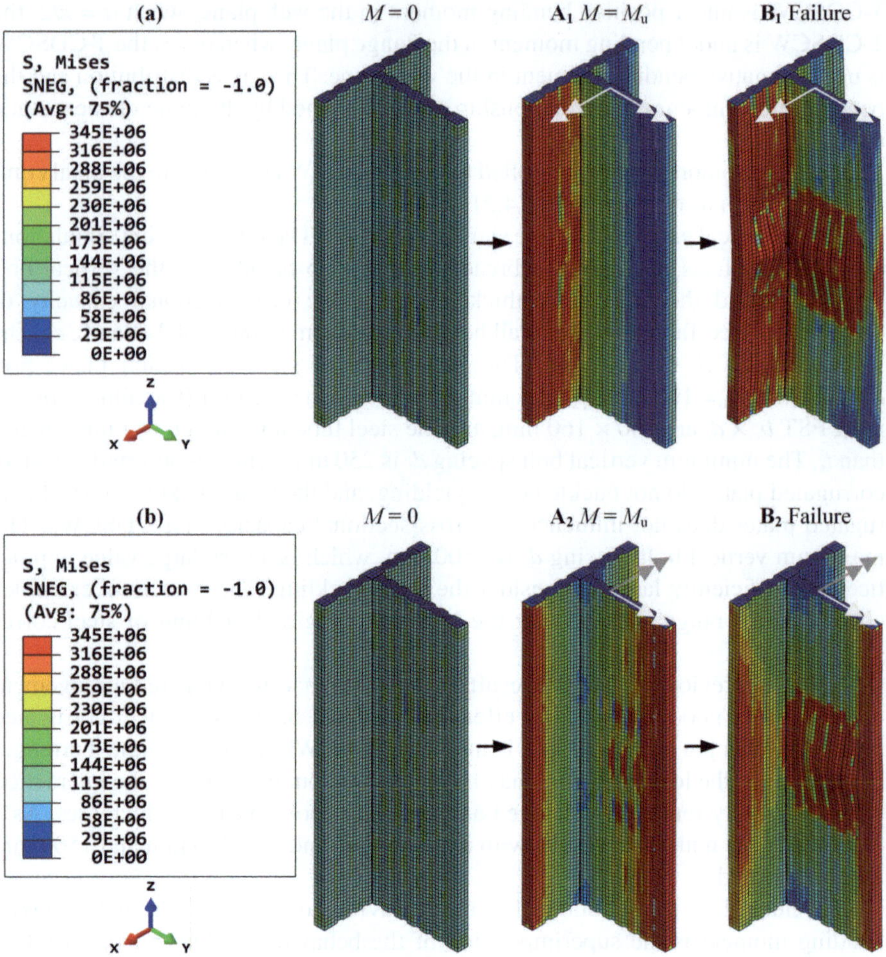

Fig. 4.58 Stress distribution (unit of stress: Pa) and failure modes of typical numerical examples (with aligned intermediate bolts, d_v = 1500 mm, n = 0.5): (**a**) $\alpha = \pi/4$; (**b**) $\alpha = 3\pi/4$

curvature ϕ_0, which is the squared root of the squared sum of ϕ_{0x} and ϕ_{0y}. The two normalised variables have no real physical meaning.

As the axial compression ratio n increases, the slope of the initial stage of the load–displacement curves do not change significantly for positive M_x ($0 \leq \alpha < \pi/2$); while, for negative M_x ($\pi/2 < \alpha \leq \pi$), the slope increases with the axial compression ratio n. This is because, for positive M_x, the area of the compression zone at the flange wall is sufficiently large so that the change in the compression zone of concrete caused by axial compression is not large, and the stiffness of the whole cross-section varies less. Whereas, for negative M_x, the area of the compression zone at the web wall is so small that the change in the compression zone of concrete caused by axial compression is large and the stiffness of the whole cross-section varies more.

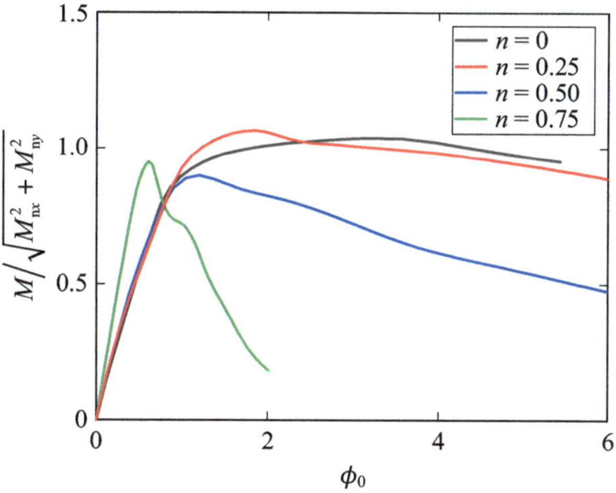

Fig. 4.59 Bending moment–curvature relationship of typical numerical examples: with aligned intermediate bolts, $b_f = 1200$ mm, $b_w = 1200$ mm, $d_v = 1500$ mm, $\alpha = \pi/8$

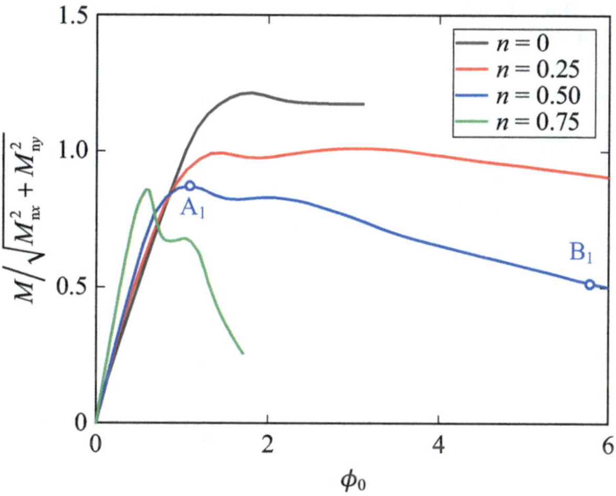

Fig. 4.60 Bending moment–curvature relationship of typical numerical examples: with aligned intermediate bolts, $b_f = 1200$ mm, $b_w = 1200$ mm, $d_v = 1500$ mm, $\alpha = \pi/4$

4.4.2 Design Method of Cross-Sectional Capacity

The interaction equations of bending moment in the web and flange planes both have axial compression term and bending moment term. If the exponent of axial compression term is the same in the interaction equations, it may be possible to obtain the interaction equation of biaxial bending moment by directly superimposing the interaction equations obtained above. However, the interaction equation of bending moment in the web plane is linear with 1 as the exponent of axial compression, while the interaction equation of bending moment in the flange plane is parabolic with 2 as the exponent of axial compression. Therefore, the interaction equations of bending moment in the web and flange planes cannot be simply

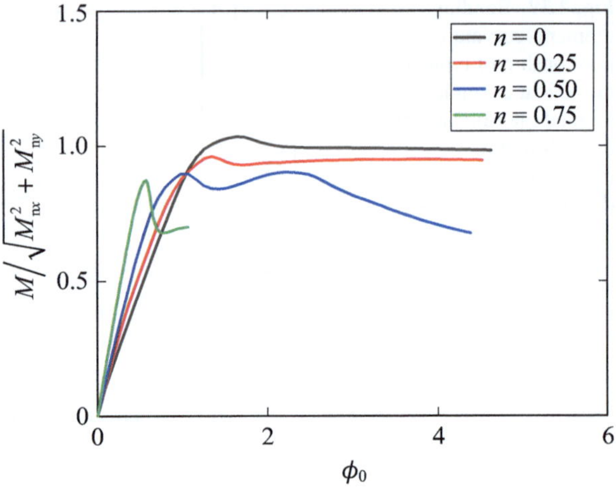

Fig. 4.61 Bending moment–curvature relationship of typical numerical examples: with aligned intermediate bolts, $b_f = 1200$ mm, $b_w = 1200$ mm, $d_v = 1500$ mm, $\alpha = 3\pi/8$

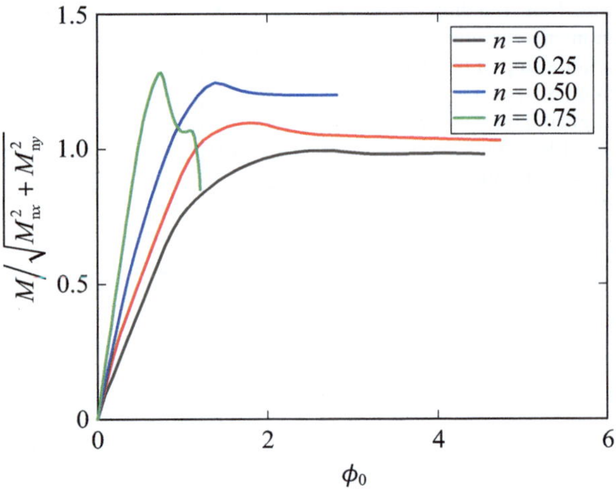

Fig. 4.62 Bending moment–curvature relationship of typical numerical examples: with aligned intermediate bolts, $b_f = 1200$ mm, $b_w = 1200$ mm, $d_v = 1500$ mm, $\alpha = 5\pi/8$

superimposed. The moment capacities M_{nx} and M_{ny} under axial compression N can be evaluated first and compared with the finite element simulation results.

To obtain the interaction equations of biaxial bending moment, we will focus on the biaxial moment capacity subjected to certain axial compression. Figure 4.14 shows the moment capacity of T-CDSCWs under biaxial bending moment.

The interaction curves of biaxial bending moment are almost exclusively related to the axial compression ratio n (see Figs. 4.65, 4.66, 4.67, 4.68, and 4.69). For positive M_x (the first quadrant). the numerical results are basically symmetric about $x = y$. The interaction curves of pure bending moment are very convex, and the degree of the bulge decreases with the increasing axial compression ratio n. For negative M_x (the second quadrant), the maximum of M_y is not obtained at $M_x = 0$. Moreover, small M_y has almost no effect on M_x, and the moment capacity of M_x is

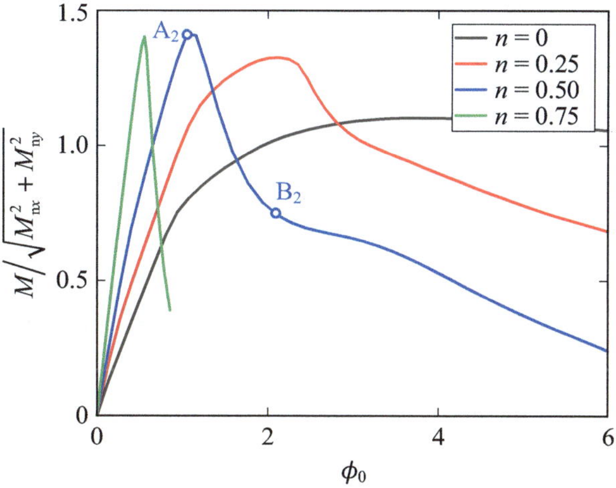

Fig. 4.63 Bending moment–curvature relationship of typical numerical examples: with aligned intermediate bolts, $b_f = 1200$ mm, $b_w = 1200$ mm, $d_v = 1500$ mm, $\alpha = 3\pi/4$

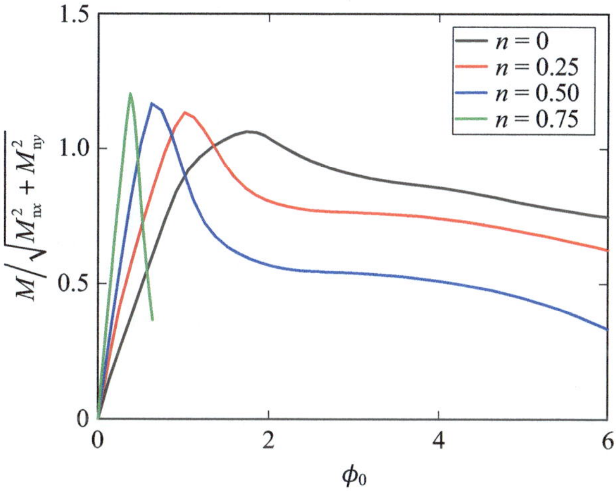

Fig. 4.64 Bending moment–curvature relationship of typical numerical examples: with aligned intermediate bolts, $b_f = 1200$ mm, $b_w = 1200$ mm, $d_v = 1500$ mm, $\alpha = 7\pi/8$

almost M_{nx}; when M_y/M_{ny} exceeds a certain threshold value, the increase of M_y makes M_x decrease, and this threshold value increases with the axial compression ratio n.

The shapes of the interaction curves can be explained by the previous analysis of stress distribution (see Figs. 4.70 and 4.71). For positive M_x, the flange wall is already compressed to yield under M_x. To balance the addition of M_y, a part of the flange wall cannot be compressed to yield when the ultimate bending moment is reached. While, for negative M_x, the flange wall is not tensioned to yield, or is under compression. In such cases, the addition of M_y has an insignificant effect on M_x. As for the maximum of M_y is not obtained at $M_x = 0$, this can be explained by the

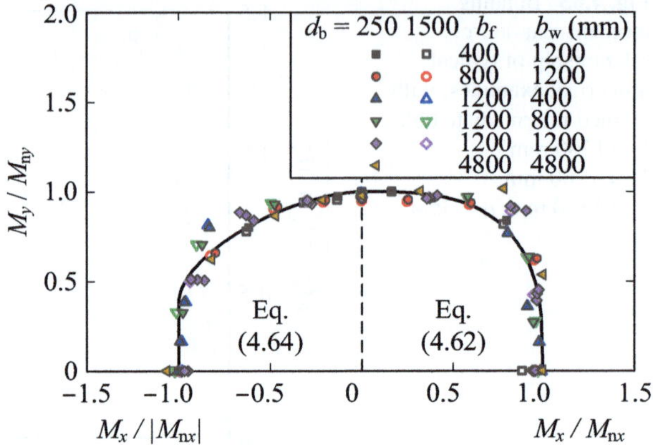

Fig. 4.65 Interaction curves of typical numerical examples: $n = 0$

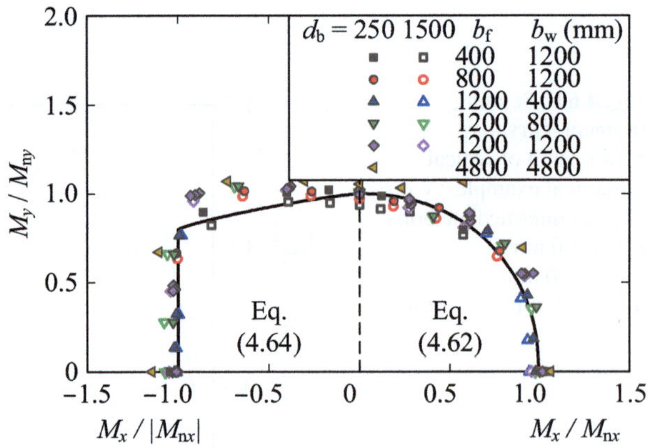

Fig. 4.66 Interaction curves of typical numerical examples: $n = 0.25$

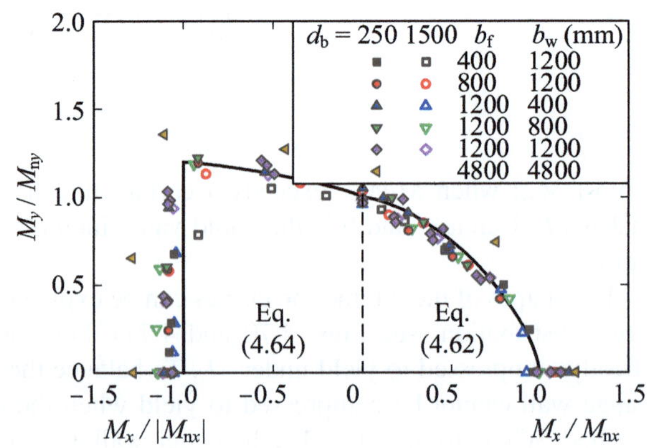

Fig. 4.67 Interaction curves of typical numerical examples: $n = 0.50$

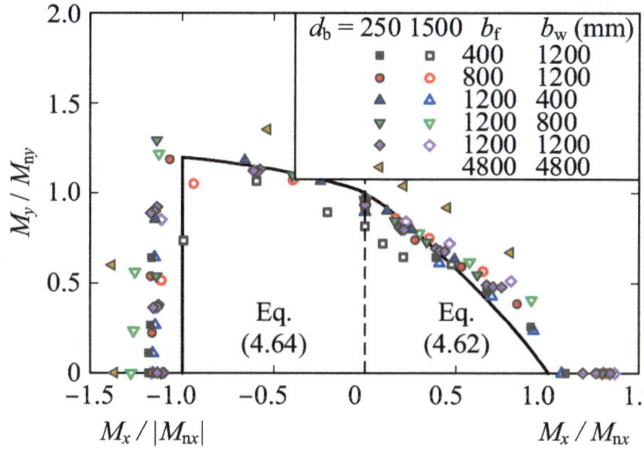

Fig. 4.68 Interaction curves of typical numerical examples: $n = 0.75$

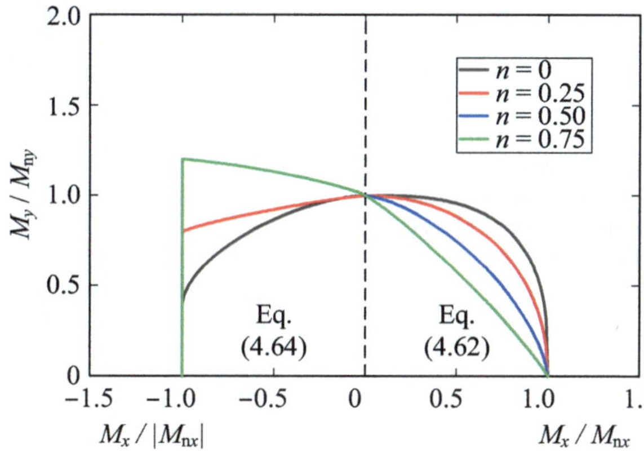

Fig. 4.69 Interaction curves of all numerical examples

violation of the assumption (3). Only if the bending direction is not parallel to the y-axis, the neutral axis is possible to be parallel to the y-axis, and M_y reaches the maximum.

Based on the numerical results, the simplified interaction equations of M_x–M_y are proposed as Eqs. (4.62)–(4.66).

The interaction equation of M_y and positive M_x:

$$\left(\frac{M_x}{M_{nx}}\right)^{n_1} + \left(\frac{M_y}{M_{ny}}\right)^{n_1} \leq 1 \tag{4.62}$$

$$n_1 = 1 + 2(1-n)^2 \tag{4.63}$$

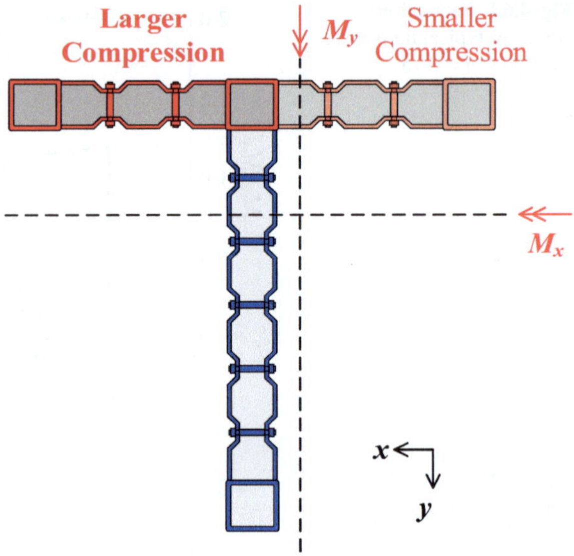

Fig. 4.70 Stress distribution under biaxial bending moment: M_y and positive M_x

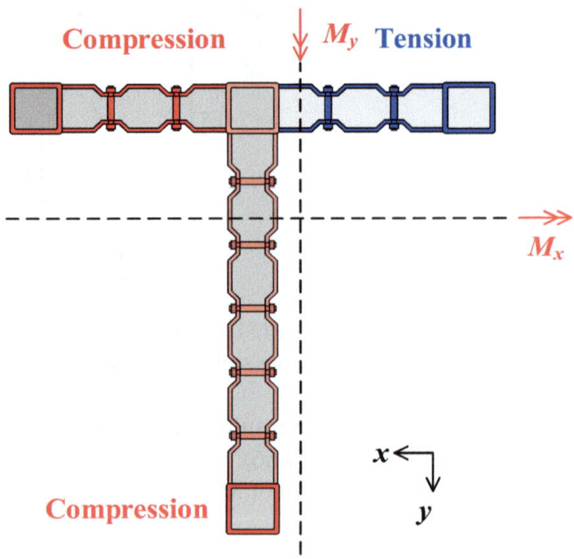

Fig. 4.71 Stress distribution under biaxial bending moment: M_y and negative M_x

The interaction equation of M_y and negative M_x:

$$\begin{cases} M_x = M_{ux} & M_y < M_{cy} \\ \left(\dfrac{M_x}{M_{nx}}\right)^{n_2} + \left(\dfrac{M_y - M_{cy}}{M_{ny} - M_{cy}}\right)^{n_2} = 1 & \min\left(M_{cy}, M_{ny}\right) \le M_y \le \max\left(M_{cy}, M_{ny}\right) \end{cases} \tag{4.64}$$

$$n_2 = 1.6(1-n) \ge 0.8 \tag{4.65}$$

$$m_{cy} = \frac{M_{cy}}{M_{ny}} = 1.6n + 0.4 \le 1.2 \tag{4.66}$$

Where M_{cy} denotes the threshold value below which M_y does not influence M_x. The exponents n_1 and n_2 can reflect the trends of the interaction curves. When $n = 0$, $n_1 = 3$, the interaction curves bulge much; when $n = 1$, $n_1 = 1$, the interaction curves are reduced to straight lines. When $n > 0.5$, $n_2 = 0.8$, $m_{cy} = 1.2$, and the interaction curves do not change anymore with the increase of the axial compression ratio n.

Compared with the numerical results of uniaxial bending moment, the numerical results of biaxial bending moment are more scattered. This is because the moment capacity under biaxial bending moment is evaluated based on the uniaxial moment capacities. The interaction Eq. (4.64) are conservative with large axial compression ratio ($n = 0.75$). The conservativeness comes from the interaction equations of uniaxial bending moment. However, in the structural design, the composite walls are usually governed by seismic design, and the requirement of ductility restricts the axial compression ratio. This reduces the impact of the conservativeness. For some numerical results, the moment capacity may be overestimated by Eqs. (4.62) and (4.64). On the one hand, the safety can be ensured by the partial factors in the design; on the other hand, the moment capacity is overestimated evidently only for the numerical examples with small web walls ($b_w = 400$ mm), which is too small for common design. Overall, the interaction Eqs. (4.62) and (4.64) agree reasonably with the numerical results.

When the wen breadth b_w and the flange breadth b_f are small, the T-CDSCW is close to the CDSCW with boundary elements, and the simplified interaction equation Eqs. (4.62) and (4.64) may overestimate the moment capacity. When the wen breadth b_w and the flange breadth b_f are large, the simplified interaction equation Eqs. (4.62) and (4.64) may underestimate the moment capacity. The application range of Eqs. (4.62) and (4.64) is limited with reference to JGJ 3 (MOHURD 2010). When both the ratio of the overall web breadth B_w to the wall thickness d_w and the ratio of the overall flange breadth B_f to the wall thickness d_w is not less than 4, Eqs. (4.62) and (4.64) can be used to design the cross-sectional capacity of T-CDSCWs under axial compression and biaxial bending moment.

Up to now, the interaction equations of uniaxial and biaxial bending moment have been obtained, and the moment capacity of T-CDSCWs under axial compression and biaxial bending moment can be designed (see Fig. 4.72).

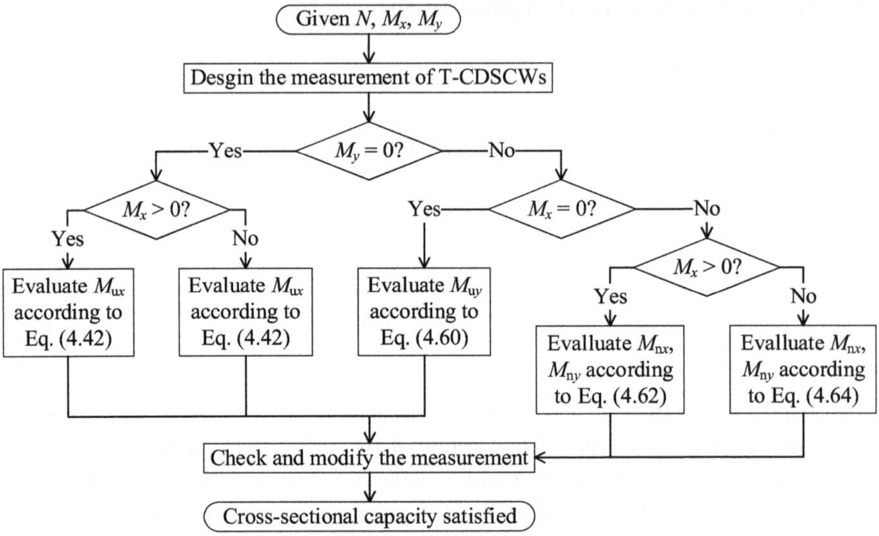

Fig. 4.72 Design flow of the cross-sectional capacity of T-CDSCWs under axial compression and biaxial bending moment

4.5 Chapter Summary

Based on the solid–shell finite element model, this study analyses the local buckling performance of steel corrugated plates under axial compression and the cross-sectional capacity of T-CDSCWs under axial compression and bending moment in the web plane, bending moment in the flange plane, and biaxial bending moment. Based on the finite element simulation results and theoretical analyses, the following conclusions are drawn.

This chapter proposes the stability design method of steel corrugated plates, and thus the local buckling of steel corrugated plates can be considered in the cross-sectional capacity of T-CDSCWs. This chapter proposes the formulas of bilateral buckling of steel corrugated plates with only intermediate bolt constraints and contact buckling of steel corrugated plates with intermediate bolt and infilled concrete constraints. Based on the normalised slenderness ratio, the stability design curve of steel corrugated plates is developed and agrees well with finite element simulation results. Based on the stability design curve, the upper limit of the normalised slenderness ratio for contact buckling is proposed to be 0.4, below which the effect of the local buckling of steel corrugated plates can be neglected on the cross-sectional capacity of T-CDSCWs.

This chapter proposes the design method of the cross-sectional capacity of T-CDSCWs under axial compression and bending moment in the web plane. When T-CDSCWs sustain bending moment in the web plane, the direction of bending moment influences the shape of the interaction curves. For positive bending moment in the web plane, the interaction curve is convex, and the the convexity becomes significant as the web breadth increases and the steel plate thickness decreases. For negative bending moment in the web plane, the interaction curve is basically linear, and the overall shape is independent on the cross-section. The local buckling of steel corrugated plates does not influence the overall shape of the interaction curve. By reducing the full-section plastic bending moment, the pure moment capacity in the web plane of T-CDSCWs can be evaluated, and thus the cross-sectional capacity under axial compression and bending moment in the web plane can be designed according to the bilinear interaction equation for positive bending moment and the linear interaction equation for negative bending moment. The design formulas agree well with the finite element simulation results.

This chapter proposes the design method of the cross-sectional capacity of T-CDSCWs under axial compression and bending moment in the flange plane. When T-CDSCWs sustain bending moment in the flange plane, the overall shape of the interaction curve is basically independent on the cross-section. The local buckling of steel corrugated plates does not influence the overall shape of the interaction curve. By reducing the full-section plastic bending moment, the pure moment capacity in the flange plane of T-CDSCWs can be evaluated, and thus the cross-sectional capacity under axial compression and bending moment in the flange plane can be designed according to the parabolic interaction equation. The design formulas agree well with the finite element simulation results.

This chapter proposes the design method of the cross-sectional capacity of T-CDSCWs under axial compression and biaxial bending moment. When T-CDSCWs sustain biaxial bending moment, the overall shape of the interaction curve primarily depends on the axial compression ratio. The local buckling of steel corrugated plates does not influence the overall shape of the interaction curve. In the design of the cross-sectional capacity of T-CDSCWs, the corresponding moment capacities are evaluated in the web and flange plane according to the given axial compression ratio, and then the cross-sectional capacity of T-CDSCWs is designed according to the interaction curve of biaxial bending moment. The design formulas agree well with the finite element simulation results.

This chapter focuses on the design method of the cross-sectional capacity of T-CDSCWs. The next chapter will use the finite element model to further analyse the global stability of T-CDSCWs and propose the corresponding design method.

References

Huang Z, Liew JYR (2016) Compressive resistance of steel-concrete-steel sandwich composite walls with J-hook connectors. J Constr Steel Res 124:142–162. https://doi.org/10.1016/j.jcsr.2016.05.001

MOHURD (2010) Technical specification for concrete structures of tall building (in Chinese). China Architecture and Building Press, Beijing

MOHURD and AQSIQ (2003) Code for design of steel structures (in Chinese). China Planning Press, Beijing

MOHURD and AQSIQ (2010) Code for design of concrete structures (in Chinese). China Architecture and Building Press, Beijing
MOHURD and AQSIQ (2017) Standard for design of steel structures (in Chinese). China Architecture and Building Press, Beijing
Song Y, Li J, Chen Y (2019) Local and post-local buckling of normal/high strength steel sections with concrete infill. Thin Wall Struct 138:155–169. https://doi.org/10.1016/j.tws.2019.02.004
Timoshenko SP, Gere JM (2009) Theory of elastic stability, 2nd edn. Dover Publications, New York
Timoshenko SP, Woinowsky-Krieger S (1959) Theory of plates and shells. McGraw-hill, New York
Yang Y, Liu J, Fan J (2016) Buckling behavior of double-skin composite walls: an experimental and modeling study. J Constr Steel Res 121:126–135. https://doi.org/10.1016/j.jcsr.2016.01.019
Zhu J (2020) Theoretical and experimental study on design methods for I-section concrete-infilled double steel corrugated-plate walls (in Chinese). Ph.D. Thesis, Tsinghua University, Beijing
Zhu J, Guo Y, Wang M, Yang X, Zhu B (2019) Strength design of concrete-infilled double steel corrugated-plate walls under uniform compressions. Thin Wall Struct 141:153–174. https://doi.org/10.1016/j.tws.2019.02.021

Chapter 5
Global Stability of T-CDSCWs

Due to the excellent cross-sectional capacity, composite walls can be designed slender and thus the failure mode is governed not by cross-section capacity, but by global instability. Based on the solid–shell finite element model, this chapter studies the global stability of T-CDSCWs under axial compression and under axial compression and bending moment in the web plane.

This chapter first analyses the global stability of axially loaded T-CDSCWs. For the member with single-symmetric section, such as T-section, two forms of elastic buckling modes may occur under axial compression. With short web walls, the flexural stiffness about the asymmetric axis is relatively small, and the first-order elastic buckling mode of the axially loaded T-CDSCW turn to be flexural buckling about the asymmetric axis. With wide web walls, the flexural stiffness about the asymmetric axis is relatively large, and the first-order elastic buckling mode of the axially loaded T-CDSCW turn to be flexural-torsional buckling about the symmetric axis. This chapter studies the flexural and flexural-torsional buckling of T-CDSCWs. The elastic buckling loads are derived, and the load-carrying performance is investigated. To evaluate the load-carrying capacity of axially loaded T-CDSCWs, the curves of the strength reduction factor are developed based on the normalised slenderness ratio of flexural and flexural-torsional buckling.

Based on the buckling analysis of axially loaded T-CDSCWs, this chapter then analyses the global stability of T-CDSCWs under axial compression and bending moment in the web plane. Due to the secondary effect, the failure occurs at the mid-span section of T-CDSCWs. Combined with the interaction curves of cross-section capacity, the interaction curves of global stability are proposed.

M.-Z. Wang, *Design Method of Concrete-Infilled Double Steel Corrugated-Plate Walls with T-section*, Springer Theses,
https://doi.org/10.1007/978-981-95-7789-7_5

5.1 Flexural Buckling About the Asymmetric Axis Under Axial Compression

For the two forms of buckling modes of axially loaded T-CDSCWs, this section focuses on the flexural buckling about the asymmetric axis, including the elastic buckling performance and load-carrying capacity. This section first analyses the elastic flexural buckling mode and proposes the corresponding formulas to evaluate the elastic buckling load. The borderline between the elastic flexural and flexural-torsional buckling is proposed to distinguish the first-order elastic buckling mode. Then, the effect of the initial geometric imperfection is considered, and the load-carrying performance of axially loaded T-CDSCWs is investigated, including the stress distribution and the load–displacement relationship. Based on the numerical results and theoretical analyses, the strength reduction factor curve is proposed to evaluate the load-carrying capacity of flexural instability.

5.1.1 *Elastic Buckling*

Based on the solid–shell finite element model, the elastic flexural buckling of axially loaded T-CDSCWs is analysed. The material properties of steel are as follows: the elastic modulus $E_s = 2.06 \times 10^5$ MPa, and the Poisson's ratio $\upsilon_s = 0.3$ (MOHURD and AQSIQ 2017). The material properties of concrete are as follows: the elastic modulus $E_c = 3.25 \times 10^4$ MPa, and the Poisson's ratio $\upsilon_c = 0.2$ (MOHURD and AQSIQ 2010). The other material properties are the same as Chap. 3.

The boundary condition of the T-CDSCW is pin ended. The top and bottom sections are coupled with the reference points at the centroid, which is weighted with the elastic modulus of the materials. The T-CDSCW is loaded at the reference points. The displacement and the rotation about the z-axis of the bottom reference point are restrained. The displacement along x- and y-axis of the top reference point are restrained. The axial compression N is applied at the top reference point.

The finite element simulations analyse the effects of geometric parameters, including the flange breadth b_f, the web breadth b_w, the wall height H_w, the wall thickness d_w, the steel plate thickness t_s and the geometric parameters of boundary elements $b_c \times d_c \times t_c$, on the flexural buckling behaviour (see Table 5.1). The

Table 5.1 Numerical examples of elastic buckling of axially loaded T-CDSCWs

NO.	Flange breadth b_f (mm)	Web breadth b_w (mm)	Wall height H_w (m)	Wall thickness d_w (mm)	Steel plate thickness t_s (mm)	CFST b_c and d_c (mm)	Steel tube thickness t_c (mm)
A-1	720	180–4320	3–9	130–200	3–7	140–210	7–11
A-2	1440	180–2160	3–9	130–200	3–7	140–210	7–11
A-3	2160	180–2160	3–9	130–200	3–7	140–210	7–11
A-4	2880	180–2160	3–9	130–200	3–7	140–210	7–11

parameters of numerical examples in Table 5.1 are within the range commonly used in engineering practice. The size of boundary elements corresponds with that of wall elements: the steel tube breadths b_c and d_c are 10 mm larger than d_w, and the steel tube thickness t_c is 4 mm larger than the steel plate thickness t_s. The parameters of steel corrugated plates are: $q = 180$ mm, $q_1 = 85$ mm, $q_2 = 27.5$ mm, $q_3 = 40$ mm, $a = 35$ mm. Intermediate bolts are arranged in the staggered way, and the vertical bolt spacing d_v is 350 mm. The normalised slenderness ratio of contact buckling of steel corrugated plates λ_{pc} is not greater than 0.4, the upper limit given by Eq. (4.18), so the effect of local buckling of steel corrugated plates can be neglected.

The elastic buckling mode and corresponding buckling load of the axially loaded T-CDSCW are obtained by the finite element buckling analysis. The elastic flexural buckling of the T-CDSCW shows different deformation modes in accordance with the first-order elastic buckling mode (see Fig. 5.1). The deformation mode of

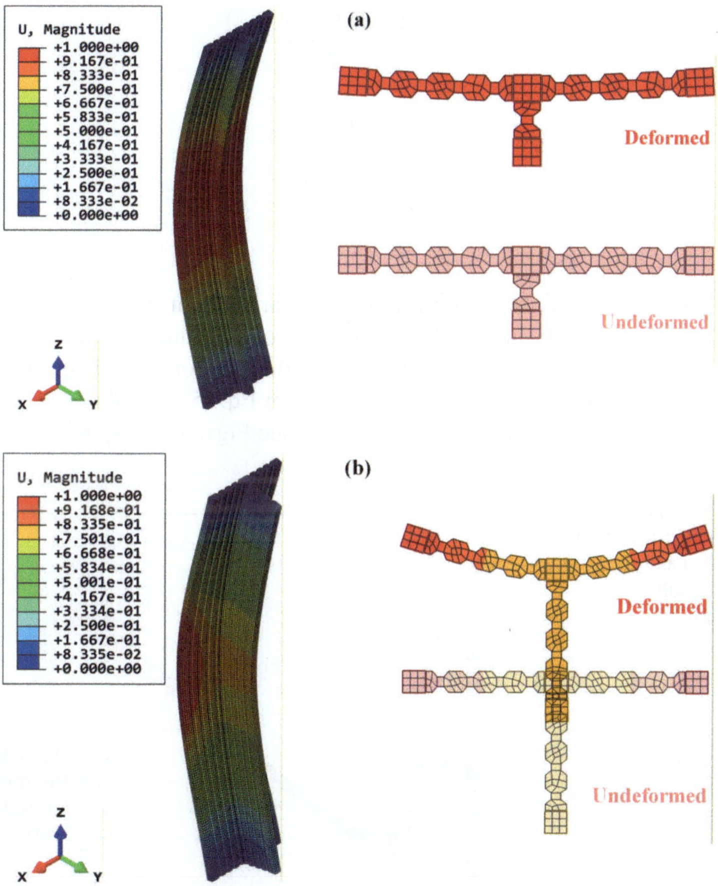

Fig. 5.1 Flexural buckling mode of T-CDSCWs ($H_w = 9000$ mm, $b_f = 1440$ mm): (**a**) $b_w = 180$ mm; (**b**) $b_w = 360$ mm

flexural buckling depends on the ratio of the overall web breadth B_w to the overall flange breadth B_f. When B_w/B_f is small and the first-order mode is flexural buckling, the deformation of the T-CDSCW is mainly global flexure about the asymmetric axis without the distortion of the cross-section. As B_w/B_f gets larger, the flexural stiffness of the T-CDSCW about the asymmetric axis increases. The flexural buckling occurs in a higher-order mode, and the first-order mode is transformed into flexural-torsional buckling. In such a case, the flexural buckling in the higher-order mode shows not only the global flexure of the T-CDSCW but also the distortion of the flange.

Based on the theory of elastic stability (Timoshenko and Gere 2009), the elastic flexural buckling load $N_{cr,x}$ can be evaluated by Eq. (5.1). The yield strength N_y can be evaluated by Eq. (5.2). To evaluate the flexural buckling behaviour of T-CDSCWs, a dimensionless variable, the normalised slenderness ratio of flexural buckling λ_x as defined in Eq. (5.3), is introduced.

$$N_{\mathrm{cr},x} = \frac{\pi^2 \left(E_{\mathrm{s}} I_{x\mathrm{s}} + E_{\mathrm{c}} I_{x\mathrm{c}}\right)}{H_{\mathrm{w}}^2} \tag{5.1}$$

$$N_{\mathrm{y}} = f_{\mathrm{y}} A_{\mathrm{s}} + f_{\mathrm{c}} A_{\mathrm{c}} \tag{5.2}$$

$$\lambda_x = \sqrt{\frac{N_{\mathrm{y}}}{N_{\mathrm{cr},x}}} \tag{5.3}$$

Where I_{xs} and I_{xc} denote the second moment of area about the asymmetric axis of steel and concrete, and A_s and A_c denote the area of steel and concrete.

No matter the first-order mode of flexural or flexural-torsional buckling, the normalised flexural slenderness ratio λ_x derived from Eq. (5.3) is always less than $\lambda_{x,n}$ obtained from finite element numerical results (see Figs. 5.2, 5.3, 5.4, 5.5, and 5.6),

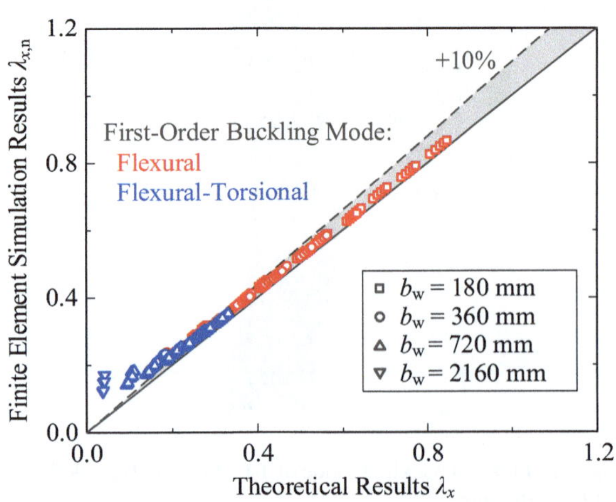

Fig. 5.2 Normalised flexural slenderness ratio of theoretical and numerical results: b_f = 720 mm

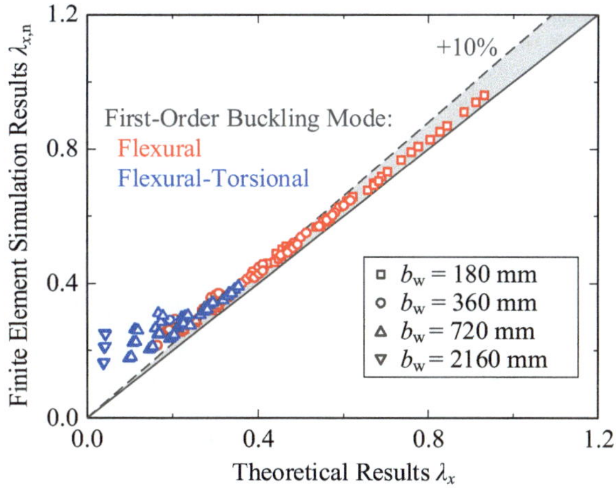

Fig. 5.3 Normalised flexural slenderness ratio of theoretical and numerical results: $b_f = 1440$ mm

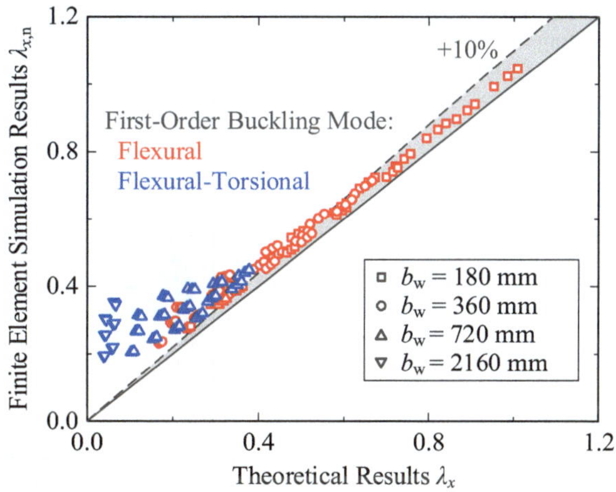

Fig. 5.4 Normalised flexural slenderness ratio of theoretical and numerical results: $b_f = 2160$ mm

which indicates that the Euler load is always larger than the elastic flexural buckling load of numerical results. The deviation comes from the difference between the derivation of the Euler load and the deformation in the finite element analysis. In the derivation of the Euler load, it is assumed that there is no distortion of the cross-section during buckling. On the contrary, due to the refinement of the finite element model, the distortion can occur in the finite element analysis, which reduces the elastic flexural buckling load. Especially for the numerical examples with the first-order mode of flexural-torsional buckling, the value of λ_x is generally less than 0.4, the numerical results $\lambda_{x,n}$ can even be many times greater than λ_x, and the deviation increases with the web breadth b_f. For T-CDSCWs with the first-order mode of flexural-torsional buckling, λ_x is small and the load-carrying capacity is not governed by flexural buckling, so the deviation has insignificant effects. For T-CDSCWs

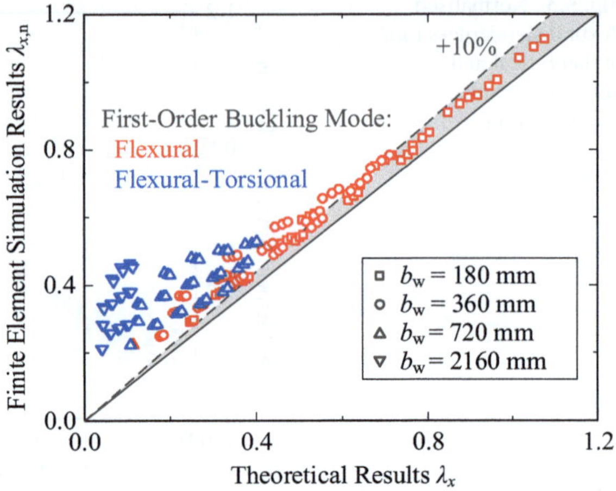

Fig. 5.5 Normalised flexural slenderness ratio of theoretical and numerical results: $b_f = 2880$ mm

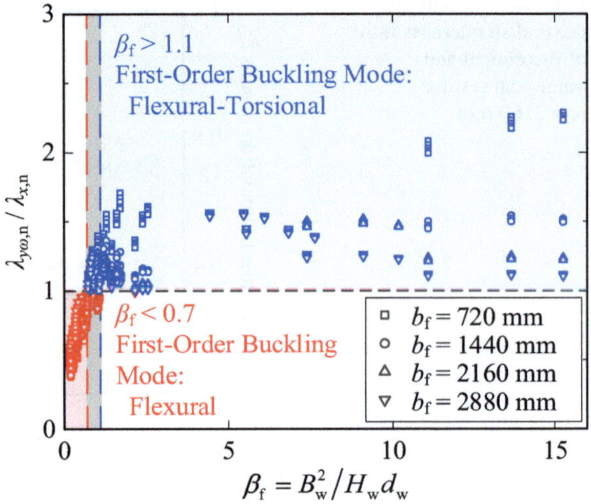

Fig. 5.6 First-order elastic buckling modes of axially loaded T-CDSCWs

with the first-order mode of flexural buckling, λ_x is large and the load-carrying capacity is governed by flexural buckling, and the deviation is small.

The design of axially loaded T-CDSCWs differs on the flexural and flexural-torsional buckling modes and would be simplified if the buckling mode that governs the failure mode is determined prior to evaluating the load-carrying capacity of different buckling modes. The first-order mode of T-CDSCWs is determined in accordance with the overall geometry, which is reflected in the dimensionless parameter β_f defined in Eq. (5.4). The parameter β_f is induced from the ratios representing the flexural and flexural-torsional behaviours of T-CDSCWs. The flexural stiffness of T-CDSCWs about the asymmetric axis is directly related to the ratio of the wall height to the overall web breadth H_w/B_w, and the flexural buckling generally corresponds with a larger height-breadth ratio H_w/B_w. The flexural-torsional behaviour about the symmetric axis is related to the ratio of the overall web breadth to the wall thickness B_w/d_w, and the flexural-torsional buckling generally corresponds with a larger breadth-thickness ratio B_w/d_w. In essence, the parameter β_f is the ratio of the height-breadth ratio H_w/B_w to the breadth-thickness ratio B_w/d_w. Therefore, large β_f indicates flexural buckling as the first-order mode, and small β_f indicates flexural-torsional buckling as the first-order mode.

$$\beta_f = \frac{B_w^2}{H_w d_w} \tag{5.4}$$

$$\lambda_{y\omega} = \sqrt{\frac{N_y}{N_{cr,y\omega}}} \tag{5.5}$$

The parameter β_f reflects the transformation trend of the first-order mode well (see Fig. 5.6). Similar to flexural buckling, the elastic flexural-torsional buckling load $N_{cr,y\omega}$ also lead to the normalised flexural-torsional slenderness ratio $\lambda_{y\omega}$ defined as Eq. (5.5). With the ratio of the numerical results $\lambda_{y\omega,n}/\lambda_{x,n}$ less than 1.0, the first-order mode of the numerical result is flexural buckling, and with the ratio of the numerical results $\lambda_{y\omega,n}/\lambda_{x,n}$ greater than 1.0, the first-order mode of the numerical result is flexural-torsional buckling. Based on the numerical results, the first-order mode of the axially loaded T-CDSCW is always flexural buckling with β_f less than 0.7, and flexural-torsional buckling with β_f greater than 1.1.

The parameter β_f does not take into consideration the overall flange breadth B_f, but such exclusion has insignificant influence on determining the borderline between different buckling modes. The first-order mode is transformed from flexural to flexural-torsional buckling as the web gets wider. With β_f around the borderline, the web is generally narrower than the flange and solely influences the buckling mode. As β_f exceeds the borderline, the influence of the overall flange breadth B_f is revealed gradually: the wider the flange extends, the less $\lambda_{y\omega,n}/\lambda_{x,n}$ becomes. This is because the increasing of β_f transforms the deformation of the flexural buckling from global flexure to the distortion of the flange, and leads to the buckling load of flexural buckling approaching that of flexural-torsional buckling. However, the web is so wide that $\lambda_{y\omega,n}/\lambda_{x,n}$ cannot go below the lower limit of 1.0 and the first-order mode is always flexural-torsional buckling. From the perspective of engineering design, despite no consideration of the flange breadth, the parameter β_f represents the borderline between flexural and flexural-torsional buckling properly.

In accordance with the parameter β_f, the elastic buckling mode of the axially loaded T-CDSCW is determined, and the corresponding stability design can be carried out. It is understandable that T-CDSCWs with β_f less than 0.7 shall be designed in accordance with flexural buckling and T-CDSCWs with β_f greater than 1.1 shall be designed in accordance with flexural-torsional buckling. For T-CDSCWs with β_f between 0.7 and 1.1, the buckling mode of stability design is uncertain merely based on the value of β_f and the design method should be further studied.

5.1.2 Load-Carrying Capacity

Based on the elastic buckling analysis, the solid–shell finite element model is used to analyse the flexural instability of axially loaded T-CDSCWs. The material properties of steel are as follows: the yield strength f_y = 345 MPa, the elastic modulus $E_s = 2.06 \times 10^5$ MPa, and the Poisson's ratio $\upsilon_s = 0.3$ (MOHURD and AQSIQ 2017). The material properties of concrete are as follows: the axial compressive strength f_c = 26.8 MPa, the elastic modulus $E_c = 3.25 \times 10^4$ MPa, and the Poisson's ratio $\upsilon_c = 0.2$ (MOHURD and AQSIQ 2010). The other material properties are the same as Chap. 3. The initial geometric imperfection of T-CDSCWs is in the form of the first-order elastic buckling, and the amplitude of the imperfection is $H_w/500$.

The finite element simulations analyse the effects of geometric parameters, including the flange breadth b_f, the web breadth b_w, the wall height H_w, the wall thickness d_w, the steel plate thickness t_s, and the geometric parameters of boundary elements $b_c \times d_c \times t_c$, on the load-carrying behaviour (see Table 5.2). The parameters of numerical examples are within the range commonly used in engineering practice. The size of boundary elements corresponds with that of wall elements: the steel tube breadths b_c and d_c are 10 mm larger than d_w, and the steel tube thickness t_c is 4 mm larger than the steel plate thickness t_s. The parameters of steel corrugated plates are: q = 180 mm, q_1 = 85 mm, q_2 = 27.5 mm, q_3 = 40 mm, a = 35 mm. Intermediate bolts are arranged in the staggered way, and the vertical bolt spacing d_v is 350 mm. The normalised slenderness ratio of contact buckling of steel corrugated plates λ_{pc} is not greater than 0.4, the upper limit given by Eq. (4.18), so the effect of local buckling of steel corrugated plates can be neglected.

From the aspect of elastic buckling load, the buckling directions are equivalent with the flange or the web on the convex side. However, from the aspect of load-carrying capacity, the buckling directions are different. With the flange on the convex side, the flange is under tension and the web is under compression. With the web on the convex side, the flange is under compression and the web is under tension. The failure modes of T-CDSCWs are related to the direction of initial geometric imperfection. The stress distribution of T-CDSCWs (see Fig. 5.6) corresponds to the points on the vertical load–displacement curves (see Fig. 5.7). The direction of initial geometric imperfection has no effect on the initial stiffness, but has a significant effect on the load-carrying capacity (see Fig. 5.8). Compared with the direction with the web on the convex side, the direction with the flange on the convex side

Table 5.2 Numerical examples of flexural instability of axially loaded T-CDSCWs

NO.	Flange breadth b_f (mm)	Web breadth b_w (mm)	Wall height H_w (m)	Wall thickness d_w (mm)	Steel plate thickness t_s (mm)	CFST b_c and d_c (mm)	Steel tube thickness t_c (mm)
B-1	1440	360	3–9	130–200	5	140–210	9
B-2	1440	360	6–9	130–200	3–7	140–210	7–11
B-3	720–2880	180–720	6	150	5	160	9
B-4	720–2880	180–720	3–9	130	3	140	7

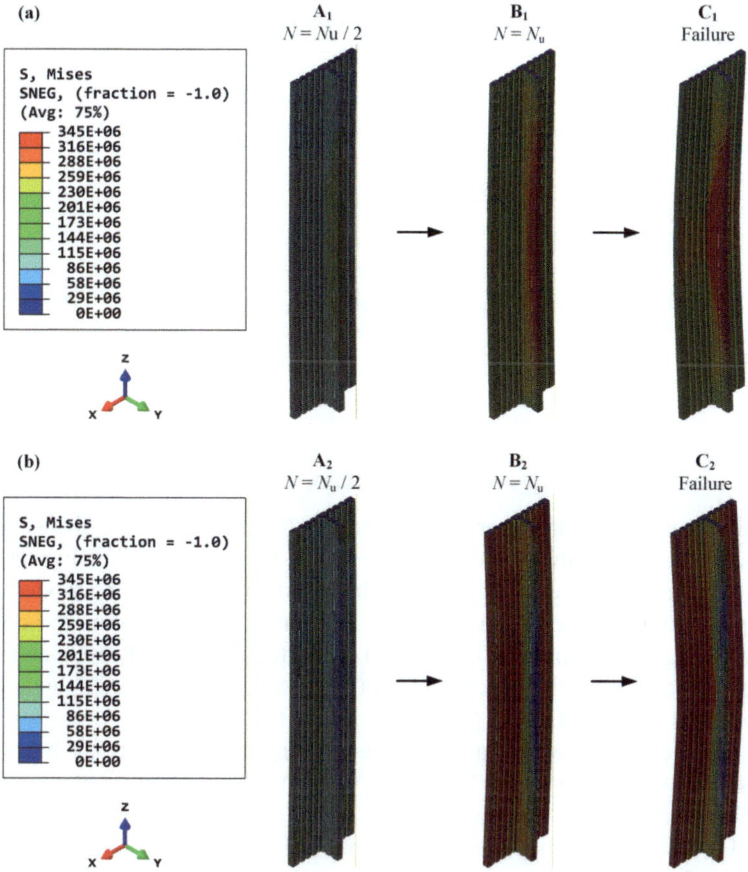

Fig. 5.7 The effect of the direction of initial geometric imperfection on the load-carrying performance ($H_w = 9000$ mm, $b_f = 1440$ mm, $b_w = 360$ mm, $d_w = 130$ mm, $t_s = 3$ mm, unit = Pa): (**a**) the initial geometric imperfection with the flange on the convex side; (**b**) the initial geometric imperfection with the web on the convex side

leads to smaller load-carrying capacity. With the web on the convex side, the plastic zone concentrate in the flange with the larger area, and more compression is shared by the concrete. With the flange on the convex side, the plastic zone concentrates in the web with the smaller area, and the load is less shared by the concrete and thus the load-carrying capacity is reduced. Therefore, the direction with the flange on the convex side is the most disadvantageous direction of flexural buckling and considered in the following finite element analysis.

Generally, the load-carrying behaviour of axially loaded members is mainly influenced by the initial geometric imperfection of the consistent first-order bucking mode. However, the first-order buckling mode of axially loaded T-CDSCWs is not merely in one form, so the imperfection of combined flexural and flexural-torsional buckling is investigated to propose a reliable design method of flexural buckling. Since the unconstrained flange breadth $B_f/2$ of T-CDSCWs with β less than 1.1 is

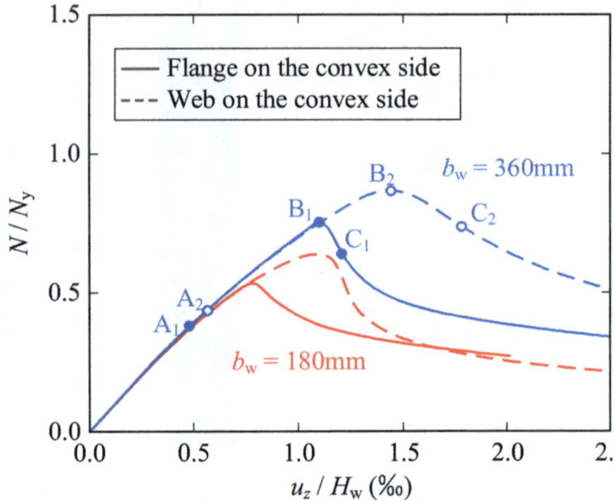

Fig. 5.8 The effect of the direction of initial geometric imperfection on the load–displacement relationship

generally greater than the overall web breadth B_w, the maximum displacement of both buckling modes always occurs at the flange tip of the mid-height section (see Fig. 5.6). Therefore, to equally consider the effects of two buckling modes, the linear superposition of flexural and flexural-torsional buckling forms the combined imperfection. The amplitude of flexural imperfection is $a_1H_w/500$, and that of flexural-torsional imperfection is $a_2H_w/500$. The sum of the factors a_1 and a_2 equals one, which is expressed in Eq. (5.6).

As expected, the flexural imperfection governs the load-carrying behaviour of T-CDSCWs with β less than 0.7 (see Fig. 5.9). As the decreasing of a_1 transforms flexural imperfection to flexural-torsional imperfection, the load-carrying capacity of T-CDSCWs approaches the yield load N_y. However, for T-CDSCWs with β between 0.7 and 1.1, the combined imperfection has no significant effect on the load-carrying behaviour no matter what mode is the first-order, and, in fact, the load-carrying capacity is close to N_y, and the strength reduction factor is generally greater than 0.9 (see Figs. 5.10, and 5.11). This is because the overall web breadth B_w is not so small that the load-carrying capacity of flexural buckling is small and not so large that the load-carrying capacity of flexural-torsional buckling is small. No matter the first-order buckling mode is flexural or flexural-torsional buckling, the initial geometric imperfection of the consistent first-order bucking mode is the most disadvantage imperfection combination or close to the most disadvantage imperfection combination. Therefore, for T-CDSCWs with β less than 1.1, the first-order bucking mode can be selected as the most disadvantageous buckling mode for initial geometric imperfection and considered in the following finite element analysis.

The load-carrying behaviour of axially loaded T-CDSCWs is obtained by finite element nonlinear analysis. The failure of T-CDSCWs occurs at the mid-height section (see Fig. 5.7). In the initial loading stage, the stress on the concave becomes greater than that on the convex side. When the load-carrying capacity is reached, steel corrugated plates and steel tubes yield, and the T-CDSCW deflects without the distortion of the cross-section in the same direction with the initial imperfection. As

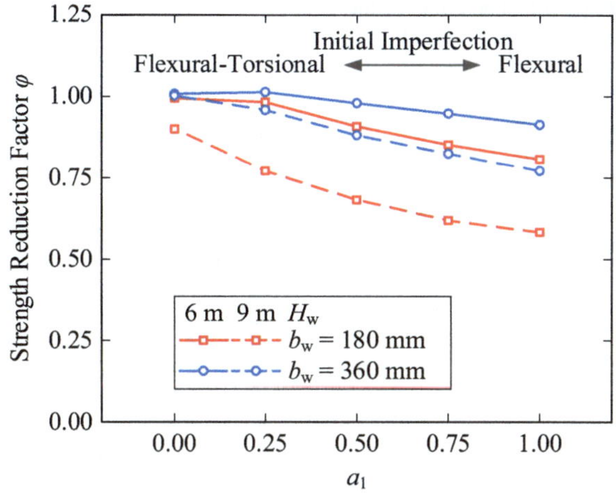

Fig. 5.9 The effect of the combination of initial geometric imperfection on the load-carrying performance: flexural buckling of typical numerical examples with $\beta < 0.7$

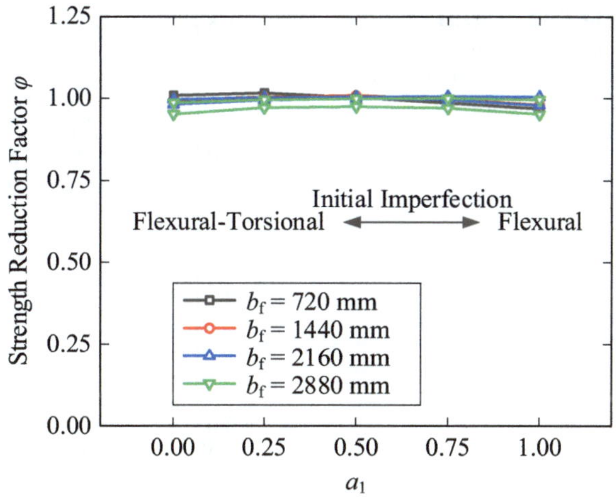

Fig. 5.10 The effect of the combination of initial geometric imperfection on the load-carrying performance: T-CDSCWs with flexural buckling as the first-order elastic buckling mode and $0.7 < \beta < 1.1$

the loading continues, the deflection and yielding develop, leading to the load decreases.

$$a_1 + a_2 = 1 \tag{5.6}$$

Various geometric parameters have different effects on the load–displacement relationship of T-CDSCWs (see Figs. 5.12, 5.13, 5.14, 5.15, and 5.16). The initial slope of normalised load–displacement curves is basically the same, and the difference mainly focuses on the load-carrying capacity and the corresponding strength reduction factor φ_x, which is the ratio of the load-carrying capacity N_u to the yield load N_y. As the wall height H_w decreases or the wall thickness d_w increases, the elastic buckling load of T-CDSCWs increases, the normalised flexural slenderness ratio λ_x

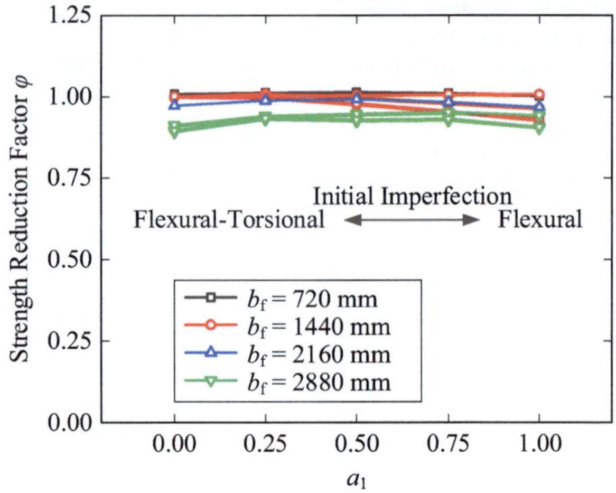

Fig. 5.11 The effect of the combination of initial geometric imperfection on the load-carrying performance: T-CDSCWs with flexural-torsional buckling as the first-order elastic buckling mode and $0.7 < \beta < 1.1$

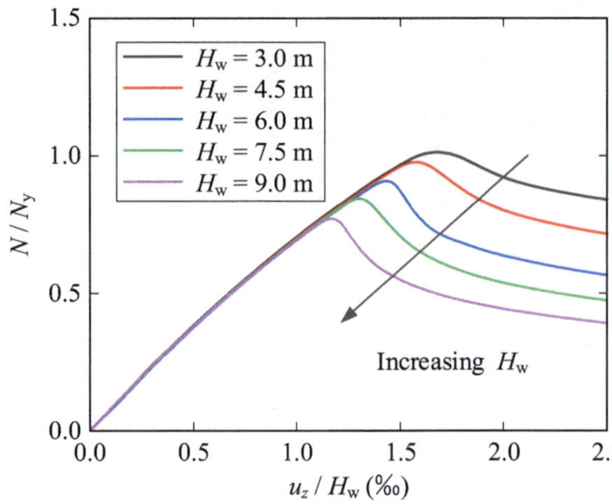

Fig. 5.12 Vertical load–displacement relationship of numerical examples B-1: d_w = 150 mm

decreases, so the strength reduction factor φ_x increases. As the steel plate thickness t_s increases, the strength reduction factor φ_x increases slightly.

The change of the web breadth b_w may transform the first-order buckling mode. Take numerical examples B-3 as an example, as the web breadth b_w increases, the strength reduction factor φ_x increases first and decreases then. The first increase results from the plastic zone expands as the web breadth b_w develops, while the following decrease is caused by that the significant stiffening effect of the web transforms the first-order buckling mode from flexural buckling to flexural-torsional buckling, and thus the failure mode following the transformation is not related to flexural buckling. In such a situation, continuing to increase the web breadth b_w leads to a smaller torsional stiffness and correspondingly a smaller load-carrying capacity of flexural-torsional instability.

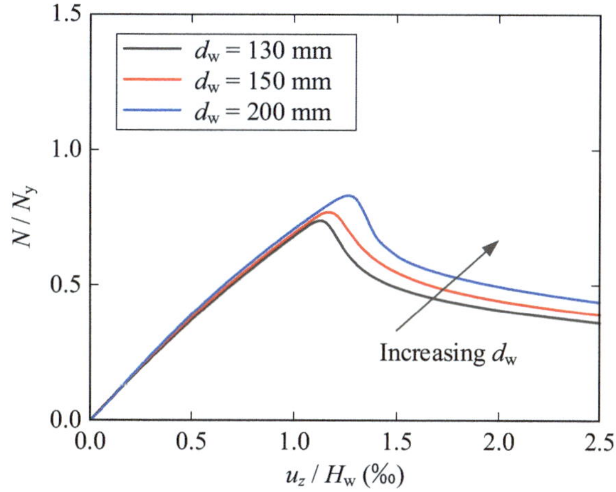

Fig. 5.13 Vertical load–displacement relationship of numerical examples B-2: $t_s = 5$ mm

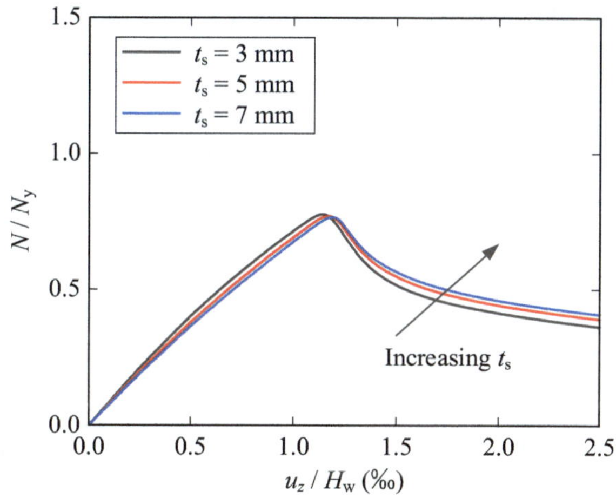

Fig. 5.14 Vertical load–displacement relationship of numerical examples B-2: $d_w = 150$ mm

The trend of the strength reduction factor φ_x is delineated pretty well by the normalised flexural slenderness ratio λ_x (see Figs. 5.17, 5.18, 5.19, 5.20, 5.21, and 5.22). The numerical results of φ_x and λ_x almost lie around one curve, no matter the variation of the flange breadth b_f, the web breadth b_w, the wall thickness d_w or the steel plates thickness t_s. From the numerical results, the lower envelope, average and upper envelope curves of the strength reduction factor φ_x are fitted in the form of Eqs. (5.7) and (5.8).

$$\varphi_x = \begin{cases} 1-\alpha_1\lambda_x^2 & \lambda_x \le 0.3 \\ \dfrac{1}{\Phi+\sqrt{\Phi^2-\lambda_x^2}} & \lambda_x > 0.3 \end{cases} \tag{5.7}$$

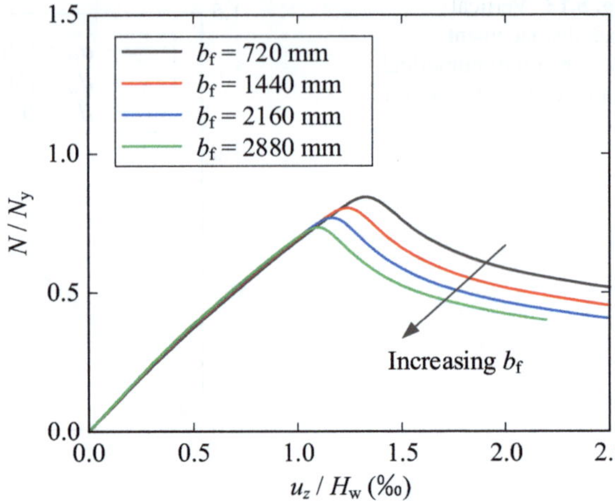

Fig. 5.15 Vertical load–displacement relationship of numerical examples B-3: $b_w = 180$ mm

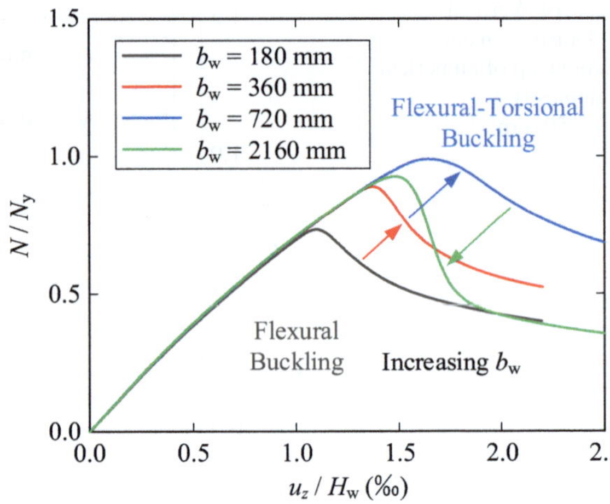

Fig. 5.16 Vertical load–displacement relationship of numerical examples B-3: $b_f = 2880$ mm

$$\Phi = 0.5\left[1+\alpha_2\left(\lambda_x-\alpha_3\right)+\lambda_x^2\right] \tag{5.8}$$

Where α_1, α_2, and α_3 are the shape parameters. For the lower envelope curve, which is the design curve, $\alpha_1 = 0.652$, $\alpha_2 = 0.884$, $\alpha_3 = 0.236$. For the average curve, $\alpha_1 = 0.171$, $\alpha_2 = 0.782$, $\alpha_3 = 0.282$. For the upper envelope curve, $\alpha_1 = 0$, $\alpha_2 = 0.696$, $\alpha_3 = 0.333$. Considering the range of parameters, Eq. (5.7) is applicable with the normalised flexural slenderness ratio λ_x no greater than 1.0. From the finite element simulation results, when the normalised flexural slenderness ratio λ_x is no greater than 0.2, the strength reduction factor φ_x approaches to 1.0, and the flexural instability can be neglected. No matter what the geometric parameters of T-CDSCWs are, Eq. (5.7) can reflect the $\varphi_x - \lambda_x$ relationship accurately (see Figs. 5.17, 5.18, 5.19, 5.20, 5.21, and 5.22).

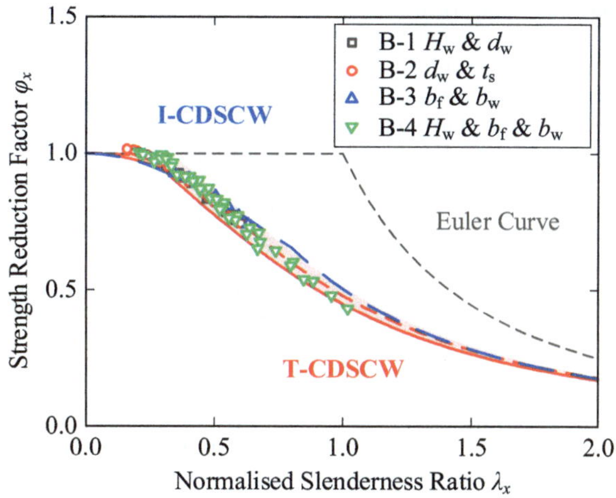

Fig. 5.17 Numerical results of the strength reduction factor and the normalised slenderness ratio of flexural instability $\varphi_x - \lambda_x$: all numerical examples

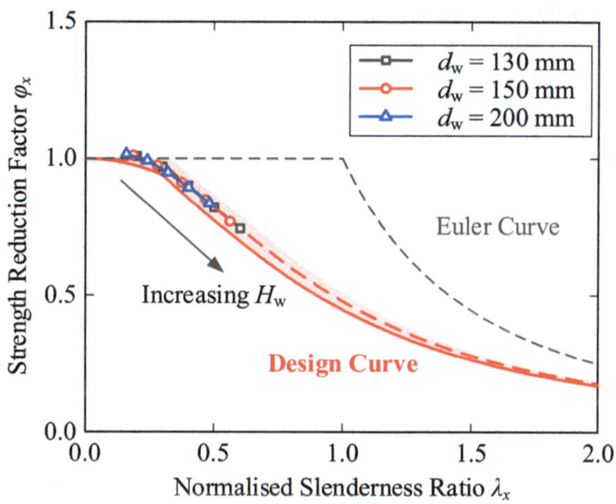

Fig. 5.18 Numerical results of the strength reduction factor and the normalised slenderness ratio of flexural buckling $\varphi_x - \lambda_x$: numerical examples B-1

Compared with the design curve of CDSCWs with I-section (I-CDSCWs) (Zhu 2020; Guo et al. 2018), the design curve of the flexural instability of T-CDSCWs underlies that of I-CDSCWs (see Fig. 5.17). This is because the T-CDSCW can be viewed as the superposition of the I-CDSCW and the web wall. The elastic flexural buckling of T-CDSCWs is significantly improved by the stiffening effect of the web, while the load-carrying capacity is not improved that much due to the small area of the web. The decreasing of the slenderness ratio overwhelms the increasing of the strength reduction factor, and thus the design curve of T-CDSCWs transfer to the left below that of I-CDSCWs.

Compared with the experimental results, the design curve Eq. (5.7) is conservative. For specimen T-2 of the global stability test, the strength reduction factor φ_x is 0.933. According to the theory of elastic buckling, the effective length ratio of the

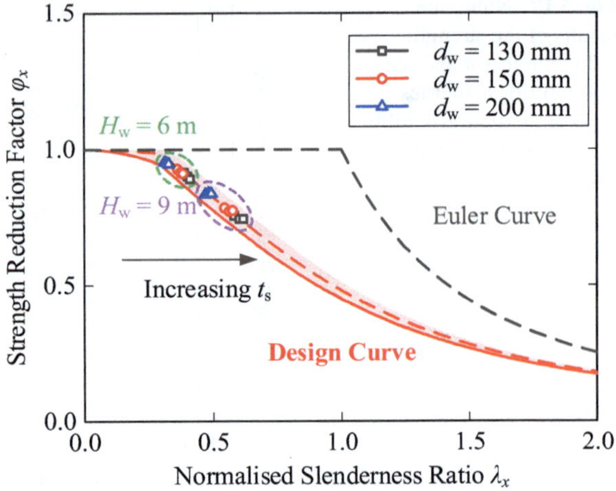

Fig. 5.19 Numerical results of the strength reduction factor and the normalised slenderness ratio of flexural buckling $\varphi_x - \lambda_x$: numerical examples B-2

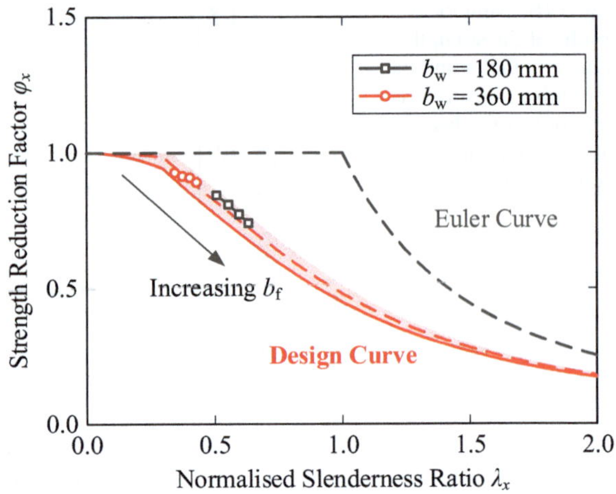

Fig. 5.20 Numerical results of the strength reduction factor and the normalised slenderness ratio of flexural buckling $\varphi_x - \lambda_x$: numerical examples B-3

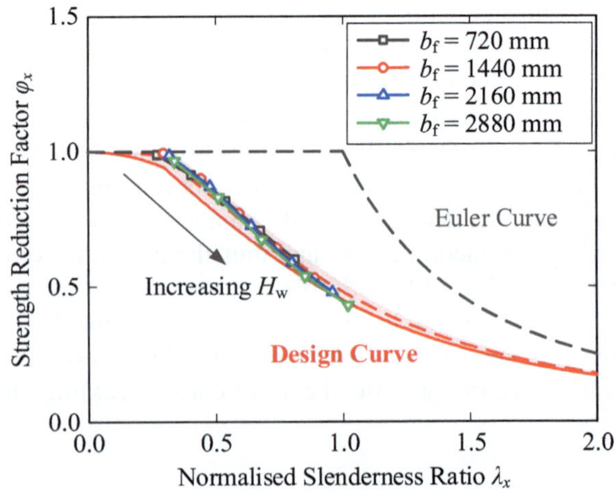

Fig. 5.21 Numerical results of the strength reduction factor and the normalised slenderness ratio of flexural buckling $\varphi_x - \lambda_x$: numerical examples B-4, b_w = 180 mm

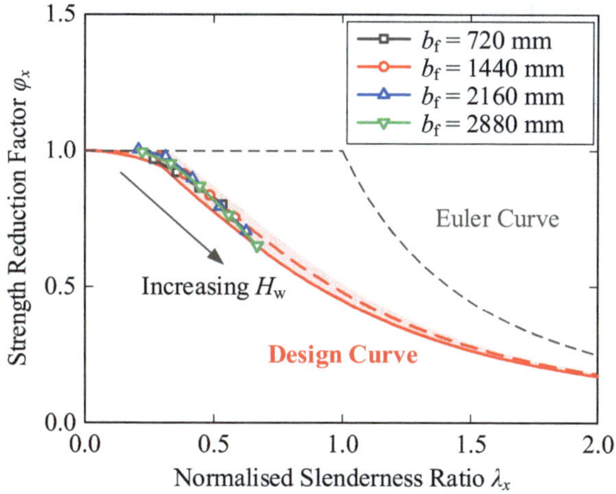

Fig. 5.22 Numerical results of the strength reduction factor and the normalised slenderness ratio of flexural buckling $\varphi_x - \lambda_x$: numerical examples B-4, $b_w = 360$ mm

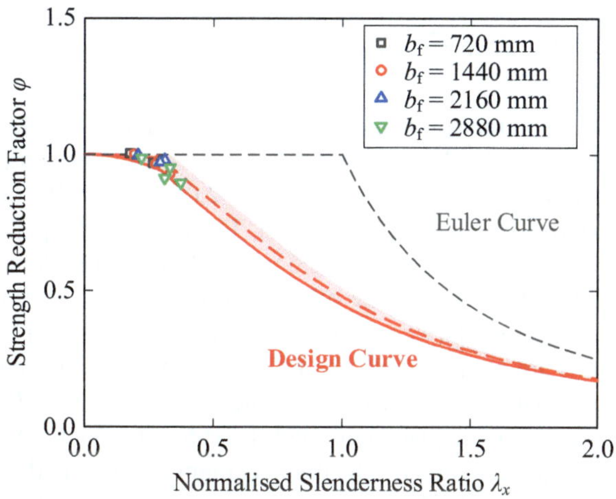

Fig. 5.23 Stability design of axially loaded T-CDSCWs with $0.7 < \beta < 1.1$

test specimen is 0.840, and the normalised flexural slenderness ratio λ_x is 0.587. Thus, derived from Eq. (5.7), the design value of strength reduction factor φ_x is 0.725, which is less than the experimental result and proves the conservation of the design formulas.

The design curve Eq. (5.7) can be used as the design criteria of T-CDSCWs with β between 0.7 and 1.1. No matter the first-order buckling mode is flexural or flexural-torsional buckling, the normalised slenderness ratio is evaluated by Eq. (5.3) in accordance with flexural buckling. Moreover, the numerical results of the strength reduction factor are the minimum value of the numerical results with combined initial imperfection. Since almost all of the numerical results lie above the design curve (see Fig. 5.23), it is proved that T-CDSCWs with β less than 1.1 can be designed by Eq. (5.7) in accordance with flexural instability.

5.2 Flexural-Torsional Buckling About the Symmetric Axis Under Axial Compression

For the two forms of buckling modes of axially loaded T-CDSCWs, this section focuses on the flexural-torsional buckling about the symmetric axis, including the elastic buckling performance and load-carrying capacity. This section first analyses the elastic flexural-torsional buckling mode and proposes the corresponding formulas to evaluate the elastic buckling load. The borderline between the elastic flexural-torsional and torsional buckling is investigated to simplify the design. Then, the effect of the initial geometric imperfection is considered, the load-carrying performance of axially loaded T-CDSCWs is investigated, including the stress distribution and load–displacement relationship. Based on the numerical results and theoretical analyses, the strength reduction factor curve is proposed to evaluate the load-carrying capacity of flexural-torsional instability.

5.2.1 *Elastic Buckling*

Based on the solid–shell finite element model, the elastic flexural-torsional buckling of axially loaded T-CDSCWs is analysed. The material properties of steel are as follows: the elastic modulus $E_s = 2.06 \times 10^5$ MPa, and the Poisson's ratio $\upsilon_s = 0.3$ (MOHURD and AQSIQ 2017). The material properties of concrete are as follows: the elastic modulus $E_c = 3.25 \times 10^4$ MPa, and the Poisson's ratio $\upsilon_c = 0.2$ (MOHURD and AQSIQ 2010). The other material properties are the same as Chap. 3.

The boundary condition of the T-CDSCW is pin ended. The top and bottom sections are coupled with the reference points at the centroid, which is weighted with the elastic modulus of the materials. The T-CDSCW is loaded at the reference points. The displacement and the rotation about the z-axis of the bottom reference point are restrained. The displacement along x- and y-axis of the top reference point are restrained. The axial compression N is applied at the top reference point.

The finite element simulations analyse the effects of geometric parameters, including the flange breadth b_f, the web breadth b_w, the wall height H_w, the wall thickness d_w, the steel plate thickness t_s and the geometric parameters of boundary elements $b_c \times d_c \times t_c$, on the flexural-torsional buckling behaviour (see Table 5.1). The parameters of numerical examples in Table 5.1 are within the range commonly used in engineering practice. The size of boundary elements corresponds with that of wall elements: the steel tube breadths b_c and d_c are 10 mm larger than d_w, and the steel tube thickness t_c is 4 mm larger than the steel plate thickness t_s. The parameters of steel corrugated plates are: $q = 180$ mm, $q_1 = 85$ mm, $q_2 = 27.5$ mm, $q_3 = 40$ mm, $a = 35$ mm. Intermediate bolts are arranged in the staggered way, and the vertical bolt spacing d_v is 350 mm. The normalised slenderness ratio of contact buckling of steel corrugated plates λ_{pc} is not greater than 0.4, the upper limit given by Eq. (4.18), so the effect of local buckling of steel corrugated plates can be neglected.

The elastic buckling mode and corresponding buckling load of the axially loaded T-CDSCW are obtained by the finite element buckling analysis. The deformation of flexural-torsional buckling about the symmetric axis includes two parts (see Fig. 5.24): the flexure about the symmetric axis and the twist about the shear centre. The flexural-torsional buckling load $N_{cr,y\omega}$ and the ratios of flexure and twist depend on the values of the flexural buckling load about the symmetric axis $N_{cr,y}$ and the torsional buckling load about the shear centre $N_{cr,\omega}$. The flexural-torsional buckling load $N_{cr,y\omega}$ is closer to the smaller load of $N_{cr,y}$ and $N_{cr,\omega}$, and the deformation is governed by the corresponding mode. Within the range of numerical examples, when the breadth-to-thickness ratio $(B_f + B_w)/d_w$ is large, the elastic buckling mode is mainly twist about the shear centre, and the flexure is small. When the breadth-to-thickness ratio $(B_f + B_w)/d_w$ is small, the elastic buckling mode is the flexure about

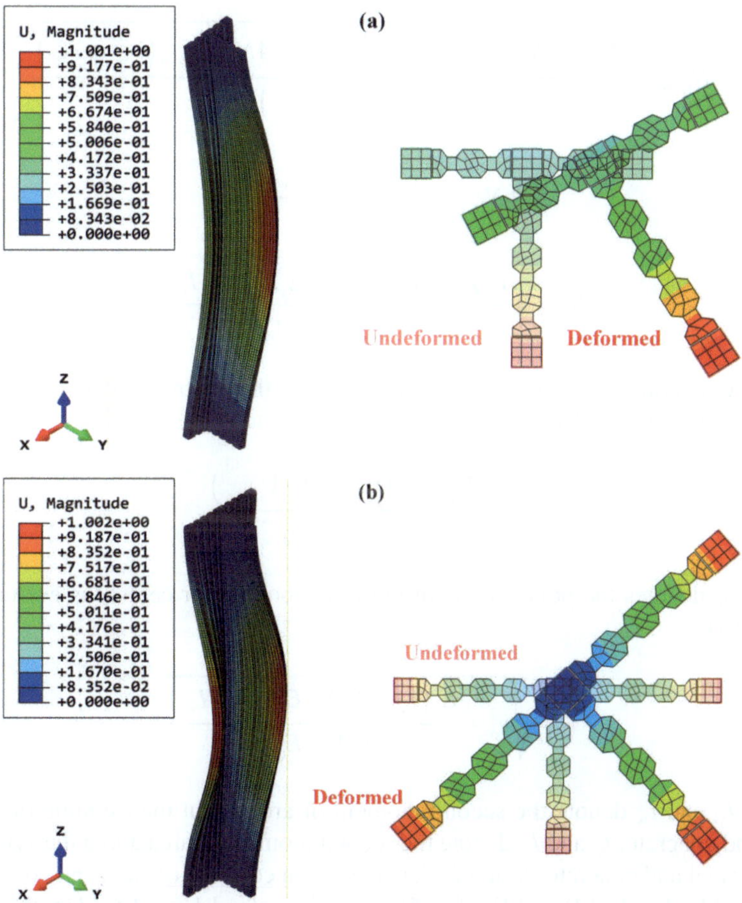

Fig. 5.24 Flexural-torsional buckling mode of T-CDSCWs (H_w = 9000 mm, b_w = 720 mm): (**a**) b_f = 720 mm; (**b**) b_f = 1440 mm

the symmetric axis and twist about the shear centre. It is noted that the torsional buckling about the shear centre is equivalent to the buckling of flange and web walls. The elastic torsional buckling load is the sum of the buckling load of wall elements with three sides simply supported and one side free. That is to say, as the breadth-to-thickness ratio $(B_f + B_w)/d_w$ increases, the flexural-torsional buckling mode is close to the buckling of wall elements.

Based on the theory of elastic stability (Timoshenko and Gere 2009), the elastic flexural-torsional buckling load $N_{cr,y\omega}$ can be evaluated by Eq. (5.9), which is determined by the flexural buckling load about the symmetric axis $N_{cr,y}$ and the torsional buckling load about the shear centre $N_{cr,\omega}$, evaluated by Eq. (5.10) and Eq. (5.11) respectively. The yield strength N_y can be evaluated by Eq. (5.2). To evaluate the flexural-torsional buckling behaviour of T-CDSCWs, a dimensionless variable, the normalised slenderness ratio of flexural buckling $\lambda_{y\omega}$ as defined in Eq. (5.5), is introduced.

$$N_{cr,y\omega} = \frac{N_{cr,y} + N_{cr,\omega} - \sqrt{\left(N_{cr,y} + N_{cr,\omega}\right)^2 - 4N_{cr,y}N_{cr,\omega}\left(1 - y_0^2 / i_0^2\right)}}{2\left(1 - y_0^2 / i_0^2\right)} \tag{5.9}$$

$$N_{cr,y} = \frac{\pi^2\left(E_s I_{ys} + E_c I_{yc}\right)}{H_w^2} \tag{5.10}$$

$$N_{cr,\omega} = \frac{G_s I_{ts} + G_c I_{tc}}{i_0^2} + \frac{\pi^2\left(E_s I_{ws} + E_c I_{wc}\right)}{\mu_w^2 H_w^2 i_0^2} \tag{5.11}$$

Where y_o denotes the shear centre distance from the centroid, which is weighted with the shear centre distance of all parts, as evaluated by Eq. (5.12).

$$y_o = \frac{\sum_i \left(E_s A_{si} y_{osi} + E_c A_{ci} y_{oci}\right)}{E_s A_s + E_c A_c} \tag{5.12}$$

Where, i_o denotes the polar radius of gyration about shear centre, as evaluated by Eq. (5.13).

$$i_o = \sqrt{y_o^2 + \frac{E_s\left(I_{xs} + I_{ys}\right) + E_c\left(I_{xc} + I_{yc}\right)}{E_s A_s + E_c A_c}} \tag{5.13}$$

Where I_{xs} and I_{xc} denote the second moment of area about the asymmetric axis of steel and concrete, I_{ys} and I_{yc} denote the second moment of area about the symmetric axis of steel and concrete, I_{ts} and I_{tc} denote torsion constant of steel and concrete, as evaluated by Eq. (5.14) and Eq. (5.15) respectively, and I_{ws} and I_{wc} denote warping constant of steel and concrete, as evaluated by Eq. (5.16) and Eq. (5.17) respectively. Since the wall elements of T-CDSCWs are not flat but corrugated, and composed of steel and concrete. In the evaluation of sectional parameters, some simplification is adopted to consider the corrugation of the section and the contribution of steel and concrete.

$$I_{\mathrm{ts}} = \frac{4\left(A_{\mathrm{s}} + A_{\mathrm{c}}\right)^2 t_{\mathrm{s}}}{p_{\mathrm{s}}} \tag{5.14}$$

$$I_{\mathrm{tc}} = \frac{1}{3}\left[\left(b_{\mathrm{f}} + 3b_{\mathrm{c1}}\right)d_{\mathrm{fc}}^3 + \left(b_{\mathrm{w}} + b_{\mathrm{c2}}\right)d_{\mathrm{wc}}^3\right] \tag{5.15}$$

$$I_{\mathrm{ws}} = \frac{t_{\mathrm{s}}}{6}\left[2\left(\frac{b_{\mathrm{f}}}{2} + b_{\mathrm{c1}}\right)^3 d_{\mathrm{f}}^2 + \left(b_{\mathrm{w}} + b_{\mathrm{c2}}\right)^3 d_{\mathrm{w}}^2\right] \tag{5.16}$$

$$I_{\mathrm{wc}} = \frac{1}{36}\left[2\left(\frac{b_{\mathrm{f}}}{2} + b_{\mathrm{c1}}\right)^3 d_{\mathrm{fe}}^3 + \left(b_{\mathrm{w}} + b_{\mathrm{c2}}\right)^3 d_{\mathrm{we}}^3\right] \tag{5.17}$$

Where p_{s} denotes the perimeter of steel, d_{fc} and d_{wc} denote the maximum thickness of the concrete in the flange and web walls, and d_{fe} and d_{we} denote the area-equivalent thickness of the concrete in the flange and web walls.

Where μ_{w} denotes the effective length coefficient of torsional buckling. In the finite element model, the twist of the T-CDSCW is restrained at the top and bottom reference points, and the top and bottom sections are coupled with the corresponding reference points and do not warp. Therefore, the value of μ_{w} is taken as 0.5.

The theoretical results of the normalised slenderness ratio $\lambda_{y\omega}$ are close to the numerical results $\lambda_{y\omega,\mathrm{n}}$ (see Figs. 5.25, 5.26, 5.27, and 5.28). Although the evaluation of the elastic buckling load adopts simplified sectional parameters, the deviation is mostly less than ±10%. For the flexural buckling of T-CDSCWs, Euler load can be used to evaluate the elastic flexural buckling load of T-CDSCWs with flexural buckling as the first-order buckling mode but show significant deviation for T-CDSCWs with flexural-torsional buckling as the first-order buckling mode. Different from flexural buckling, for flexural-torsional buckling, Eq. (5.9) shows good agreement with numerical results no matter the first-order buckling mode is flexural or flexural-torsional buckling, and theoretical results satisfy the accuracy requirement of global stability.

Since the breadth-to-thickness ratio of T-CDSCWs is large with flexural-torsional buckling as the first-order buckling mode. Therefore, the elastic flexural buckling load about the symmetric axis $N_{\mathrm{cr},y}$ is generally much greater than the elastic torsional buckling load $N_{\mathrm{cr},\omega}$. That is to say, the elastic flexural-torsional buckling load $N_{\mathrm{cr},y\omega}$ is closer to the elastic torsional buckling load $N_{\mathrm{cr},\omega}$ as the upper bound.

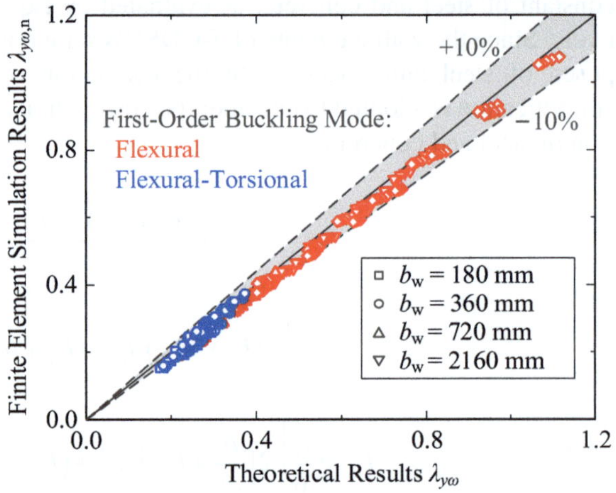

Fig. 5.25 Normalised flexural-torsional slenderness ratio of theoretical and numerical results: $b_f = 720$ mm

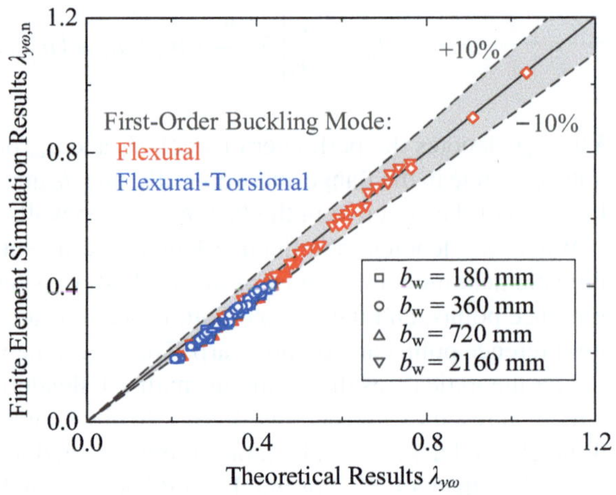

Fig. 5.26 Normalised flexural-torsional slenderness ratio of theoretical and numerical results: $b_f = 1440$ mm

The relationship among different buckling loads can be reflected by the overall geometric parameters of T-CDSCWs. The ratio of the wall height to the overall flange breadth H_w/B_f influences the flexural buckling about the symmetric axis $N_{cr,y}$: the less H_w/B_f, the greater $N_{cr,y}$. The ratio of the overall breadth to the wall thickness $(B_f + B_w)/d_w$ influences the torsional buckling about the shear centre $N_{cr,\omega}$: the greater $(B_f + B_w)/d_w$, the less $N_{cr,\omega}$. By dividing the breadth-to-thickness ratio $(B_f + B_w)/d_w$ by the height-to-breadth ratio of the flange H_w/B_f, the dimensionless parameter β_t is obtained as Eq. (5.18). As β_t increases, the flexural buckling load about the symmetric axis $N_{cr,y}$ becomes much greater than the torsional buckling load $N_{cr,\omega}$, which is closer to the flexural-torsional buckling load $N_{cr,y\omega}$.

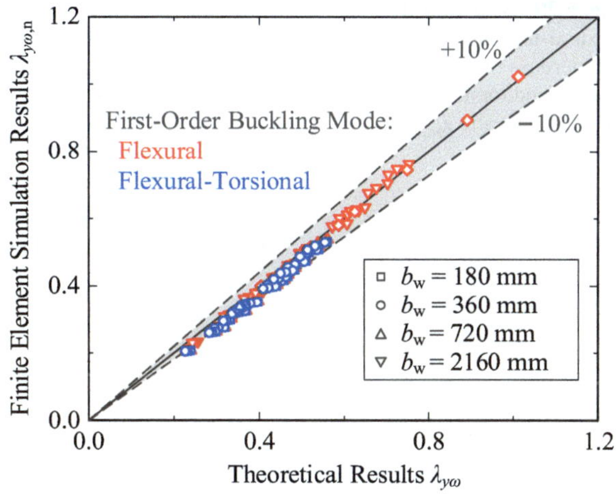

Fig. 5.27 Normalised flexural-torsional slenderness ratio of theoretical and numerical results: $b_f = 2160$ mm

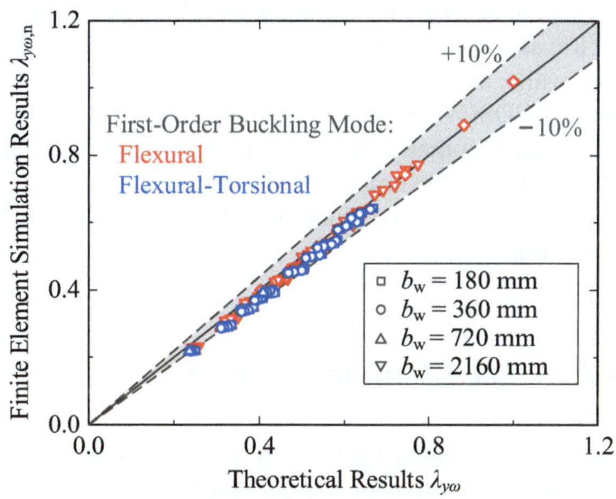

Fig. 5.28 Normalised flexural-torsional slenderness ratio of theoretical and numerical results: $b_f = 2880$ mm

$$\beta_t = \frac{B_f\left(B_f + B_w\right)}{H_w d_w} \tag{5.18}$$

$$\lambda_\omega = \sqrt{\frac{N_y}{N_{cr,\omega}}} \tag{5.19}$$

The normalised flexural-torsional slenderness ratio $\lambda_{y\omega}$ is the upper bound of the normalised torsional slenderness ratio λ_ω defined in Eq. (5.19). The increasing of β_t leads to the ratio of λ_ω to $\lambda_{y\omega}$ approaching the maximum of 1.0 (see Fig. 5.29). The difference between λ_ω and $\lambda_{y\omega}$ is significant with $\beta_t < 3.0$, while the difference is quite small with $\beta_t > 3.0$ and the flexural-torsional buckling about the symmetric

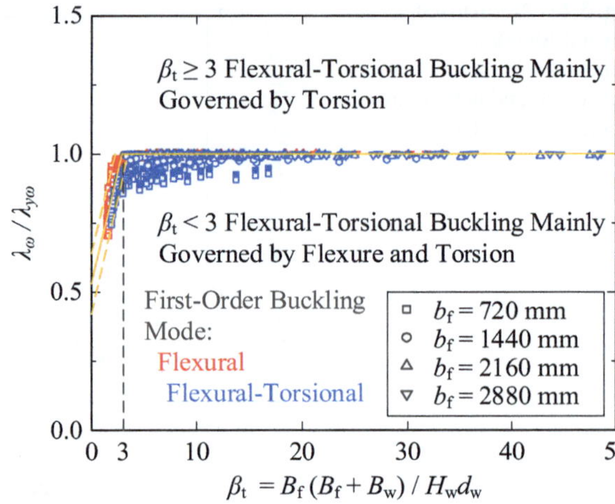

Fig. 5.29 Normalised flexural-torsional slenderness ratio $\lambda_{y\omega}$ and normalised torsional slenderness ratio λ_ω

axis could be treated as torsional buckling. In such a case, the flexure about the symmetric axis is insignificant, the deformation is governed by the torsion of flange and web walls, and flexural-torsional buckling is transformed to torsional buckling. For most T-CDSCWs with flexural-torsional buckling as the first-order buckling mode, the overall geometry meets the requirement of $\beta_t > 3.0$. Therefore, it can be considered to design flexural-torsional instability of axially loaded T-CDSCWs in accordance with torsional buckling. The feasibility will be further verified in the following.

5.2.2 *Load-Carrying Capacity*

Based on the elastic buckling analysis, the solid–shell finite element model is used to analyse the flexural-torsional instability of axially loaded T-CDSCWs. The material properties of steel are as follows: the yield strength f_y = 345 MPa, the elastic modulus E_s = 2.06 × 10^5 MPa, and the Poisson's ratio υ_s = 0.3 (MOHURD and AQSIQ 2017). The material properties of concrete are as follows: the axial compressive strength f_c = 26.8 MPa, the elastic modulus E_c = 3.25 × 10^4 MPa, and the Poisson's ratio υ_c = 0.2 (MOHURD and AQSIQ 2010). The other material properties are the same as Chap. 3. The initial geometric imperfection of T-CDSCWs is in the form of the first-order elastic buckling, and the amplitude of the imperfection is H_w/500.

The finite element simulations analyse the effects of geometric parameters, including the flange breadth b_f, the web breadth b_w, the wall height H_w, the wall thickness d_w, the steel plate thickness t_s, and the geometric parameters of boundary elements $b_c \times d_c \times t_c$, on the load-carrying behaviour (see Table 5.3). The parameters of numerical examples are within the range commonly used in engineering practice. The size of boundary elements corresponds with that of wall elements: the steel tube

Table 5.3 Numerical examples of flexural-torsional instability of axially loaded T-CDSCWs

NO.	Flange breadth b_f (mm)	Web breadth b_w (mm)	Wall height H_w (m)	Wall thickness d_w (mm)	Steel plate thickness t_s (mm)	CFST b_c and d_c (mm)	Steel tube thickness t_c (mm)
C-1	720	2160	3–9	130–200	5	140–210	9
C-2	720	2160	6	130–200	3–7	140–210	7–11
C-3	720–2880	360–2880	6	150	5	160	9
C-4	720–2880	360–2880	3–9	130	3	140	7

breadths b_c and d_c are 10 mm larger than d_w, and the steel tube thickness t_c is 4 mm larger than the steel plate thickness t_s. The parameters of steel corrugated plates are: $q = 180$ mm, $q_1 = 85$ mm, $q_2 = 27.5$ mm, $q_3 = 40$ mm, $a = 35$ mm. Intermediate bolts are arranged in the staggered way, and the vertical bolt spacing d_v is 350 mm. The normalised slenderness ratio of contact buckling of steel corrugated plates λ_{pc} is not greater than 0.4, the upper limit given by Eq. (4.18), so the effect of local buckling of steel corrugated plates can be neglected.

The load-carrying behaviour of axially loaded T-CDSCWs is obtained by finite element nonlinear analysis. The failure of T-CDSCWs occurs at the mid-height section (see Fig. 5.7). In the initial loading stage, the stress on the concave becomes greater than that on the convex side. When the load-carrying capacity is reached, steel corrugated plates and steel tubes yield, and the T-CDSCW deflects without the distortion of the cross-section in the same direction with the initial imperfection. As the loading continues, the deflection and yielding develop, leading to the load decreases.

Besides the elastic buckling behaviour of T-CDSCWs, the load-carrying capacity and failure mode are influenced significantly by the ratio of the flange breadth to the web breadth b_f/b_w. With the web breadth b_w remaining unchanged, the increase of the flange breadth b_f changes the buckling mode of T-CDSCWs. The T-CDSCW with a small flange (see Fig. 5.30a) shows not only twist about the shear centre but also flexure about the symmetric axis, while the T-CDSCW with a large flange (see Fig. 5.30b) shows almost merely twist. Although the buckling modes of the two examples are different, there is no significant difference between their strength reduction factors.

The failure mode of T-CDSCWs also changes when the relative breadth of the flange and web changes with the same cross-sectional area of T-CDSCWs. The T-CDSCW with a larger relative web breadth (see Fig. 5.31a) has a large unrestrained breadth of wall elements. The failure concentrates on the web wall, and the free side of the web yields first and then deforms out-of-plane. The T-CDSCW with a larger relative flange breadth (see Fig. 5.31b) has half of the flange breadth as the unrestrained breadth of wall elements, which is half of that in the previous numerical example. Due to the effect of initial geometric imperfection, the flange sustains different compression on the two sides. The side sustains more compression yields first and then the out-of-plane instability occurs. Despite the same cross-sectional area of the numerical examples, the strength reduction factor $\varphi_{y\omega}$ of the latter

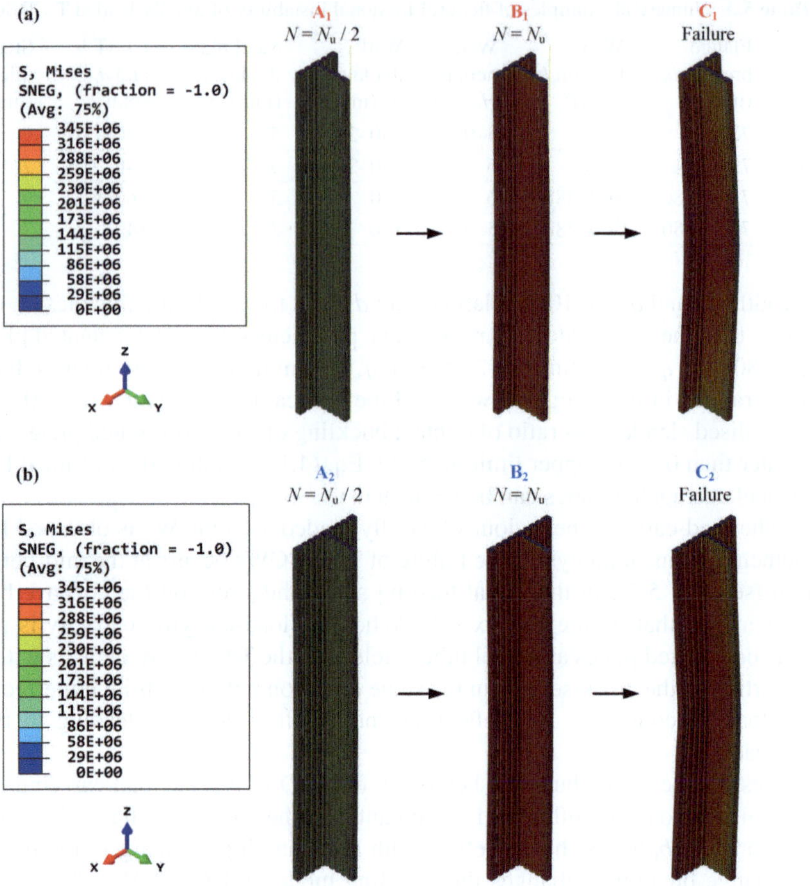

Fig. 5.30 The effect of the flange breadth b_f on the load-carrying performance ($H_w = 9000$ mm, $b_w = 720$ mm, $d_w = 130$ mm, $t_s = 3$ mm, unit = Pa): (**a**) $b_f = 720$ mm; (**b**) $b_f = 1440$ mm

improves significantly. The T-CDSCW with a larger relative flange breadth has higher load-bearing efficiency than that with a larger relative web breadth (see Figs. 5.32 and 5.33).

Various geometric parameters have different effects on the load–displacement relationship of T-CDSCWs (see Figs. 5.34, 5.35, 5.36, 5.37, and 5.38). The initial slope of normalised load–displacement curves is basically the same, and the difference mainly focuses on the load-carrying capacity and the corresponding strength reduction factor $\varphi_{y\omega}$, which is the ratio of the load-carrying capacity N_u to the yield load N_y.

Some geometric parameters have insignificant effects on the strength reduction factor $\varphi_{y\omega}$, including detail parameters, such as the steel plate thickness t_s, and overall parameters, such as the flange breadth b_f. It is understandable that detail parameters hardly influence global instability. The reason why the flange breadth b_f as an overall parameter has no significant effect either is that the unrestrained breadth is

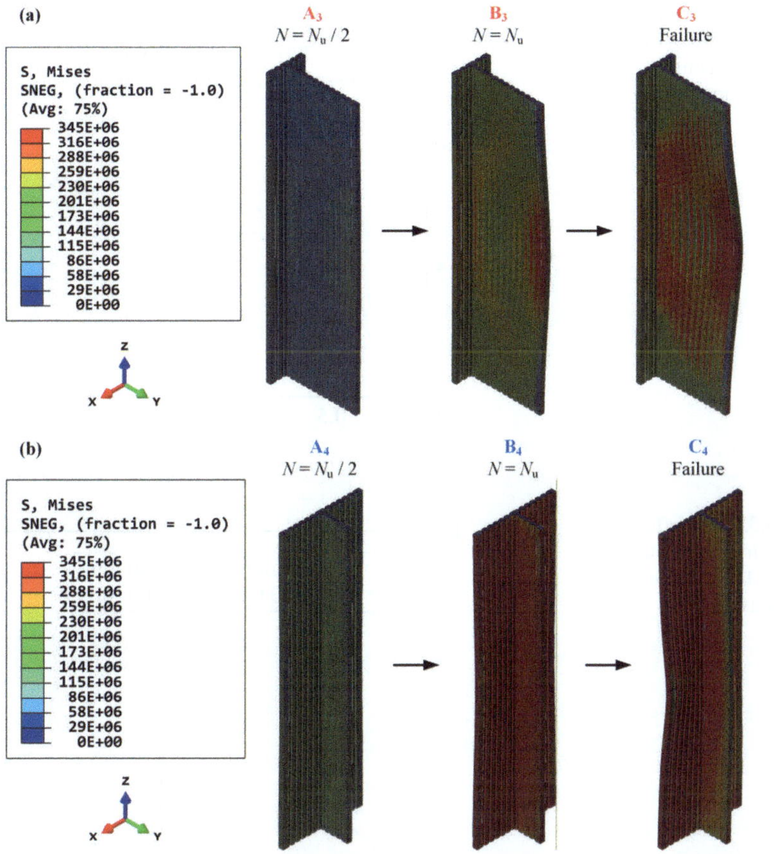

Fig. 5.31 The effect of geometric parameters on the load-carrying performance (H_w = 9000 mm, d_w = 130 mm, t_s = 3 mm, unit = Pa): (**a**) b_f = 720 mm, b_w = 2160 mm; (**b**) b_f = 2160 mm, b_w = 720 mm

half of the flange breadth b_f, and thus the flexural-torsional instability is governed by the failure of the web wall.

The wall height H_w and thickness d_w have significant effects on the normalised load–displacement curves. However, the sensitivity of flexural-torsional buckling behaviour to the wall height H_w and thickness d_w is different from that of flexural buckling behaviour about the asymmetric axis. The strength reduction factor φ_x of flexural instability about the asymmetric axis decreases rapidly as the wall height H_w increases while varies a little due to the varying wall thickness d_w since the flexural stiffness is almost proportional to the wall thickness d_w. For flexural-torsional instability about the symmetric axis, the sensitivity to the wall height H_w and thickness d_w is reversed: the strength reduction factor $\varphi_{y\omega}$ decreases a little as the wall height H_w continues to increase but increases rapidly as the wall thickness d_w increases. That is because the relative change of the wall height H_w decreases and the wall height H_w only influences the restrained torsional buckling load. The effect of a high height is limited and the torsional buckling load is mainly determined by

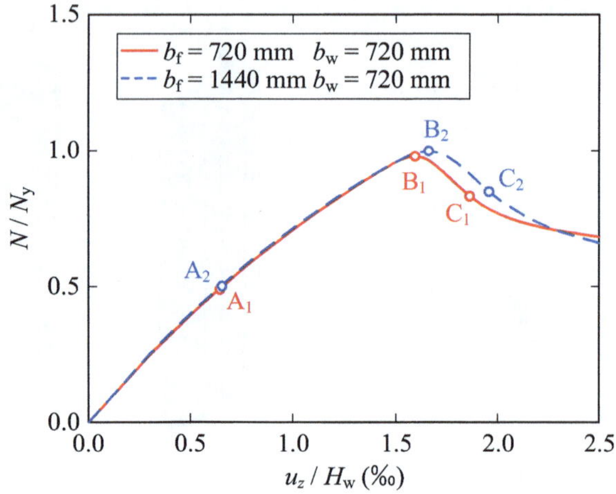

Fig. 5.32 The effect of the flange breadth b_f on the load–displacement relationship

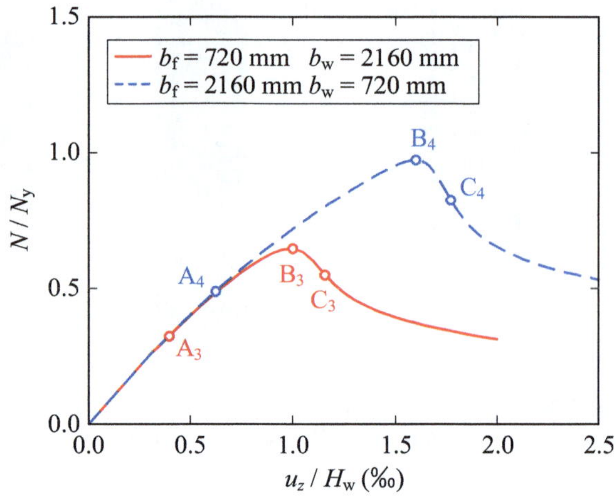

Fig. 5.33 The effect of geometric parameters on the load–displacement relationship

the free torsional buckling load, which governs the flexural-torsional buckling behaviour. As the wall thickness d_w increases, since the torsion and warping constants are proportional to the third power of the wall thickness d_w, the elastic torsional buckling load increases significantly together with the strength reduction factor $\varphi_{y\omega}$.

The varying of the web breadth b_w can change the buckling mode of T-CDSCWs, and the strength reduction factor $\varphi_{y\omega}$ does not change with the web breadth b_w monotonously. For T-CDSCWs with a small web, the first-order elastic buckling mode is flexural buckling about the asymmetric axis, and the increase of the web breadth b_w reduces the normalised flexural slenderness ratio λ_x and improves the strength reduction factor φ_x of flexural instability. However, as the web exceeds a certain breadth, the first-order elastic buckling mode transforms to flexural-torsional buckling about the symmetric axis. In such a case, the increasing of the web breadth

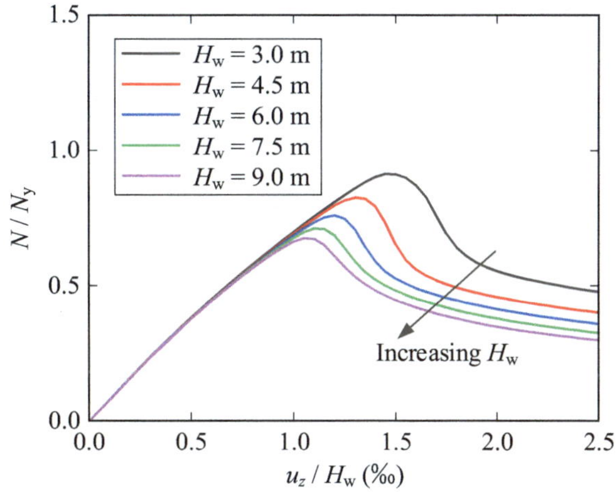

Fig. 5.34 Vertical load–displacement relationship of numerical examples C-1: $d_w = 150$ mm

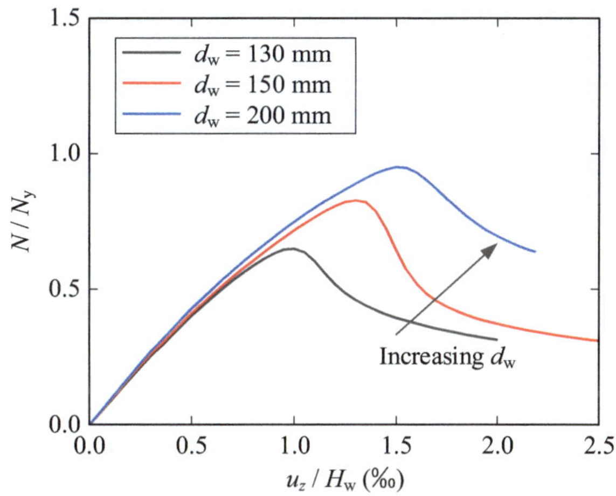

Fig. 5.35 Vertical load–displacement relationship of numerical examples C-2: $t_s = 5$ mm

b_w leads to a larger normalised flexural-torsional slenderness ratio $\lambda_{y\omega}$ and reduces the strength reduction factor $\varphi_{y\omega}$ of flexural-torsional instability.

The strength reduction factor $\varphi_{y\omega}$ of flexural-torsional instability of T-CDSCWs basically decreases as the normalised flexural-torsional slenderness ratio $\lambda_{y\omega}$ increases (see Figs. 5.39, 5.40, 5.41, 5.42, 5.43, and 5.44). Some methods design flexural-torsional instability based on the principle of equivalent elastic buckling load (MOHURD and AQSIQ 2017): treating the normalised flexural-torsional slenderness ratio $\lambda_{y\omega}$ as the normalised flexural slenderness ratio λ_x, and obtaining the corresponding strength reduction factor in accordance with the design curve of flexural instability. For T-CDSCWs, when the normalised flexural-torsional slenderness ratio is small ($\lambda_{y\omega} < 0.4$), the design curve of flexural instability Eqs. (5.7) and (5.8) ensures the safety of flexural-torsional instability. However, when the normalised flexural-torsional slenderness ratio is relatively large ($\lambda_{y\omega} > 0.4$), the numerical

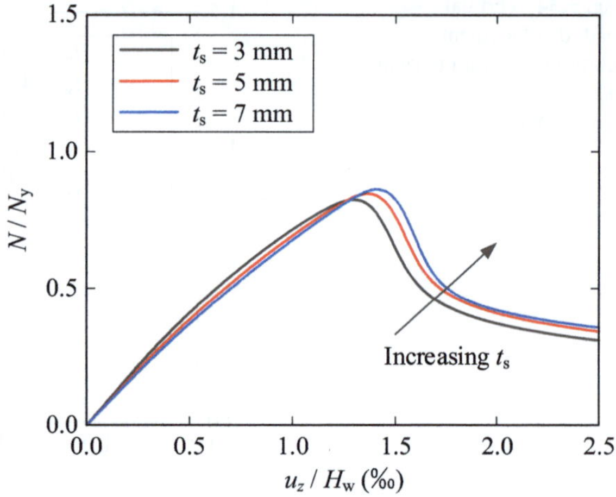

Fig. 5.36 Vertical load–displacement relationship of numerical examples C-2: d_w = 150 mm

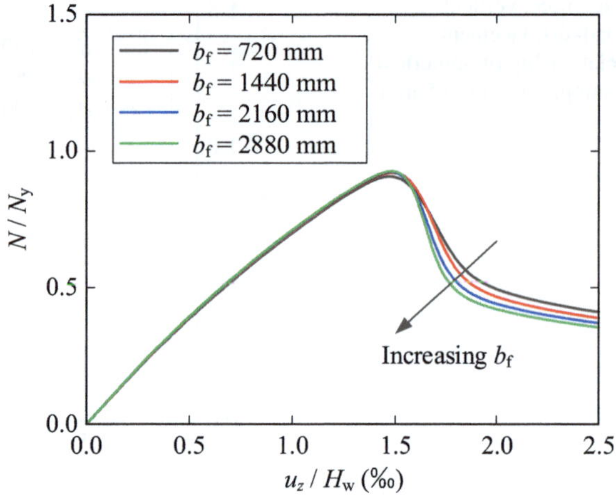

Fig. 5.37 Vertical load–displacement relationship of numerical examples C-3: b_w = 2160 mm

results of the strength reduction factor $\varphi_{y\omega}$ go below the design curve of flexural instability. Therefore, flexural-torsional instability is not proper to be designed based on the principle of equivalent elastic buckling load, and in the need of additional design curves.

Since the flexural-torsional buckling is governed by torsional buckling for most T-CDSCWs with flexural-torsional buckling as the first-order buckling mode, it is feasible to delineate the strength reduction factor $\varphi_{y\omega}$ of flexural-torsional buckling with the normalised torsional slenderness ratio λ_ω. On the one hand, the design of flexural-torsional buckling of T-CDSCWs is reduced to the buckling of flange and web walls, which is equivalent to the buckling of wall panels with one side simply supported under axial compression, and such equivalent coincides with the concept of Chinese design code (MOHURD 2010). On the other hand, the substitution of the normalised flexural-torsional slenderness ratio $\lambda_{y\omega}$ with the normalised torsional

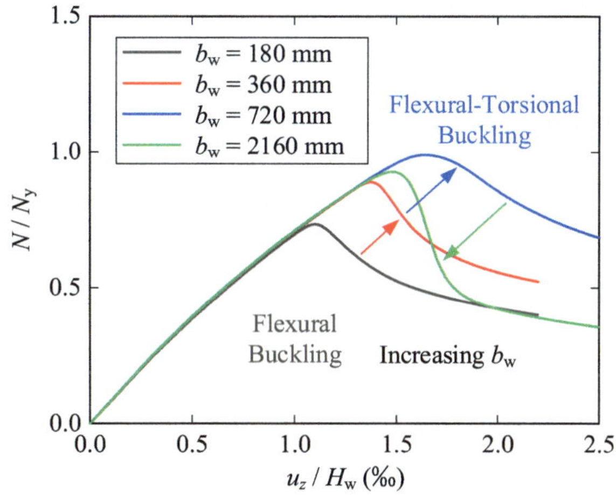

Fig. 5.38 Vertical load–displacement relationship of numerical examples C-3: $b_f = 720$ mm

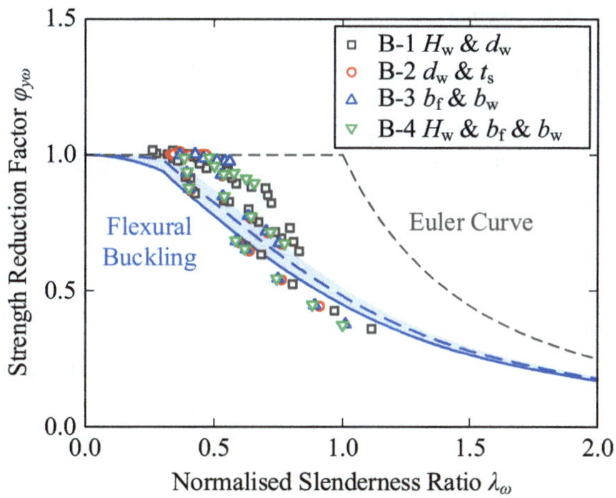

Fig. 5.39 Numerical results of the strength reduction factor and the normalised slenderness ratio of flexural-torsional buckling $\varphi_{y\omega} - \lambda_{y\omega}$: all numerical examples

slenderness ratio λ_ω agrees with the discussion on the elastic bucking, simplifies the calculation of the normalised slenderness ratio and leads to a more concentrated distribution of $\varphi_{y\omega} - \lambda$.

Compared with the results of flexural instability $\varphi_x - \lambda_x$, the distribution of $\varphi_{y\omega} - \lambda_\omega$ is more dispersed. If a single design curve is used to evaluate the strength reduction factor $\varphi_{y\omega}$, the design value would be too conservative and lead to more unnecessary cost. To attain an economic design, the design formulas of flexural-torsional instability should take into consideration not only the effect of the normalised slenderness ratio but also the effects of other parameters.

The steel plate thickness t_s and the flange breadth b_f have limited effects on the distribution of $\varphi_{y\omega} - \lambda_\omega$. As the steel plate thickness t_s varies, the normalised torsional slenderness ratio λ_ω almost keeps unchanged, and the strength reduction factor $\varphi_{y\omega}$ changes a bit only when the wall thickness d_w is small because of a large

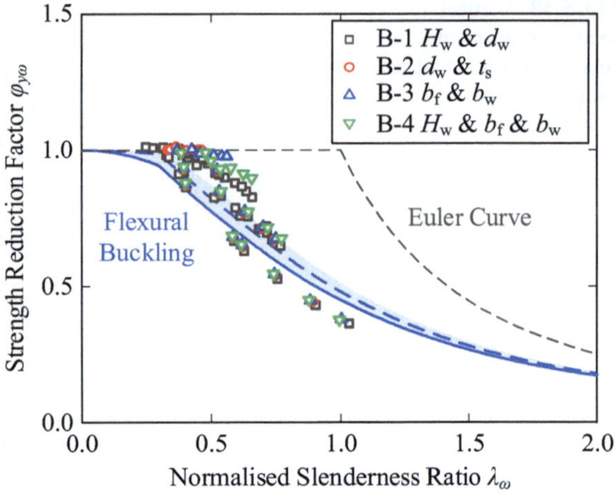

Fig. 5.40 Numerical results of the strength reduction factor and the normalised slenderness ratio of torsional buckling $\varphi_{y\omega} - \lambda_\omega$: all numerical examples

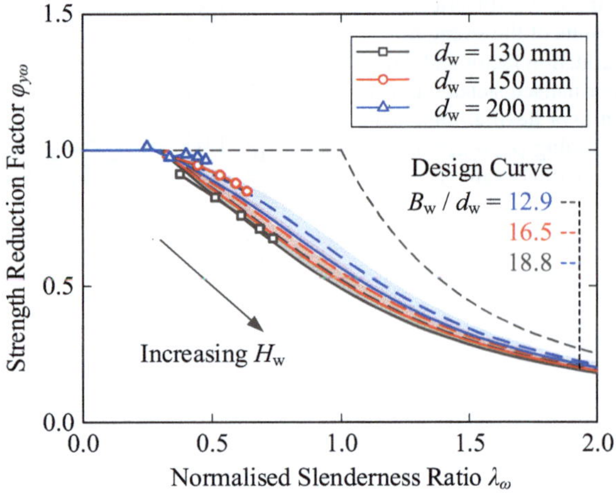

Fig. 5.41 Numerical results of the strength reduction factor and the normalised slenderness ratio of torsional buckling $\varphi_{y\omega} - \lambda_\omega$: numerical examples C-1

change of the steel ratio. As the flange breadth b_f varies, the strength reduction factor $\varphi_{y\omega}$ almost keeps unchanged, and the normalised torsional slenderness ratio λ_ω changes a bit only when the web breadth b_w is small because the maximum unrestrained breadth of wall elements is generally governed by the web breadth b_w. With the same web breadth b_w, the strength reduction factors $\varphi_{y\omega}$ with different flange breadth b_f are located on the same curve.

By contrast, the distribution of $\varphi_{y\omega} - \lambda_\omega$ is influenced by the wall thickness d_w and the web breadth b_w significantly. The smaller the wall thickness d_w and the wider the web breadth b_w, the more the curve of strength reduction factor $\varphi_{y\omega}$ of flexural-torsional instability goes downward. That is to say, the trend of strength reduction factor $\varphi_{y\omega}$ does not depend on merely the normalised torsional slenderness ratio λ_ω.

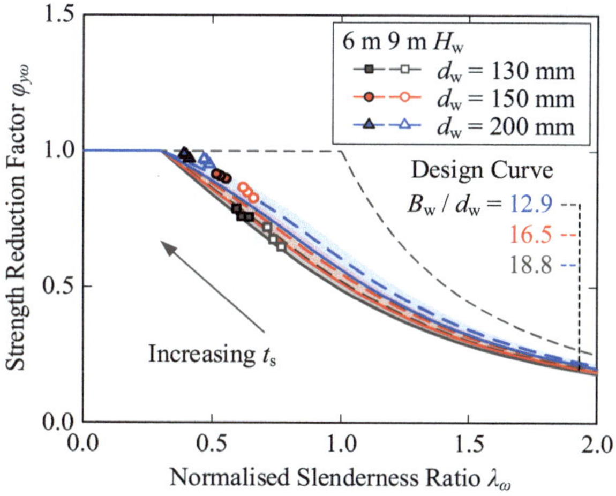

Fig. 5.42 Numerical results of the strength reduction factor and the normalised slenderness ratio of torsional buckling $\varphi_{y\omega} - \lambda_\omega$: numerical examples C-2

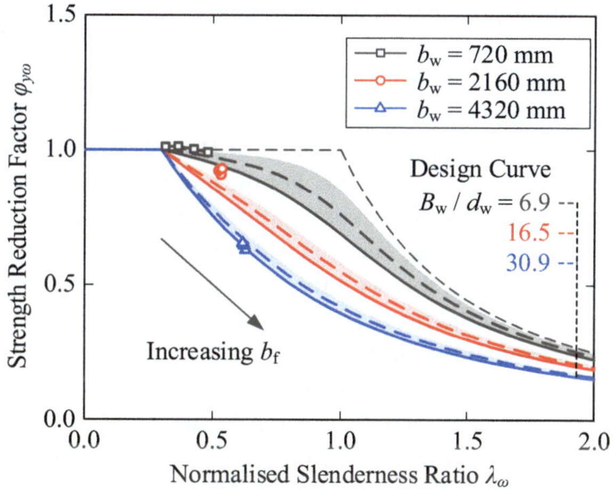

Fig. 5.43 Numerical results of the strength reduction factor and the normalised slenderness ratio of torsional buckling $\varphi_{y\omega} - \lambda_\omega$: numerical examples C-3

To consider the effects of the wall thickness d_w and the web breadth b_w, Eqs. (5.20) and (5.21) on the strength reduction factor $\varphi_{y\omega}$ of flexural-torsional buckling are proposed.

$$\varphi_{y\omega} = \begin{cases} 1 - \alpha_1 \lambda_\omega^2 & \lambda_\omega \le 0.3 \\ \dfrac{1}{\Phi + \sqrt{\Phi^2 - \lambda_\omega^2}} & \lambda_\omega > 0.3 \end{cases} \tag{5.20}$$

$$\Phi = 0.5\left[1 + \alpha_2 \left(\lambda_\omega - \alpha_3\right) + \lambda_\omega^2\right] \tag{5.21}$$

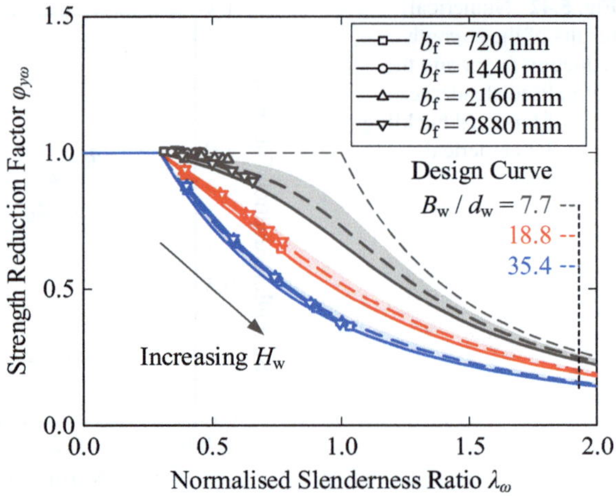

Fig. 5.44 Numerical results of the strength reduction factor and the normalised slenderness ratio of torsional buckling $\varphi_{y\omega} - \lambda_\omega$: numerical examples C-4

Where α_1, α_2, and α_3 are the shape parameters. $\alpha_1 = 0$ and $\alpha_3 = 0.3$, that is to say, when $\lambda_\omega \leq 0.3$, the failure mode is always governed by the cross-section capacity for T-CDSCWs with flexural-torsional buckling as the first-order buckling mode, and the flexural-torsional instability can be neglected. For the lower envelope curve, which is the design curve, $\alpha_2 = 0.0466\ B_w/d_w - 0.121$. For the average curve, $\alpha_2 = 0.0438\ B_w/d_w - 0.196$. For the upper envelope curve, $\alpha_2 = 0.0406\ B_w/d_w - 0.243$. As the breadth-to-thickness ratio of the web B_w/d_w increases, α_2 becomes larger and the corresponding curve go downward. Considering the range of parameters, Eq. (5.20) is applicable with the normalised flexural slenderness ratio λ_x no greater than 1.0. Within the range of practical parameters, α_2 is always positive, which ensures the reasonability of design formulas. Equation (5.20) reflects the trend of the curve of strength reduction factor $\varphi_{y\omega}$ with the changes of the normalised torsional slenderness ratio λ_ω and the breadth-to-thickness ratio of the web B_w/d_w. In the stage of preliminary design, Eq. (5.20) could be reduced to a certain curve with the maximum breadth-to-thickness ratio of the web B_w/d_w, and the design values would be more conservative.

Generally, the curve of strength reduction factor adopts a single form, while Eq. (5.20) consider the effect of the breadth-to-thickness ratio of the web B_w/d_w. The design values $\varphi_{y\omega}$ obtained from Eq. (5.20) is conservative to the numerical results $\varphi_{y\omega,n}$, and the deviation ranges within +15% (see Fig. 5.45). It is reliable to design flexural-torsional instability of axially loaded T-CDSCWs by Eq. (5.20).

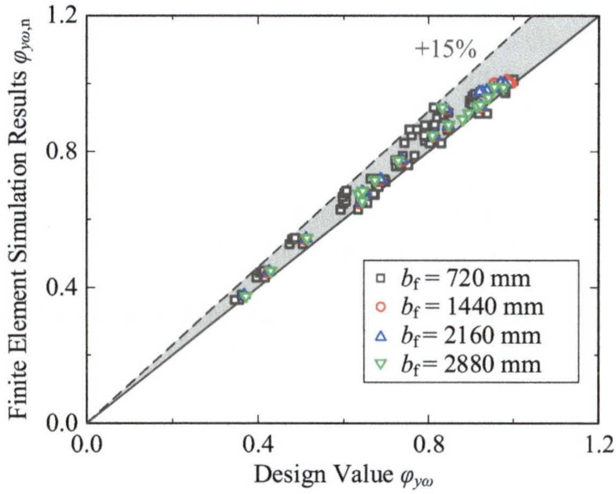

Fig. 5.45 Design value and numerical results of the strength reduction factor $\varphi_{y\omega}$

5.3 Global Stability Under Axial Compression and Bending Moment in the Web Plane

In addition to axial compression, T-CDSCWs are also subjected to bending moment. Based on the research of the global stability under axial compression, this section focuses on the global stability of T-CDSCWs under axial compression and bending moment in the web plane. The effect of the initial geometric imperfection is considered, the load-carrying capacity of T-CDSCWs under axial compression and bending moment in the web plane is investigated. Based on the numerical results, the theoretical analyses, and the interaction equations of cross-sectional capacity, the interaction equations are proposed to evaluate the load-carrying capacity of global instability under axial compression and bending moment in the web plane.

5.3.1 Load-Carrying Capacity

Based on the previous analysis, the solid–shell finite element model is used to analyse the global instability of T-CDSCWs under axial compression and bending moment in the web plane. The material properties of steel are as follows: the yield strength f_y = 345 MPa, the elastic modulus $E_s = 2.06 \times 10^5$ MPa, and the Poisson's ratio $\upsilon_s = 0.3$ (MOHURD and AQSIQ 2017). The material properties of concrete are as follows: the axial compressive strength f_c = 26.8 MPa, the elastic modulus $E_c = 3.25 \times 10^4$ MPa, and the Poisson's ratio $\upsilon_c = 0.2$ (MOHURD and AQSIQ 2010). The other material properties are the same as Chap. 3. The initial geometric imperfection of T-CDSCWs is in the form of the first-order elastic buckling, and the amplitude of the imperfection is $H_w/500$.

The boundary condition of the T-CDSCW is pin ended. The top and bottom sections are coupled with the reference points at the centroid, which is weighted with the elastic modulus of the materials. The T-CDSCW is loaded at the reference points. The displacement and the rotation about the z-axis of the bottom reference point are restrained. The displacement along x- and y-axis of the top reference point are restrained. The axial compression N is first applied at the top reference point, and then the bending moment M_x in the web plane is applied at both reference points. The load-carrying capacity can be obtained by the finite element static analysis.

The axial compression N is applied to the T-CDSCW according to the axial compression ratio n as defined in Eq. (5.22).

$$n = \frac{N}{N_{\mathrm{u}}} \tag{5.22}$$

Where N_{u} denotes the load-carrying capacity of flexural instability determined by the strength reduction factor φ_x, which is evaluated by Eq. (5.7).

The finite element simulations analyse the effects of geometric parameters, including the flange breadth b_{f}, the web breadth b_{w}, the wall thickness d_{w}, the steel plate thickness t_{s}, and the geometric parameters of boundary elements $b_{\mathrm{c}} \times d_{\mathrm{c}} \times t_{\mathrm{c}}$, on the load-carrying behaviour (see Table 5.4). The parameters of numerical examples are within the range commonly used in engineering practice. The wall height H_{w} is taken as 9 mm. The size of boundary elements corresponds with that of wall elements: the steel tube breadths b_{c} and d_{c} are 10 mm larger than d_{w}, and the steel tube thickness t_{c} is 4 mm larger than the steel plate thickness t_{s}. The parameters of steel corrugated plates are: $q = 180$ mm, $q_1 = 85$ mm, $q_2 = 27.5$ mm, $q_3 = 40$ mm, $a = 35$ mm. Intermediate bolts are arranged in the staggered way, and the vertical bolt spacing d_{v} is 350 mm. The normalised slenderness ratio of contact buckling of steel corrugated plates λ_{pc} is not greater than 0.4, the upper limit given by Eq. (4.18), so the effect of local buckling of steel corrugated plates can be neglected.

Similar to flexural buckling under axial compression, the direction of initial geometric imperfection also influences the load-carrying performance of T-CDSCWs under axial compression and bending in web plane (see Figs. 5.46 and 5.47). For axial compression N, the direction of initial geometric imperfection with the flange

Table 5.4 Numerical examples of instability of T-CDSCWs under axial compression and bending moment in the web plane

NO.	Flange breadth b_{f} (mm)	Web breadth b_{w} (mm)	Wall thickness d_{w} (mm)	Steel plate thickness t_{s} (mm)	CFST b_{c} and d_{c} (mm)	Steel tube thickness t_{c} (mm)	Axial compression ratio n
D-1	720–2880	360	130	3	140	5	0–0.75
D-2	1440	180–720	130	3	140	5	0–0.75
D-3	1440	360	130–200	3	140–210	5	0–0.75
D-4	1440	360	130	3–7	140	5–9	0–0.75

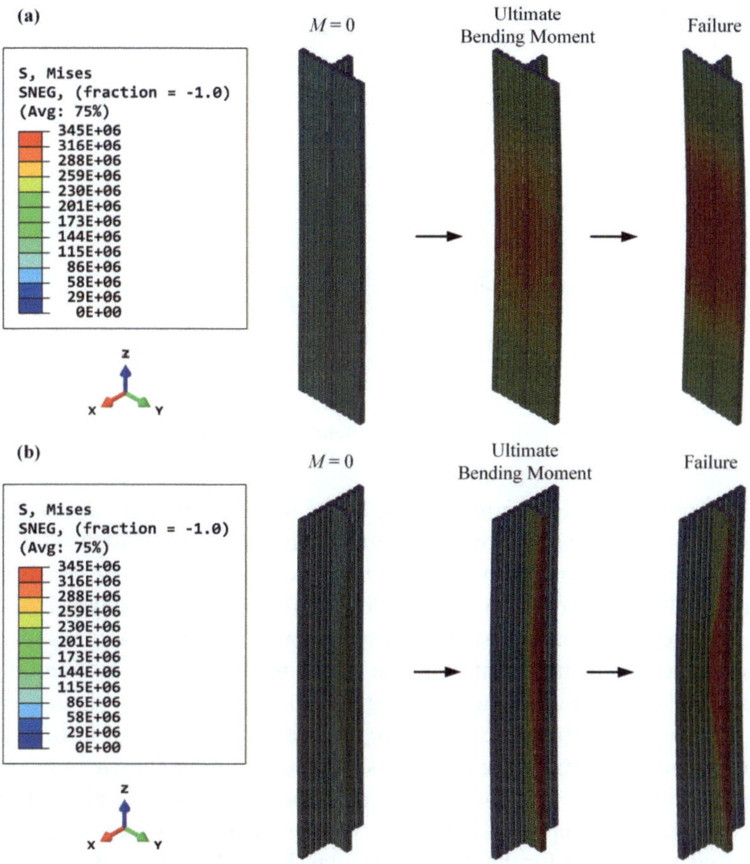

Fig. 5.46 The effect of the direction of initial geometric imperfection on the in-plane instability ($H_w = 9000$ mm, $b_f = 1440$ mm, $b_w = 180$ mm, $d_w = 130$ mm, $t_s = 3$ mm, $n = 0.5$, unit = Pa): (**a**) the T-CDSCW under positive bending and the initial geometric imperfection with the web on the convex side; (**b**) the T-CDSCW under negative bending moment and the initial geometric imperfection with the flange on the convex side

on the convex side is the most disadvantageous direction. For the bending moment M_x in the web plane, the direction of initial geometric imperfection corresponding to the bending direction is the most disadvantageous direction. When T-CDSCWs are subjected to positive bending moment M_x, the most disadvantageous direction corresponding to the bending moment M_x is the direction with the web on the convex side, which is opposite to the most disadvantageous direction corresponding to the axial compression N. However, in most cases, the bending capacity under the same axial compression N corresponding to the direction with the web on the convex side is smaller. Only when the axial compression N approaches the ultimate compression capacity N_u, the bending capacity under the same axial compression N corresponding to the direction with the flange on the convex side is smaller. When T-CDSCWs are subjected to negative bending moment M_x, the most

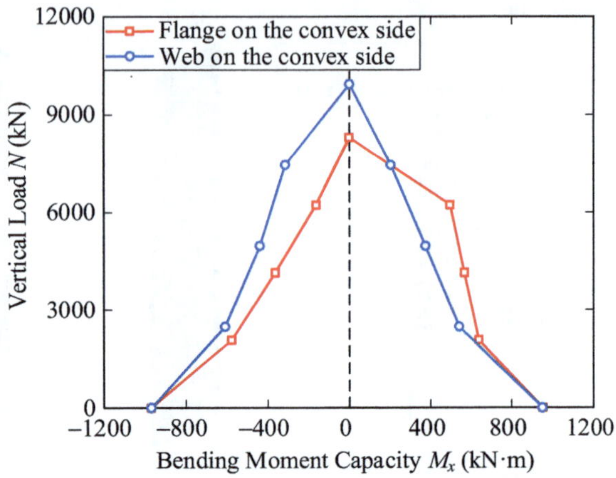

Fig. 5.47 The effect of the direction of initial geometric imperfection on the interaction curve

disadvantageous direction corresponding to the bending moment M_x is the same as that corresponding to the axial compression *N*. Therefore, the most disadvantageous direction of initial geometric imperfection can be determined to be consistent with the bending direction and used in the following finite element analyses.

The failure mode of global instability under axial compression and bending moment in the web plane is similar to the failure modes of cross-section failure and flexural instability, which is the mid-height failure influenced by the second-order effect. The stress distribution is similar to the stress distribution of flexural instability, and the moment–curvature curve is similar to the moment–curvature curve of cross-sectional failure, which will not be described repeatedly in this section.

The load-carrying capacity of T-CDSCWs obtained by finite element simulation can be used as the basis for the design formulas. From the perspective of the overall trend, the interaction curves of axial compression and positive bending moment protrude outwards, while the interaction curves of axial compression and negative bending moment are concave (see Figs. 5.48, 5.49, 5.50, 5.51, 5.52, 5.53, 5.54, 5.55, 5.56, and 5.57). For the interaction curves of axial compression and positive bending moment, the lower part is no longer a linear distribution, but an internal concave distribution. The degree of convex and concave is significantly influenced by the change of the web breadth b_w, but the change of the flange breadth b_f, the wall thickness d_w, and the steel plate thickness t_s have insignificant effects on the interaction curves. As the increase of the web breadth b_w, for the interaction curves of axial compression and positive bending moment, the degree of convex increases, while for the interaction curves of axial compression and negative bending moment, the degree of concave decreases.

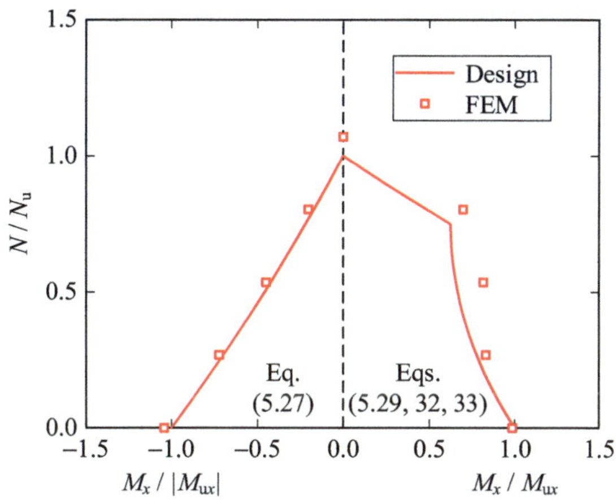

Fig. 5.48 Interaction curves of typical numerical examples: Group D-1, $b_f = 720$ mm, $b_w = 360$ mm

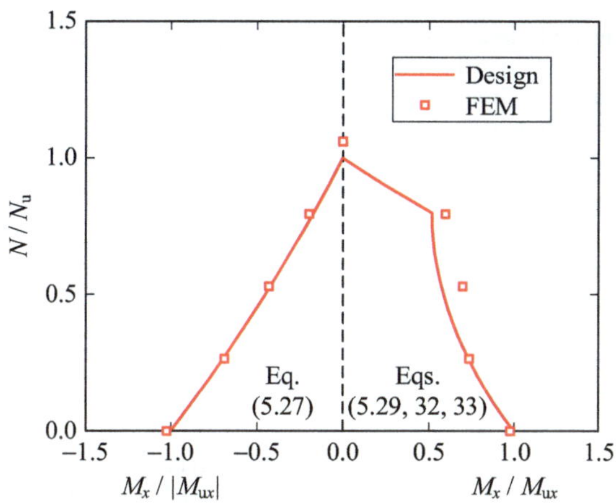

Fig. 5.49 Interaction curves of typical numerical examples: Group D-1, $b_f = 1440$ mm, $b_w = 360$ mm

5.3.2 Design Method

By analogy with the interaction equation of in-plane stability of the symmetric section member under axial compression and bending moment (Tong 2005), the interaction equation of in-plane stability of T-CDSCWs under axial compression and bending moment in the web plane can be derived. The interaction equation of the load-carrying capacity of the symmetric section member based on the yielding criterion of extreme fibre is shown in Eq. (5.23) (MOHURD and AQSIQ 2017).

$$\frac{N}{N_y} + \frac{M}{M_y} = 1 \tag{5.23}$$

Where N_y denotes the compression capacity and M_y denotes the moment capacity.

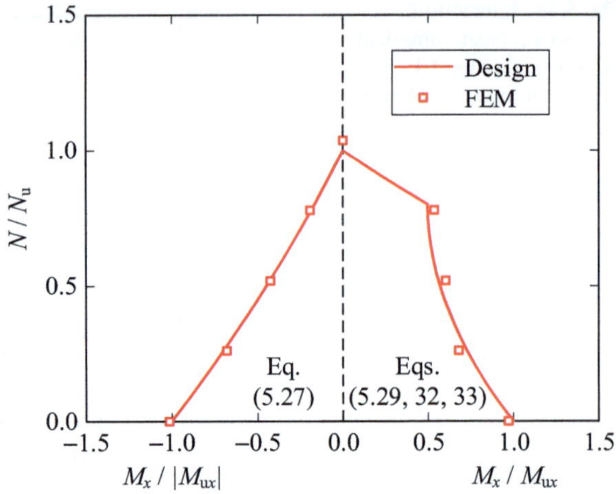

Fig. 5.50 Interaction curves of typical numerical examples: Group D-1, $b_f = 2160$ mm, $b_w = 360$ mm

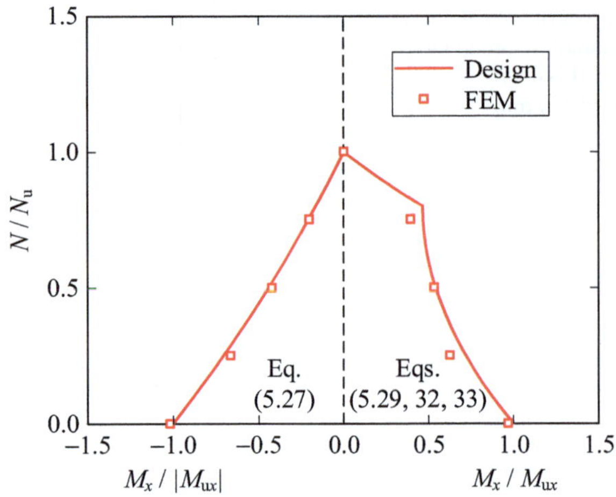

Fig. 5.51 Interaction curves of typical numerical examples: Group D-1, $b_f = 2880$ mm, $b_w = 360$ mm

Considering the initial geometric imperfection, initial loading eccentricity, and welding residual stress, the initial imperfection are equivalent to the initial curvature e_0. Considering the amplification effect of initial curvature e_0 and the axial compression N on the bending moment M, Eq. (5.24) is derived.

$$\frac{N}{N_y} + \frac{\left(1 + 0.25\dfrac{N}{N_{cr,x}}\right)M + Ne_0}{\left(1 - \dfrac{N}{N_{cr,x}}\right)M_y} = 1 \tag{5.24}$$

Where $N_{cr,x}$ denotes the elastic flexural buckling load.

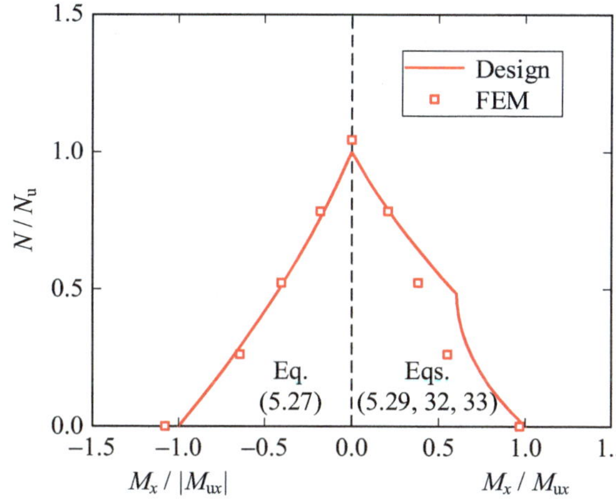

Fig. 5.52 Interaction curves of typical numerical examples: Group D-2, $b_f = 1440$ mm, $b_w = 180$ mm

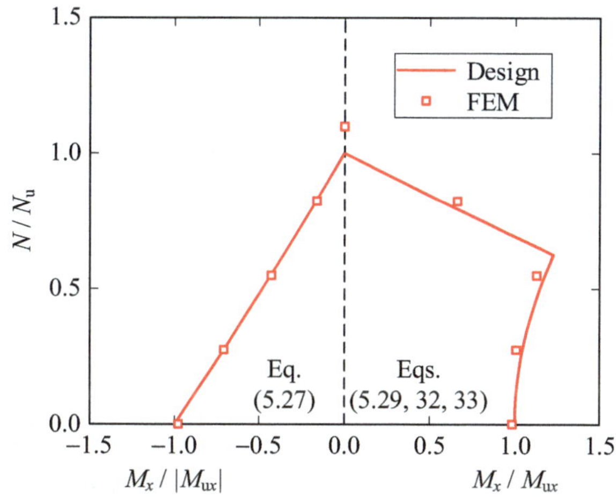

Fig. 5.53 Interaction curves of typical numerical examples: Group D-2, $b_f = 1440$ mm, $b_w = 720$ mm

When the bending moment $M = 0$, the load-carrying capacity of flexural instability N_u should equal $\varphi_x N_y$ obtained by Eq. (5.24). Introducing that to Eq. (5.24), the interaction equation under axial compression and in-plane bending moment can be obtained by simplification, which is shown in Eq. (5.25).

$$\frac{N}{N_u} + \frac{M}{\left[1 - (\varphi_x + 0.25)\dfrac{N}{N_{cr,x}}\right] M_y} = 1 \tag{5.25}$$

In the application of the interaction equation to the I-section steel member (MOHURD and AQSIQ 2017), the plastic development coefficient γ_x is introduced.

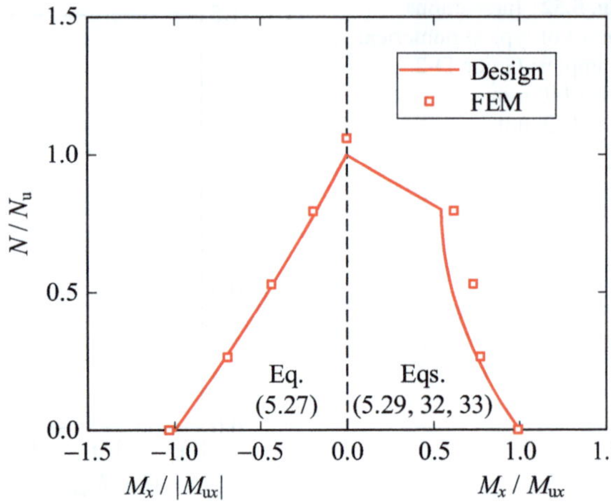

Fig. 5.54 Interaction curves of typical numerical examples: Group D-3, d_w = 150 mm

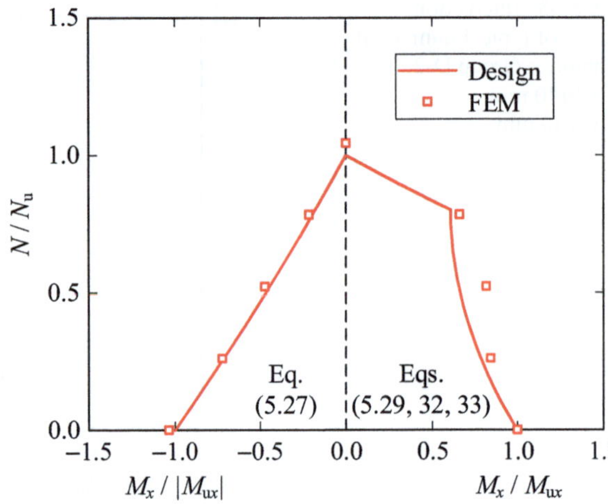

Fig. 5.55 Interaction curves of typical numerical examples: Group D-3, d_w = 200 mm

By further simplification of Eq. (5.25), Eq. (5.26) is proposed to design the interaction capacity of in-plane stability.

$$\frac{N}{N_u} + \frac{M}{\left(1 - 0.8\frac{N}{N_{cr,x}}\right)\gamma_x f_y W_x} = 1 \tag{5.26}$$

Where W_x denotes the section modulus.

Since the interaction equation of the cross-sectional capacity of T-CDSCWs under axial compression and negative bending moment is in the linear form, which is the same as Eq. (5.23), and simpler than the bilinear interaction equation under axial compression and positive bending moment. By analogy with the above

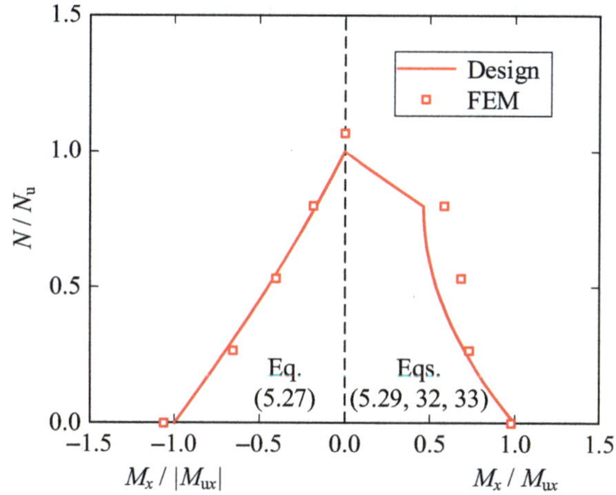

Fig. 5.56 Interaction curves of typical numerical examples: Group D-4, $t_s = 5$ mm

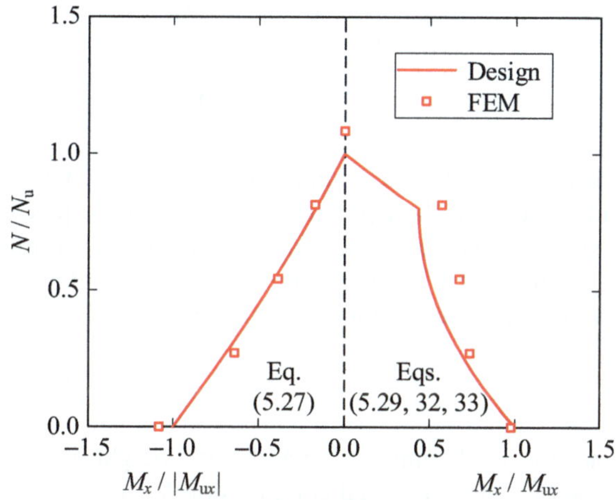

Fig. 5.57 Interaction curves of typical numerical examples: Group D-4, $t_s = 7$ mm

derivation, the interaction equation of the global stability of T-CDSCWs under axial compression and negative bending moment is derived as Eq. (5.27).

$$\frac{N}{N_u} + \frac{M_x}{\left(1 - \frac{N}{N_{cr,x}}\right) M_{ux}} = 1 \tag{5.27}$$

Where N_u denotes the load-carrying capacity of flexural instability, which is determined by the strength reduction factor φ_x obtained from Eq. (5.7); $N_{cr,x}$ denotes the elastic flexural buckling load, which is determined by Eq. (5.1); M_u denotes the negative moment capacity, which is determined by Eq. (4.45). In Eq. (5.27), the term of the bending moment M considers the secondary effect caused by the axial compression N, so the linear interaction equation of the cross-sectional capacity turns in the concave interaction equation of global stability.

By analogy with the derivation of Eq. (5.25), the amplification effect of initial curvature e_0 and the axial compression N on the bending moment M is introduced to the upper part of Eq. (4.42), and Eq. (5.28) is derived.

$$\frac{N-N_{cx}}{N_u-N_{cx}}+\frac{\left(1+0.25\frac{N}{N_{cr,x}}\right)M+Ne_0}{\left(1-\frac{N}{N_{cr,x}}\right)M_{cx}}=1 \quad N_{cx}<N\le N_u \tag{5.28}$$

By simplifying Eq. (5.28), the upper part of the interaction equation of global stability of T-CDSCWs under axial compression and positive bending moment can be derived as shown in Eq. (5.29).

$$\frac{N}{N_u}+\left(1-\frac{N_{cx}}{N_y}\right)\frac{M_x}{\left(1-\frac{N}{N_{cr,x}}\right)M_{cx}}\le 1 \quad N_{cux}<N\le N_u \tag{5.29}$$

Where N_y denotes the compression capacity, which is determined by Eq. (5.2); N_{cx} and M_{cx} denote the axial compression and the positive bending moment with the neutral axis located at the centroid, which are determined by Eqs. (4.23) and (4.24); N_{cux} denotes the axial compression corresponding to the protruding point of the interaction curve of global stability, as shown in Eq. (5.30).

$$N_{cux}=\min\left(N_{cx},0.8N_u\right) \tag{5.30}$$

Since the axial compression at the mid-height section equals the axial compression at the two ends, the axial compression N_{cux} corresponding to global stability should equal the axial compression N_{cx} corresponding to cross-sectional capacity. However, the load-carrying capacity of flexural instability N_u may be relatively small, even less than N_{cx}. By trial calculation, therefore, the upper limit of N_{cux} is taken as $0.8N_u$.

After the upper part of the interaction equation of global stability is determined, the positive bending moment M_{cux} corresponding to the protruding point can be determined subsequently, as shown in Eq. (5.31).

$$M_{cux}=\frac{\left(1-\frac{N_{cux}}{N_u}\right)\left(1-\frac{N_{cux}}{N_{cr,x}}\right)}{\left(1-\frac{N_{cx}}{N_y}\right)}M_{cx} \tag{5.31}$$

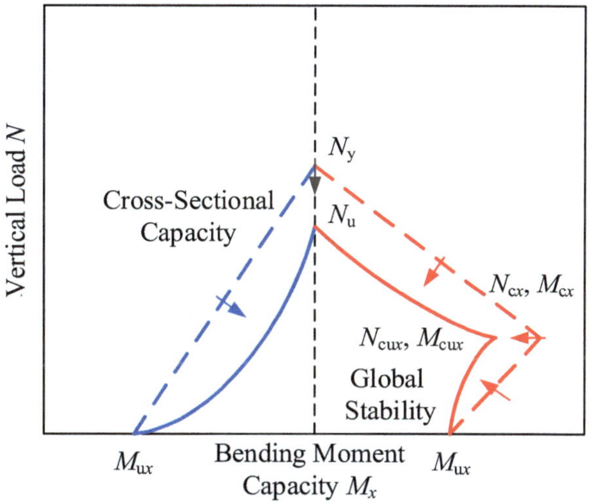

Fig. 5.58 Interaction curves of global stability under axial compression and bending moment in the web plane

According to the characteristics of concave, the lower part of the interaction equation of global stability is fitted by Eqs. (5.32) and (5.33).

When $M_{cux} > M_{ux}$:

$$\frac{M_x}{M_{ux}} - \left(\frac{M_{cux}}{M_{ux}} - 1\right)\left(\frac{N}{N_{cux}}\right)^2 = 1 \quad 0 < N \leq N_{cux} \tag{5.32}$$

When $M_{cux} \leq M_{ux}$:

$$\frac{M_x}{M_{cux}} - \left(\frac{M_{ux}}{M_{cux}} - 1\right)\left(1 - \frac{N}{N_{cux}}\right)^2 \leq 1 \quad 0 < N \leq N_{cux} \tag{5.33}$$

Taking into consideration the secondary effect of axial compression, the interaction curves of global stability of T-CDSCWs under axial compression and bending moment in the web plane are generated by reducing the interaction curves of cross-sectional capacity (see Fig. 5.58). The theoretical results obtained by Eqs. (5.27), (5.29), (5.32), and (5.33) agree well with the numerical results (see Figs. 5.48, 5.49, 5.50, 5.51, 5.52, 5.53, 5.54, 5.55, 5.56, and 5.57) and reflect the characteristics of the interaction curves of global stability. Therefore, Eqs. (5.27), (5.29), (5.32), and (5.33) can be used to evaluated the global stability of T-CDSCWs under axial compression and bending moment in the web plane. The applicable range is the same with the design formulas of flexural instability, which is the normalised flexural slenderness ratio λ_x no greater than 1.0. When the normalised flexural slenderness ratio λ_x is no greater than 0.2, the global instability can be neglected.

Up to now, the design formulas of flexural and flexural-torsional instability and the interaction equations of global stability have been obtained, and the global

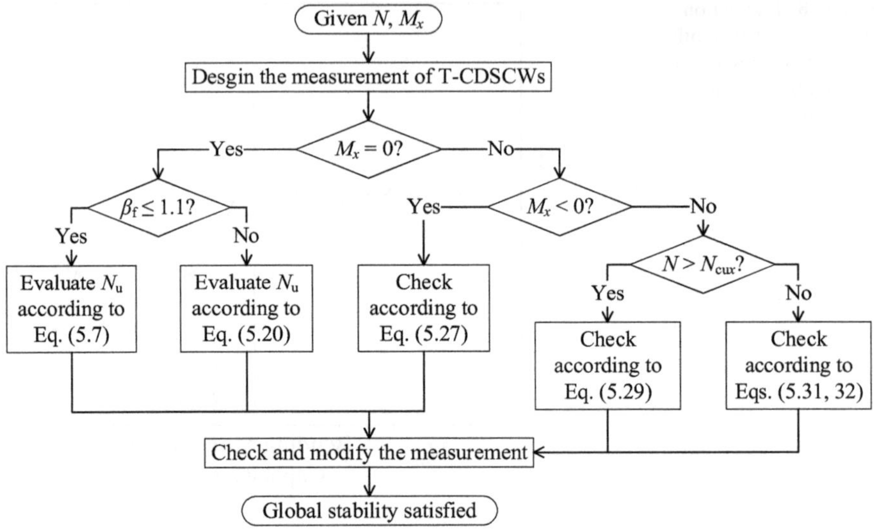

Fig. 5.59 Design flow of the global stability of T-CDSCWs under axial compression and biaxial bending

stability T-CDSCWs under axial compression and bending moment in the web plane can be designed (see Fig. 5.59).

5.4 Chapter Summary

Based on the solid–shell finite element model, this chapter analyses the flexural buckling about the asymmetric axis and the flexural-torsional buckling about the symmetric axis of T-CDSCWs under axial compression and the global instability of T-CDSCWs under axial compression and bending moment in the web plane. Based on the finite element simulation results and theoretical analyses, the following conclusions are drawn.

This chapter proposes the flexural stability design method of T-CDSCWs under axial compression. This chapter proposes the formulas of flexural buckling about the asymmetric axis, which can accurately evaluate the flexural buckling load of T-CDSCWs with flexural buckling as the first-order mode. Based on the geometry feature of T-CDSCWs, the parameter β_f is obtained from the breadth-thickness ratio and the height-breadth ratio to determine the first-order buckling mode. When β_f is no greater than 1.1, T-CDSCWs under axial compression can be design according to flexural instability. The strength reduction curve of flexural instability is developed based on the normalised flexural slenderness ratio and agrees well with the numerical results.

This chapter proposes the flexural-torsional stability design method of T-CDSCWs under axial compression. This chapter proposes the formulas of flexural-torsional buckling about the symmetric axis, which can accurately evaluate the elastic flexural-torsional buckling load of T-CDSCWs. Based on the geometry feature of T-CDSCWs, the parameter β_t is obtained from the breadth-thickness ratio and the height-breadth ratio. When β_t is no greater than 3.0, which applies to most cases, the flexural-torsional buckling is basically governed by the torsional buckling. Taking into consideration the effect of the breadth-thickness ratio of web walls, the strength reduction curve of flexural-torsional instability is developed based on the normalised torsional slenderness ratio and agrees well with the numerical results.

This chapter proposes the design method of the in-plane global stability of T-CDSCWs under axial compression and bending moment in the web plane. Based on the interaction equations of cross-sectional capacity and the design formulas of flexural instability, this chapter proposes the interaction equations of global stability of T-CDSCWs under axial compression and bending moment in the web plane. Taking into consideration the secondary effect of axial compression, the interaction curves of global stability are generated by reducing the interaction curves of cross-sectional capacity. For positive bending moment, the interaction curve is still divided into two parts, and the protruding point is determined by the interaction curve of cross-sectional capacity. The interaction equations of global stability agree well with the numerical results.

This chapter focuses on the design method of the global capacity of T-CDSCWs. The next chapter will further analyse the seismic performance of T-CDSCWs and propose the corresponding design recommendation.

References

Guo Y, Zhu J, Wang M, Yang X, Zhou P (2018) Overall instability performance of concrete-infilled double steel corrugated plate wall. Thin Wall Struct 130:372–394. https://doi.org/10.1016/j.tws.2018.05.026

MOHURD (2010) Technical specification for concrete structures of tall building (in Chinese). China Architecture and Building Press, Beijing

MOHURD and AQSIQ (2010) Code for design of concrete structures (in Chinese). China Architecture and Building Press, Beijing

MOHURD and AQSIQ (2017) Standard for design of steel structures (in Chinese). China Architecture and Building Press, Beijing

Timoshenko SP, Gere JM (2009) Theory of elastic stability, 2nd edn. Dover Publications, New York

Tong G (2005) In-plane stability of steel structures (in Chinese). China Architecture and Building Press

Zhu J (2020) Theoretical and experimental study on design methods for I-section concrete-infilled double steel corrugated-plate walls (in Chinese). Ph.D. Thesis, Tsinghua University, Beijing

Chapter 6
Seismic Performance of T-CDSCWs

Numerous seismic experiments have been conducted on various forms of concrete-infilled double steel plate walls (CDSWs) to study load-carrying capacities, stiffness, energy dissipation, and ultimate deformation. Based on existing experimental results, this chapter will analyse the deformation capacity of CDSWs, especially the CDSWs with CFSTs as boundary elements. Meanwhile, some other types of composite walls are also supplemented for comparison.

Previous seismic experiments focus on composite walls with rectangular section, and the seismic experiments of composite walls with T-section are limited. Based on the shell–beam finite element model, this chapter will analyse the seismic performance of T-CDSWs under axial compression and cyclic horizontal load, and propose the seismic requirements of CDSWs.

6.1 Deformation Analysis of CDSWs

To better evaluate the deformation capacity of CDSWs, this section summarises the previous experimental results and proposes an empirical formula for the ultimate drift ratio of CDSWs. Although circular tubes provide better confinement to the internal concrete (Alzeni and Bruneau 2017; Ji et al. 2013), rectangular tubes are more suitable for meeting the requirements of component connections and building layouts in building structures. Therefore, the main focus of this section is CDSWs with CFSTs as boundary elements.

M.-Z. Wang, *Design Method of Concrete-Infilled Double Steel Corrugated-Plate Walls with T-section*, Springer Theses,
https://doi.org/10.1007/978-981-95-7789-7_6

6.1.1 Seismic Tests of CDSWs

The typical setup for seismic experiments of steel-concrete composite walls is shown in Fig. 2.4. First, the vertical load is applied to the specimen by the hydraulic jack and remains constant. Subsequently, the cyclic in-plane lateral load is applied to the specimen by the actuator. Each step of the lateral load repeats two to three times, and the amplitude of displacement increases step by step until the specimen is obviously damaged or the load-carrying capacity drops to 85% of the maximum horizontal load (MOHURD 2015a), which corresponds to the ultimate drift ratio θ_u. In some experimental programs, in-plane or out-of-plane eccentric vertical loads are applied by changing the position of the hydraulic jack (Guo et al. 2020; Zhu et al. 2020).

The load state of CDSWs can be measured by the axial compression ratio. Compared with reinforced concrete shear walls, the steel plates in the CDSWs sustain a large number of vertical load, so the contribution of steel is included in the axial compression. The design value n_d and test value n_t of axial compression ratio are expressed as Eqs. (6.1) and (6.2), considering the contribution of steel and concrete (MOHURD and AQSIQ 2016; MOHURD 2015b, 2016), where 1.20 is the partial coefficient of load combination (MOHURD and AQSIQ 2012), 1.11 is the partial coefficient for resistance of steel (MOHURD and AQSIQ 2003), and 1.40 is the partial coefficient for resistance of concrete (MOHURD and AQSIQ 2010).

$$n_d = \frac{1.20N}{f_y A_s / 1.11 + f_c A_c / 1.40} \tag{6.1}$$

$$n_t = \frac{N}{f_y A_s + f_c A_c} \tag{6.2}$$

Where N denotes the axial compression, f_y and f_c denote the yield strength of steel and the characteristic compressive strength of concrete respectively, and A_s and A_c denote the cross-sectional area of steel and concrete respectively. If the strength of steel is different in the steel plates and tubes, Eqs. (6.1) and (6.2) are calculated according to respective strength and cross-sectional area.

Axial compression ratio n can significantly influence the load-carrying capacity and ductility of CDSWs. As axial compression ratio n increases, the moment or shear capacities of CDSWs increase first and decrease then, This is because the increasing of axial compression ratio n makes more concrete change from tension to compression and participate in sustaining load. However, the increasing of axial compression ratio n generally impairs the ductility of CDSWs. This is because the section height of compression zone increases and the damage of the extreme fibre becomes more severe.

In the seismic experiments, the typical failure modes of CDSWs includes flexural and shear modes (see Fig. 6.1). In the flexural failure mode, steel tubes and plates buckle or yield and concrete crushes under compression, and steel tubes and

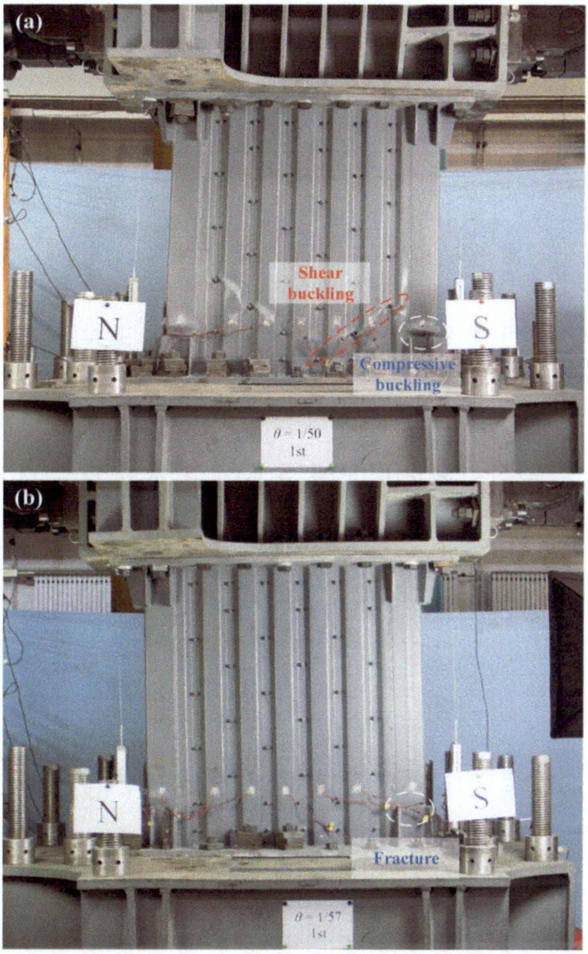

Fig. 6.1 Typical failure modes of CDSWs: (**a**) local buckling of steel corrugated plates and rectangular tubes and concrete crushing under compression and shear; (**b**) fracture of steel corrugated plates and rectangular tubes and concrete cracking under tension

plates fracture and concrete cracks under tension. In the shear failure mode, the steel plates fail by elastic shear buckling if steel plates are thin and not equipped with connectors or shear buckle after yielding if steel plates are connected with enough connectors. The shear failure mode may occur in the seismic experiments, while the flexural mode always governs the failure in the high-rise building since the aspect ratio of CDSWs is generally large.

The results of seismic experiments on steel-concrete composite walls (see Tables 6.1 and 6.2) mainly discuss the influence of variations in test parameters on deformation capacity. The test parameters include the material properties, geometric dimensions, and axial compression ratio n of steel-concrete composite walls. Among them, since the failure of steel-concrete composite walls with T-section mainly concentrates on the web walls, the listed ultimate drift ratio θ_u corresponds to the state of the web wall under compression.

Table 6.1 Design and results of seismic tests of CDSWs

Literature	NO.	f_{yw} (MPa)	f_{yc} (MPa)	f_{cu} (MPa)	H_w (mm)	B_w (mm)	ξ	ξ_w	n_t	n_d	$\theta_{u,t}$ (%)	$\theta_{u,d}$ (%)	$\theta_{u,d}/\theta_{u,t}$	Note
Nie et al. (2013)	CFSCW-1	306	306	87.5	2568	1284	0.50	0.33	0.35	0.56	1.74	1.66	0.95	CDSFWs
	CFSCW-2	306	306	86.1	2568	1284	0.51	0.33	0.35	0.56	1.60	1.64	1.03	
	CFSCW-3	306	306	86.1	2568	1284	0.51	0.33	0.35	0.56	1.53	1.64	1.07	
	CFSCW-4	351	351	89.8	2568	1284	0.44	0.29	0.35	0.56	1.59	1.63	1.02	
	CFSCW-5	443	443	88.1	2568	1284	0.42	0.28	0.36	0.56	1.67	1.60	0.96	
	CFSCW-6	306	306	65.0	2568	1284	0.69	0.45	0.35	0.55	1.44	1.59	1.10	
	CFSCW-7	306	306	102.6	2568	1284	0.42	0.27	0.33	0.52	1.49	1.79	1.21	
	CFSCW-8	351	363	88.4	2568	1284	0.72	0.38	0.35	0.56	1.75	1.88	1.07	
	CFSCW-9	306	306	83.3	2568	1284	0.53	0.35	0.37	0.58	1.39	1.60	1.15	
	CFSCW-10	443	443	83.7	1500	750	0.78	0.51	0.36	0.55	1.55	1.90	1.23	
	CFSCW-11	443	443	80.7	1125	750	0.82	0.53	0.36	0.55	1.39	1.81	1.30	
	CFSCW-12	443	443	88.0	750	750	0.74	0.48	0.35	0.54	1.84	1.74	0.94	
Chen et al. (2015)	DCHCW1	283	292	83.5	2350	900	0.58	0.42	0.10	0.16	4.21	3.17	0.75	CDSFWs
	DCHCW2	283	292	83.5	2350	900	0.58	0.42	0.00	0.00	4.19	4.41	1.05	
	DCHCW3	283	292	83.5	2350	900	0.58	0.42	0.10	0.16	4.89	3.17	0.65	
Hu et al. (2016)	SRCW-1	432	306	67.3	2568	1284	0.67	0.53	0.36	0.56	1.76	1.56	0.88	SFRCWs
	SRCW-2	432	306	87.8	2568	1284	0.50	0.39	0.36	0.56	1.75	1.64	0.94	
	SRCW-3	363	443	83.3	2568	1284	0.45	0.27	0.36	0.57	1.62	1.57	0.97	
Shen and Li (2016)	SCSW-1	410	410	57.8	2000	920	0.55	0.45	0.07	0.11	2.71	2.87	1.06	CDSFWs without boundary elements
	SCSW-2	410	410	57.8	2000	920	0.55	0.45	0.07	0.11	2.61	2.87	1.10	
	SCSW-3	410	410	57.8	2000	920	0.55	0.45	0.07	0.11	2.73	2.87	1.05	
	SCSW-4	410	410	57.8	2000	920	0.55	0.45	0.07	0.11	3.20	2.87	0.90	

Literature	NO.	f_{yw} (MPa)	f_{yc} (MPa)	f_{cu} (MPa)	H_w (mm)	B_w (mm)	ξ	ξ_w	n_t	n_d	$\theta_{u,t}$ (%)	$\theta_{u,d}$ (%)	$\theta_{u,d}/\theta_{u,t}$	Note
Zhang et al. (2016a)	BLC-C-1	433	433	47.1	2700	1324	1.67	1.08	0.25	0.38	1.90	2.42	1.27	Bundled lipped channel–concrete composite walls
	BLC-C-2	433	433	47.1	2700	1324	1.67	1.08	0.45	0.66	1.14	1.53	1.34	
	BLC-C-3	433	433	47.1	2700	1444	1.67	1.06	0.26	0.38	1.81	2.39	1.32	
	BLC-C-4	433	433	47.1	2700	1444	1.67	1.06	0.38	0.57	1.57	1.73	1.10	
	BLC-C-5	433	433	47.1	2700	1444	1.67	1.06	0.38	0.57	1.36	1.73	1.27	
	BLC-C-6	424	419	47.1	2700	1484	1.71	0.92	0.39	0.58	1.62	1.72	1.06	
	BLC-C-7	433	433	47.1	2700	1924	1.67	1.02	0.39	0.57	1.70	1.64	0.96	
Guo et al. (2017)	I-3.0b-1	440	440	34.5	3000	1420	2.62	1.66	0.37	0.53	3.09	2.98	0.96	CDSCWs
	I-3.0b-2	440	440	34.5	3000	1420	2.62	1.66	0.26	0.38	2.67	2.63	0.99	
Ji et al. (2017)	DSCW3	341	302	50.5	850	850	1.70	1.31	0.20	0.30	1.85	1.64	0.89	CDSFW
Guo et al. (2018)	CSW-7	350	406	65.6	750	400	2.48	1.37	0.44	0.64	0.98	1.40	1.43	CDSFWs
	CSW-8	350	406	65.6	750	400	2.48	1.37	0.68	0.99	1.36	1.41	1.04	
Huang et al. (2018)	CWS1	321	321	42.2	1475	900	2.06	1.21	0.39	0.57	2.08	2.12	1.02	CDSFWs
	CWS2	321	321	42.2	1475	900	2.06	1.21	0.59	0.86	1.49	1.41	0.94	
	CWS3	321	321	42.2	1475	900	2.06	1.21	0.20	0.29	3.23	3.53	1.09	
	CWS4	321	321	42.2	1025	900	2.06	1.21	0.39	0.57	2.15	2.00	0.93	
	CWS5	321	321	42.2	1925	900	2.06	1.21	0.39	0.57	2.06	2.20	1.06	
Qiao et al. (2018)	S0N2R	252	375	45.3	1110	740	1.54	0.88	0.16	0.24	3.13	3.28	1.05	CDSFWs
	S2N2R	252	375	45.3	1110	740	1.54	0.88	0.16	0.24	2.63	3.28	1.24	
	S3N2	252	375	45.3	1110	740	1.54	0.71	0.16	0.25	2.70	3.28	1.21	
	S3H2R	252	375	45.3	1110	740	1.54	0.88	0.27	0.40	2.27	2.40	1.06	
	S4N2R	252	375	45.3	1110	740	1.54	0.88	0.16	0.24	2.56	3.28	1.28	
	S5N2R	252	375	45.3	1110	740	1.54	0.88	0.16	0.24	2.13	3.28	1.54	

(continued)

Table 6.1 (continued)

Literature	NO.	f_{yw} (MPa)	f_{yc} (MPa)	f_{cu} (MPa)	H_w (mm)	B_w (mm)	ξ	ξ_w	n_t	n_d	$\theta_{u,t}$ (%)	$\theta_{u,d}$ (%)	$\theta_{u,d}/\theta_{u,t}$	Note
Yan et al. (2018)	W1	235	235	27.0	2000	1000	2.13	1.52	0.33	0.48	2.56	2.29	0.89	CDSFWs
	W2	235	235	27.0	2000	1000	2.13	1.52	0.33	0.48	3.18	2.29	0.72	
	W3	235	235	27.0	2000	1000	2.13	1.52	0.33	0.48	2.64	2.29	0.87	
	W4	235	235	27.0	2000	1000	2.13	1.52	0.33	0.48	3.41	2.29	0.67	
	W5	235	235	27.0	2000	1000	2.13	1.52	0.46	0.67	2.43	1.68	0.69	
	W6	235	235	27.0	2000	1000	2.13	1.52	0.33	0.48	2.41	2.29	0.95	
	W7	235	235	27.0	2000	1000	2.13	1.52	0.33	0.48	2.78	2.29	0.82	
Zhang et al. (2019)	CWSC-1 abc	467	467	52.3	1050	700	1.20	0.94	0.32	0.48	2.88	1.92	0.79	CDSFWs The test value takes the average value of each group
	CWSC-2 abc	467	467	52.3	1050	700	1.07	0.94	0.32	0.48	1.96	1.80	0.99	
	CWSC-3 a	467	467	37.8	1050	700	1.50	1.07	0.32	0.48	2.25	2.20	0.98	
	CWSC-4 abc	467	467	52.3	1050	700	1.31	1.05	0.32	0.48	2.57	2.00	0.83	
	CWSC-5 abc	467	467	52.3	1050	700	1.11	0.86	0.32	0.48	2.22	1.86	0.96	
Guo et al. (2020)	SC-1	325	325	46.4	1000	1000	0.87	0.67	0.11	0.16	2.02	2.71	1.34	CDSFWs without boundary elements Vertical load with out-of-plane eccentricity
	SC-2	325	325	46.4	1000	1000	0.87	0.67	0.21	0.32	1.48	1.95	1.32	
	SC-3	325	325	46.4	1000	1000	0.87	0.67	0.11	0.16	2.28	2.71	1.19	
Zhao et al. (2020a)	CW-F1	305	410	64.9	1500	1000	1.19	0.74	0.16	0.24	3.39	3.27	0.96	CDSFWs
	CW-F2	307	389	43.0	1500	1000	1.35	0.95	0.19	0.29	2.27	2.69	1.19	
	CW-C1	305	410	65.8	1500	1000	1.17	0.62	0.16	0.24	3.36	3.25	0.97	CDSCWs
	CW-C2	305	410	67.0	1500	1000	1.15	0.60	0.16	0.24	3.46	3.29	0.95	

Literature	NO.	f_{yw} (MPa)	f_{yc} (MPa)	f_{cu} (MPa)	H_w (mm)	B_w (mm)	ξ	ξ_w	n_t	n_d	$\theta_{u,t}$ (%)	$\theta_{u,d}$ (%)	$\theta_{u,d}/\theta_{u,t}$	Note
Zhao et al. (2020b)	DCW1	307	328	43.8	1000	1000	1.12	0.68	0.21	0.32	2.78	2.20	0.79	CDSCWs
	DCW2	307	328	43.8	1500	1000	1.12	0.68	0.21	0.32	2.20	2.32	1.05	
	DCW3	307	328	43.8	2000	1000	1.12	0.68	0.21	0.32	1.93	2.39	1.24	
	DCW4	307	328	43.8	2000	1000	1.12	0.68	0.21	0.32	1.97	2.39	1.21	
Zhu et al. (2020)	W-1.5-1	387	429	42.8	1500	1397	3.51	1.72	0.20	0.28	1.39	1.35	0.97	CDSCWs
	W-1.5-2	387	429	42.8	1500	1397	3.51	1.72	0.29	0.42	1.51	1.33	0.88	
	W-2.0-1	387	429	42.8	2000	1397	3.51	1.72	0.29	0.42	1.43	1.40	0.98	
	W-2.0-2	387	429	42.8	2000	1397	3.51	1.72	0.29	0.42	1.36	1.40	1.03	
	W-2.0-3	387	429	42.8	2000	1397	3.51	1.72	0.21	0.30	1.71	1.42	0.83	
	W-2.0-4	387	429	42.8	2000	1397	3.51	1.72	0.29	0.42	1.30	1.40	1.08	
	W-2.5-1	387	429	42.8	2500	1397	3.51	1.72	0.34	0.50	1.24	1.44	1.16	
	W-2.5-2	387	429	42.8	2500	1397	3.51	1.72	0.40	0.58	1.08	1.43	1.32	
	Average ± Standard Deviation												1.04 ± 0.18	

f_{yw} denotes the yield strength of steel plates, f_{yc} denotes the yield strength of steel tubes, and f_{cu} denotes the cubic compressive strength of concrete. H_w denotes the wall height, B_w denotes the overall wall breadth, ξ denotes the constraining factor of boundary elements, and ξ_w denotes the constraining factor of composite walls. n_t and n_d denote the test and design values of axial compression ratio, and $\theta_{u,t}$ and $\theta_{u,d}$ denote the test and design values of ultimate drift ratio

Table 6.2 Design and results of seismic tests of T-CDSWs

Literature	NO.	f_{yw} (MPa)	f_{yc} (MPa)	f_{cu} (MPa)	H_w (mm)	B_w (mm)	ξ	ξ_w	n_t	n_d	$\theta_{u,t}$ (%)	$\theta_{u,d}$ (%)	$\theta_{u,d}/\theta_{u,t}$	Note
Zhang et al. (2016b)	BLC-C-T1	433	433	46.0	2890	1280	1.71	1.15	0.36	0.52	1.68	2.08	1.24	Bundled lipped channel–concrete composite walls with T-section
	BLC-C-T2	433	433	46.0	2890	1280	1.71	1.15	0.36	0.52	1.56	2.08	1.34	
	BLC-C-T3	424	419	46.0	2890	1360	1.50	0.94	0.33	0.48	1.54	2.04	1.32	
	BLC-C-T4	433	433	46.0	2890	2160	1.55	1.06	0.44	0.65	2.01	1.49	0.74	
	BLC-C-T5	433	433	46.0	2890	2160	1.55	1.06	0.35	0.52	2.34	1.84	0.79	
This study	–	440	440	34.5	3000	1420	2.62	1.77	0.39	0.56	0.64	0.70	1.10	T-CDSCW
	Average ± Standard Deviation												1.09 ± 0.24	

f_{yw} denotes the yield strength of steel plates, f_{yc} denotes the yield strength of steel tubes, and f_{cu} denotes the cubic compressive strength of concrete. H_w denotes the wall height, B_w denotes the overall web breadth, ξ denotes the constraining factor of boundary elements, and ξ_w denotes the constraining factor of composite walls. n_t and n_d denote the test and design values of axial compression ratio, and $\theta_{u,t}$ and $\theta_{u,d}$ denote the test and design values of ultimate drift ratio

In the previous seismic experiments of CDSFWs, the steel flat plates of specimens are generally equipped with dense studs, and some are equipped with stiffeners inside to further improve the buckling capacity. Steel flat plates on both sides are connected by batten plates or tie bars. In some specimens, studs and U-shaped bars are set on the outside of CFSTs towards wall elements to improve the shear transfer of the interface (Chen et al. 2015; Ji et al. 2017). Nie et al. (2013) and Chen et al. (2015) conducted seismic experiments on CDSFWs with high-strength concrete, of which the cubic compressive strength ranges from 60 MPa to 100 MPa, and the test value of axial compression ratio n_t ranges from 0 to 0.35. Ji et al. (2017) conducted seismic experiments on CDSFWs with strong flange walls, and the test value of axial compression ratio n_t is 0.20. The flange walls improve the moment capacity significantly, so the CDSFWs fail in the shear mode. Huang et al. (2018) conducted seismic experiments on CDSFWs with middle CFST columns, which divide the whole steel plates into small parts and achieve more convenient modular production and welding of the inner reinforcements. The test value of axial compression ratio n_t ranges from 0.20 to 0.59. Yan et al. (2018) conducted seismic experiments on CDSFWs with inside over-headed studs, and the test value of axial compression ratio n_t ranges from 0.33 to 0.46.

In some CDSFWs, the batten plates connecting the steel plates on both sides are arranged in full length. Zhang et al. (2016a, b) conducted seismic experiments on bundled lipped channel composite walls with rectangular section and T-section, and the test value of axial compression ratio n_t ranges from 0.25 to 0.45. The composite walls are formed by welding steel channel, of which the flanges are at both sides and the web acts as stiffeners and connectors. The contribution of the channel web is ignored and regarded as safety reserves since the channel web cannot be connected vertically. Inside the steel channels, studs can be set to improve the interface. Guo et al. (2018) conducted seismic experiments on multi-partition composite walls, and the test value of axial compression ratio n_t ranges from 0.44 to 0.68. Multi-partition composite walls are similar to bundled lipped channel composite walls, except that multi-partition composite walls are formed by welding H-steel. Zhang et al. (2019) conducted seismic experiments on CDSFWs with full-length batten plates, and the test value of axial compression ratio n_t is 0.32.

In the previous seismic experiments of CDSCWs, the steel corrugated-plates on both sides are generally connected by bolts, which can be arranged at wave crest (Zhao et al. 2020a, b) or wave trough (Zhu et al. 2020). Zhu et al. (2020) conducted compressive, bending and seismic experiments on CDSCWs with rectangular section, and the test value of axial compression ratio n_t ranges from 0.20 to 0.39. Zhao et al. (2020a, b) conducted seismic experiments on both CDSFWs and CDSCWs, and the test value of axial compression ratio n_t ranges from 0.16 to 0.21.

When discussing the ultimate deformation of CDSWs, Zhao et al. (2020a) used half of the wall breadth $0.5B_w$ as the plastic hinge length, and Chen et al. (2015) modified the formula of the plastic hinge length proposed by Kazaz (2013) to Eq. (6.3).

$$H_p = 0.27B_w\left(1-\frac{N}{f_c'A_c+f_yA_s}\right)\left(1-\frac{f_y\rho_{sh}}{f_c'}\right)\left(\frac{M}{VB_w}\right)^{0.45} \tag{6.3}$$

Where f_c' denotes the cylinder compressive strength of concrete, ρ_{sh} denotes the steel content ratio, which is A_s/A_c, and M and V are the bending moment and shear force applied to CDSWs. It is noted that the formula proposed by Kazaz (2013) is applicable to reinforced concrete shear walls, and the difference between the formula and Eq. (6.3) is that the calculation of axial compression ratio does not consider the contribution of steel.

In addition, some specimens or loading are different from above typical seismic experiments, yet they are still supplemented here. Hu et al. (2016) conducted seismic experiments on SFRCWs, and the test value of axial compression ratio n_t is 0.36. Steel plates equipped with studs are embedded inside the concrete of SFRCWs and fully constrained, yet the concrete is not confined. The cubic compressive strength of the concrete f_{cu} in the specimens ranges from 60 MPa to 90 MPa. Shen and Li (2016) conducted seismic experiments on CDSFWs with only steel plates as boundary elements, and the test value of axial compression ratio n_t is 0.07. Guo et al. (2020) also conducted seismic experiments on CDSFWs with only steel plates as boundary elements, while the vertical loads were out-of-plane eccentric, and the test value of axial compression ratio n_t ranges from 0.11 to 0.21. Qiao et al. (2018) conducted seismic experiments on retrofitted specimens, which are similar to the specimens of Huang et al. (2018), and the test value of axial compression ratio n_t ranges from 0.16 to 0.27. The specimen includes three CFST columns connected by steel plates, and the steel plates are covered by concrete. After the first stage of loading, additional steel plates were welded on the specimens to cover the concrete, and then the cyclic loading was restarted.

The above-mentioned specimens have various forms and a large range of design parameters, which can fully represent the seismic performance of composite walls. The design parameters of CDSWs are numerous, and it is necessary to find key parameters that can influence the deformation capacity of composite walls, including the parameters related to material, geometry, and load, to further analysing the deformation capacity.

The boundary elements at both ends sustain a large proportion of bending moment, and thus sustain large compression and tension respectively. The load-carrying and deformation capacities of CFSTs are influenced by the constraining factor ξ (Han et al. 2001), which is determined by the geometric dimension and material properties as expressed in Eq. (6.4).

$$\xi = \frac{f_yA_{sc}}{f_cA_{cc}} \tag{6.4}$$

Where A_{sc} and A_{cc} denote the cross-sectional area of steel and concrete in the CFST respectively.

Analogous to the CFST, the constraining factor of the composite wall ξ_w is determined by the geometric dimension and material properties of the wall as expressed in Eq. (6.5).

$$\xi_w = \frac{f_y A_s}{f_c A_c} \tag{6.5}$$

When composite walls are tall, composite walls sustain large bending moment at the bottom section and fail in the flexural mode. Otherwise, composite walls sustain large shear and fail in the shear mode. The lateral load is applied at the top of the specimens in the seismic experiments, so the ratio of bending moment to shear force is the wall height. Therefore, the aspect ratio H_w/B_w can reflect the relative values of bending moment and shear force, and the wall height H_w is calculated as the distance from the loading point to the bottom section.

The axial compression ratio n has a great influence on composite walls. The greater the axial compression N, the more obvious the second-order effect. Meanwhile, the large axial compression may prevent the steel in the tensile zone from yielding, and composite walls fail in the mode similar to the reinforced concrete compression members with small eccentricity. The large axial compression ratio leads to the poor deformation capacity directly.

Above all, the key parameters that influence the deformation capacity of composite walls include the constraining factor of boundary elements ξ, the constraining factor of composite walls ξ_w, the aspect ratio H_w/B_w and the axial compression ratio n, which are already considered in Eq. (6.3). For the specimens, the constraining factor of boundary elements ξ ranges from 0.5 to 3.0, the constraining factor of composite walls ξ_w is about half the constraining factor of boundary elements, the aspect ratio of composite walls H_w/B_w ranges from 0.5 to 3.0, and the axial compression ratio n ranges from 0 to 0.7.

If only each single relationship is examined between the above four parameters and the ultimate drift ratio θ_u of composite walls (see Figs. 6.2, 6.3, 6.4, and 6.5), the ultimate drift ratio θ_u has no obvious relationship with the constraining factor of boundary elements ξ, the constraining factor of composite walls ξ_w, and the aspect ratio H_w/B_w, yet it is mainly influenced by the axial compression ratio n_t. As the axial compression ratio n_t increases by 0.1, the ultimate drift ratio θ_u decreases by 0.34% (1/300). The ultimate drift ratio θ_u of the supplementary experiments (grey scatters in Fig. 6.5) is basically located below the regression line, which demonstrates the unfavorable effects of the detailing or loading on the deformation capacity of composite walls. If the data of the supplementary experiments are ignored, as the axial compression ratio n_t increases by 0.1, the ultimate drift ratio θ_u decreases by 0.40% (1/250).

Regarding the limit values of inter-story drift ratio, GB 50011 (MOHURD and AQSIQ 2016) specifies that for reinforced concrete shear walls, the limit value of the elastic inter-story drift ratio is 1/1000, and the limit value of the elastic-plastic inter-story drift ratio is 1/120; for steel structures, the limit value of the elastic

Fig. 6.2 Relationship between the ultimate drift ratio of composite walls θ_u and the constraining factor of boundary elements ξ

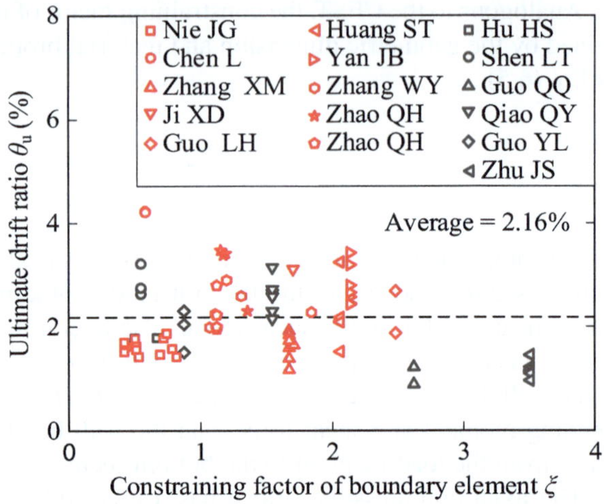

Fig. 6.3 Relationship between the ultimate drift ratio of composite walls θ_u and the constraining factor of composite walls ξ_w

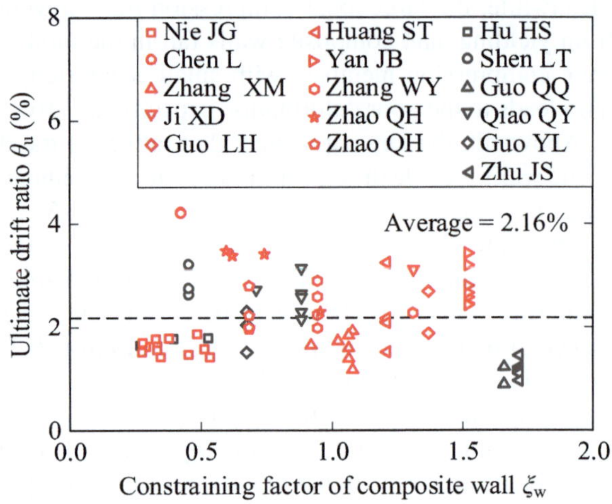

inter-story drift ratio is 1/250, and the limit value of the elastic-plastic inter-story drift ratio is 1/50. JGJ/T 380 (MOHURD 2015b) specifies that for CDSFWs, the limit value of the elastic inter-story drift ratio is 1/400, and the limit value of the elastic-plastic inter-story drift ratio is 1/80. The limit value of the elastic inter-story drift ratio of CDSFWs refers to the inter-story drift ratio when steel plates yield, and the limit value of the elastic-plastic inter-story drift ratio refers to the ultimate drift ratio θ_u.

Regarding the limit value of axial compression ratio, as mentioned above, GB 50011 (MOHURD and AQSIQ 2016) specifies that when the seismic design intensity is Grade 9 for the first seismic design level, the axial compression ratio n should not exceed 0.4; when the seismic design intensity is Grade 7 or 8, the axial compression ratio n should not exceed 0.5; for seismic design levels 2 and 3, the axial

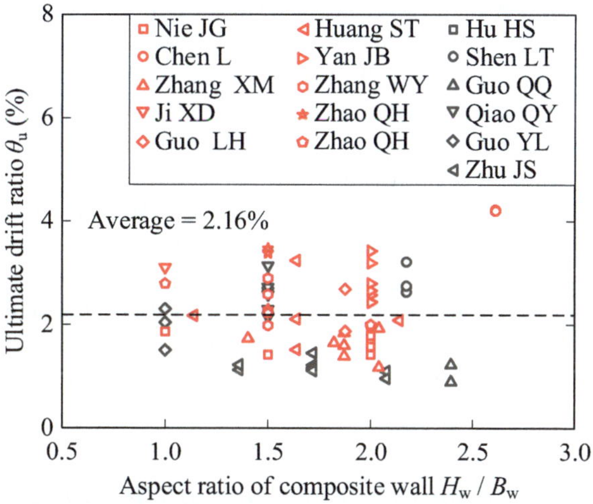

Fig. 6.4 Relationship between the ultimate drift ratio of composite walls θ_u and the aspect ratio of composite walls H_w/B_w

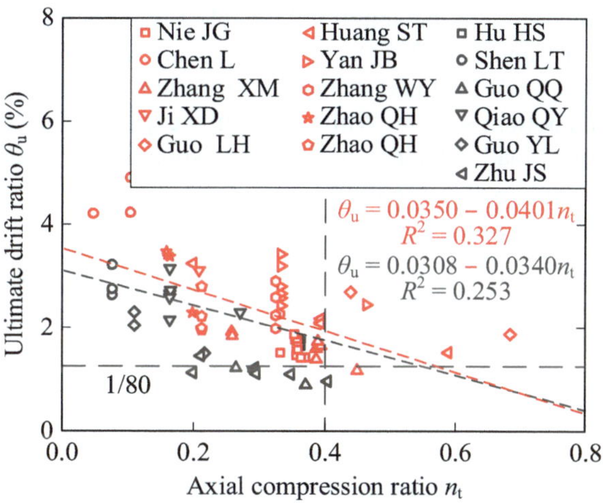

Fig. 6.5 Relationship between the ultimate drift ratio of composite walls θ_u and the axial compression ratio n_t

compression ratio n should not exceed 0.6. The specifications in JGJ/T 380 (MOHURD 2015b) on the limit value of axial compression ratio and seismic design level for steel-concrete composite walls are similar to those in GB 50011 (MOHURD and AQSIQ 2016), but the calculation of axial compression ratio n considers the contribution of steel plates.

From the experimental results, the deformation capacities of composite walls can be ensured with the axial compression ratio n_t less than 0.4, while the ultimate drift ratio θ_u can still be greater than 1/80 with the axial compression ratio n_t greater than 0.4. Considering that the design value of the axial compression ratio n_d is approximately 1.5 times the test value, that is to say, the deformation performance of composite walls may still be good with the design value of the axial compression ratio

n_d greater than 0.6. It can be seen from the comparison with the experimental results that the requirements of JGJ/T 380 (MOHURD 2015b) are reasonable on the limits of elastic-plastic drift ratio θ_u and axial compression ratio n.

However, the data in Figs. 6.2, 6.3, 6.4, and 6.5 are very scattered. This demonstrates that although the axial compression ratio n is the main factor, it is not the only factor that influences the deformation capacity of composite walls. The reason why the ultimate drift ratio θ_u has no obvious relationship with other parameters, including the constraining factor of boundary elements ξ, the constraining factor of composite walls ξ_w, and the aspect ratio H_w/B_w, is that it is influenced by many parameters together. Therefore, to evaluate the deformation capacity more accurately and optimise the seismic performance design of composite walls, it is necessary to propose more precise prediction of the ultimate drift ratio θ_u.

6.1.2 Empirical Formulas of Ultimate Displacement

According to the simplified model of shear walls (see Fig. 6.6) (Wang et al. 2020; Kang and Kim 2014), only the flexural deformation of shear walls is considered, horizontal load is distributed in inverted triangular, and plastic hinge is formed as the bottom section. While in seismic experiments, composite walls sustain concentrated load at the top, and the corresponding ultimate drift of the cantilever wall δ_u is expressed as Eq. (6.6).

$$\delta_u = \frac{1}{3}\phi_y H_w^2 + (\phi_u - \phi_y) H_p \left(H_w - \frac{1}{2} H_p \right) \tag{6.6}$$

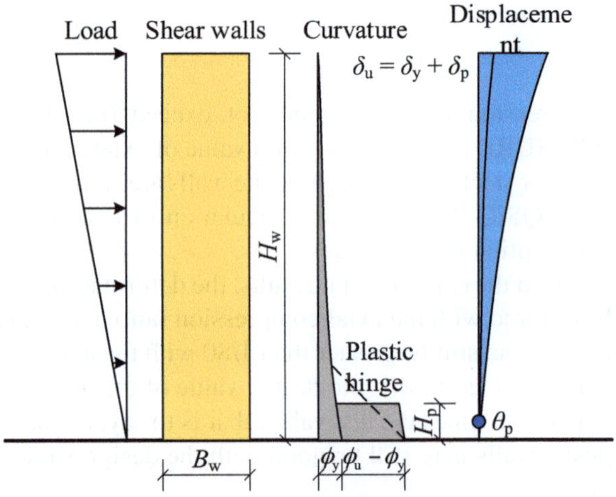

Fig. 6.6 Simplified curvature and displacement of shear walls

Where ϕ_y denotes the yield curvature, which is approximately $2\varepsilon_y/B_w$, ϕ_u denotes the ultimate curvature, which is ε_u/B_c, B_c denotes the height of the plastic compressive zone, and H_p denotes the plastic hinge length.

After algebraic manipulation, the ultimate drift ratio θ_u is expressed as Eq. (6.7).

$$\theta_u = \frac{\delta_u}{H_w} = \frac{1}{3}\phi_y H_w + \left(\phi_u - \phi_y\right) H_p \left(1 - \frac{1}{2}\frac{H_p}{H_w}\right) \tag{6.7}$$

For the plastic hinge length H_p, Hu et al. (2016) and Zhao et al. (2020a, b) assume $H_p = 0.5B_w$, which is the same as reinforced concrete shear walls (Kowalsky 2001; Wallace and Moehle 1992). Some studies suggest other values, such as $0.33B_w$ (Thomsen and Wallace 2004) and empirical values determined by wall height and reinforcement (Kazaz 2013). However, such values may not be applicable to steel–concrete composite walls. Take Eq. (6.3) as an example, the formula induce unreasonable negative values due to the large constraining factor of composite walls ξ_w.

There is another assumption that may be applicable: the height of the plastic compressive zone changes linearly along the wall height, and the effective plastic hinge length can be taken as half of the height of the plastic compression zone $0.5B_c$ (Kazaz 2013). This converts the evaluation of the plastic hinge length H_p to the height of the plastic compression zone B_c.

The height of the plastic compression zone B_c is directly determined by the axial load ratio n and the constraining ratio of composite walls ξ_w. If the cross-section of boundary elements is neglected and composite walls are taken as composite panels, the height of the plastic compression zone B_c can be derived from Eq. (6.8) based on axial load equilibrium and the full-section plasticity.

$$N = f_y\left(2B_c t_{se} - A_s\right) + f_c B_c t_{ce} \tag{6.8}$$

Where t_{se} and t_{ce} denote the equilibrium thickness of steel and concrete respectively.

After algebraic manipulation, the height of the plastic compression zone Bc is simplified as Eq. (6.9).

$$\frac{B_c}{B_w} = n + \left(1 - n\right)\frac{\xi_w}{1 + 2\xi_w} \tag{6.9}$$

From the perspective of the height of the plastic zone B_c, when the plastic neutral axis is located at the centroid, the moment capacity of the composite wall reaches the maximum. At this time, $B_c = 0.5B_w$. If the constraining factor of the composite wall ξ_w is taken as 1.0, then the corresponding axial compression ratio n is about 0.4, and the corresponding design value n_d is 0.6. This is consistent with the limit value of axial compression ratio required by JGJ/T 38037 (MOHURD 2015b). When the axial compression ratio n is less than 0.4, B_c is less than $0.5B_w$, the load-carrying capacity of the composite wall is below the protruding point of the interaction curve, which leads to good seismic performance.

For T-CDSWs, Eq. (6.9) is not applicable due to the large cross-sectional area of flange walls. After considering the cross-sectional area of flange walls, the height of the plastic compressive zone B_c with the web wall under compression is expressed as Eq. (6.10).

$$\frac{B_c}{B_w} = n + \frac{B_f / B_w + \xi_w + n\left(B_f / B_w - \xi_w + B_f \xi_w / B_w\right)}{1 + 2\xi_w} \tag{6.10}$$

Where B_f and B_w denote the overall breadth of the flange and web walls. In the calculation, the constraining factor of the composite wall ξ_w is evaluated according to the web wall. When the T-CDSW is reduced to the rectangular wall by removing the flange wall, $B_f = 0$, and Eq. (6.10) is reduced to Eq. (6.9). After the height of the plastic zone B_c is determined, the half of B_c is taken as the plastic hinge length H_p.

The only variable remaining uncertain is the ultimate curvature ϕ_u. The determination of the ultimate curvature ϕ_u is based on the ultimate strain ε_u of the concrete at the extreme fibre, which is governed by the concrete in the CFST. It is noted that the ultimate curvature ϕ_u corresponds to the state that the load of composite walls drops to 85% of the maximum load. The ultimate strain ε_u is related to the stress of concrete and may be different from the ultimate strain of axially loaded CFST columns. The compressive stress–strain relationship of confined concrete in the CFST has been proposed by Han et al. (2001) and Tao and Nie (2015), and the descending stage of the curve is expressed as Eq. (3.16). Combined with the seismic tests of CDSWs, the ultimate stress of the concrete in the boundary elements σ_u when CDSWs reaches the ultimate displacement can be derived, as shown in Eq. (6.11).

$$\frac{\sigma_u}{\sigma_c} = 0.143 + 0.238n \tag{6.11}$$

The fitting formula, Eq. (6.11), agrees well with the ultimate stress σ_u derived from the experimental results ($R^2 = 0.673$, which means the axial compression ratio n explains 67.3% of the change of the ultimate stress σ_u). Other influencing factors include the uncertainty of material properties and the detailing of composite walls. Even considering the supplementary experiments, Eq. (6.11) can still evaluate the ultimate stress σ_u well. It is noted that although the constraining factor of boundary elements ξ of Zhu et al. (2020) is relatively large, yet due to the limitation of the welding quality, the ultimate stress σ_u cannot reach Eq. (6.11). After trial and error, the ultimate strain of concrete ε_u is 0.3 in the rectangular specimens, and 0.15 in the specimens with T-section, of which the web wall suffered more damage. For the specimens of Shen and Li (2016) and Guo et al. (2018), there are no CFST columns as boundary elements, so the wall thickness is taken as the breadth of boundary elements to calculate the constraining factor of boundary elements ξ.

The greater the axial compression ratio n, the greater the ultimate stress σ_u, and the less the corresponding strain and deformation. The corresponding ultimate strain ε_u obtained from the ultimate stress σ_u ranges from 0.2 to 0.6, which is close to the ultimate strain of axially loaded CFST columns (Susantha et al. 2001). Based on the ultimate strain ε_u, the ultimate curvature ϕ_u can be obtained.

After the plastic hinge length H_p and the ultimate curvature ϕ_u are determined, the ultimate drift ratio θ_u can be obtained according to Eq. (6.7). (see Figs. 6.7 and 6.8). Although Eq. (6.7) is derived based on the flexural deformation of cantilever walls, Eq. (6.7) is still applicable to the composite walls with small aspect ratio ($H_w/B_w < 2$, red scatters in Fig. 6.9). This is because the shear capacity of composite walls is high, and the failure modes are always flexural. Moreover, Eq. (6.7) is also applicable to SPRCWs, retrofitted CDSWs, and even composite walls subjected to both in- and out-of-plane bending. For rectangular composite walls, the average ratio of design values obtained by Eq. (6.7) to experimental results $\theta_{u,d}/\theta_{u,t}$ is 1.04 with a standard deviation of 0.18; for T-CDSWs, the average ratio of design values to experimental results $\theta_{u,d}/\theta_{u,t}$ is 1.09 with a standard deviation of 0.24. To ensure a guarantee rate of 95% in the design, the ultimate drift ratio θ_u obtained by Eq. (6.7) can be multiplied

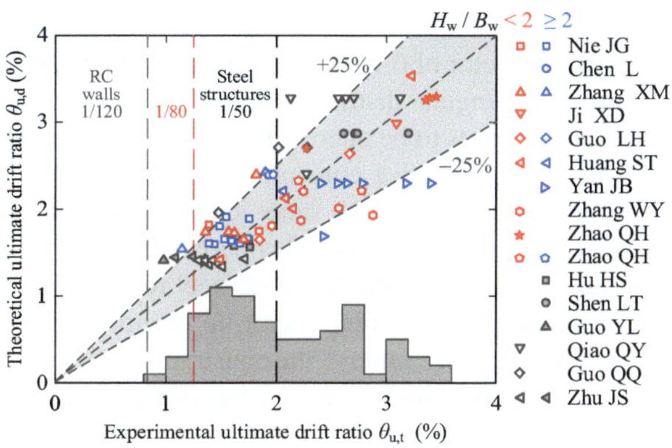

Fig. 6.7 Theoretical and experimental values of the ultimate drift ratio θ_u of CDSWs with rectangular section

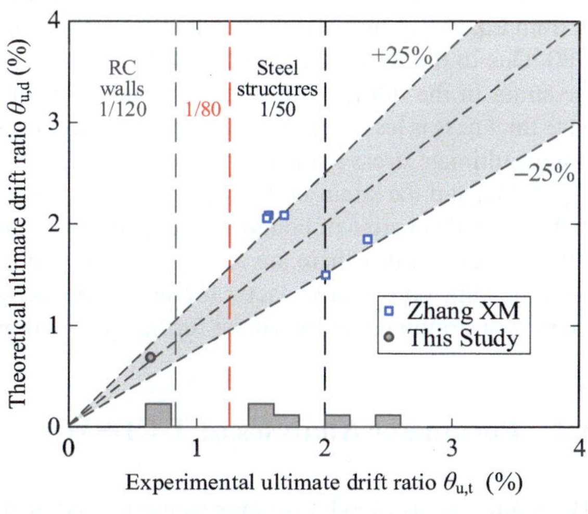

Fig. 6.8 Theoretical and experimental values of the ultimate drift ratio θ_u of CDSWs with T-section

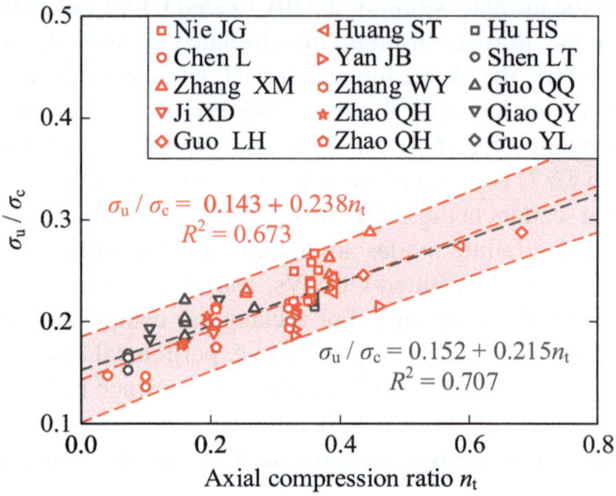

Fig. 6.9 Relationship between the ratio of the ultimate stress to the peak strength and the axial compression ratio

by a partial factor of 0.75. If the load distribution is not concentrated at the top, such as inverted triangular distributed loads, only the coefficient of the elastic deformation in Eq. (6.7) needs to be altered according to the load distribution.

Due to the effective constraint provided by steel rectangular tubes and plates, CDSWs do not require the same length requirements for boundary elements along the cross-section as reinforced concrete shear walls or SPRCWs. However, there is a requirement for the thickness of steel rectangular tubes and plates, as well as the corresponding constraining factor, to ensure effective confinement of infilled concrete. The threshold value of the constraining factor ξ, which guarantees that CDSWs reaches the drift ratio limit, can be evaluated based on Eq. (6.7). Parameter analysis of Eq. (6.7) reveals that for higher concrete strength, the constraining factor ξ required is smaller. For the concrete strength grade of C40, when the constraining factor ξ of boundary elements reaches 1.4, the design value of the ultimate drift ratio θ_u from Eq. (6.7) can meet the specified elastic-plastic inter-storey drift ratio limit of 1/80. Due to the lower strain in the infilled concrete of wall elements compared to the strain in the infilled concrete of boundary elements, the requirement for steel plate thickness is less stricter than the requirement for steel tube thickness. Assuming that the ultimate stress in the infilled concrete of wall elements is also determined by Eq. (6.11), and the strain in the edge portions of wall elements is 90% of the strain in the infilled concrete of boundary elements, the constraining factor ξ should reach 1.0. Considering that there are steel plates only on two sides of wall elements, the corresponding constraining factor for the composite wall ξ_w is 0.5. The requirements for the constraining factors will be further analysed by finite element simulations.

6.2 Parameter Analyses of T-CDSWs

The requirements of axial compression ratio and drift ratio are mainly applicable to rectangular composite walls (MOHURD 2015b), but not to other sections. Since the experimental programs of T-CDSWs are limited, a parametric analysis based on the shell–beam finite element model is conducted to further analyse the influence of the axial compression ratio on the seismic performance of T-CDSWs.

Based on the shell–beam finite element model, the seismic performance of T-CDSWs is investigated. The material properties of steel are as follows: the yield strength $f_y = 345$ MPa, the elastic modulus $E_s = 2.06 \times 10^5$ MPa, and the Poisson's ratio $\upsilon_s = 0.3$ (MOHURD and AQSIQ 2017), the failure strain $\varepsilon_f = 50\varepsilon_y$, the tangent modulus $E_t = 0.02E_s$, and the tangent modulus after failure $E_f = -0.1E_s$. The material properties of concrete are as follows: the axial compressive strength $f_c = 26.8$ MPa, the elastic modulus $E_c = 3.25 \times 10^4$ MPa, and the Poisson's ratio $\upsilon_c = 0.2$ (MOHURD and AQSIQ 2010). The other material properties are the same as Chap. 3.

The finite element analyses investigates the influences of steel plate thickness t_s and axial compression ratio n on the seismic performance of T-CDSWs. The steel tube thickness t_c and steel plate thickness t_s are the same, and the breadth of the boundary element $b_c \times d_c$ and the wall thickness d_w are the same. The parameters of numerical examples are shown in Table 6.3.

The hysteretic curves of T-CDSWs can be obtained by the finite element static analysis. The increment of drift ratio is 1/400, and the loading of each step repeats 2 times until the drift ratio reaches 1/50. If T-CDSWs still does not reach the ultimate deformation at this time, it is considered that the deformation capacity of T-CDSWs is good enough.

As the axial compression ratio n increases, the ultimate drift ratio θ_u of T-CDSWs gradually decreases, the positive maximum horizontal load increases, while the negative maximum horizontal load decreases (see Figs. 6.10, 6.11, 6.12, 6.13, 6.14, 6.15, 6.16, 6.17, 6.18, 6.19, 6.20, 6.21, 6.22, 6.23, 6.24, 6.25, 6.26, 6.27, 6.28, 6.29, 6.30, 6.31, 6.32, 6.33, and 6.34). For the numerical examples with the constraining factor of boundary elements ξ less than 1.0, the ultimate drift ratio θ_u of T-CDSWs is less than 1/80 when the axial compression ratio n reaches 0.3.

The deformation capacity of T-CDSWs improves as the constraining factor of boundary elements ξ increases, and the ultimate drift ratio θ_u under the same axial compression ratio gradually increases. As the constraining factor ξ reaches 1.5, the

Table 6.3 Numerical examples of seismic performance of T-CDSWs

NO.	Overall flange breadth B_f (mm)	Overall web breadth B_w (mm)	Wall height H_w (mm)	Wall thickness d_w (mm)	Steel plate thickness t_s (mm)	Constraining factor ξ	Axial compression ratio n
1	1000	1000	3000	150	1.4	0.5	0~0.4
2	1000	1000	3000	150	2.8	1.0	0~0.4
3	1000	1000	3000	150	4.9	1.5	0~0.4
4	1000	1000	3000	150	5.2	2.0	0~0.4
5	1000	1000	3000	150	7.5	3.0	0~0.4

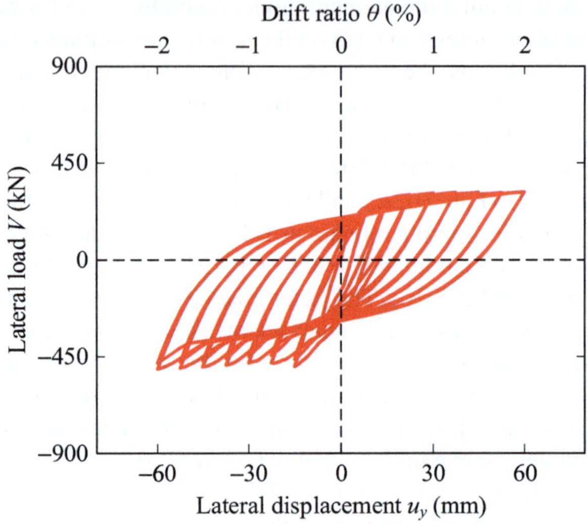

Fig. 6.10 Load–displacement curves of the T-CDSW: $\xi = 0.5$, $n_t = 0$

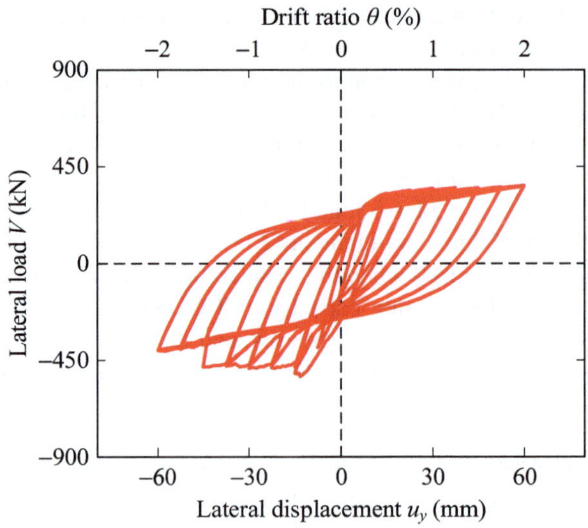

Fig. 6.11 Load–displacement curves of the T-CDSW: $\xi = 0.5$, $n_t = 0.1$

ultimate drift ratio θ_u reaches 1/60. As the constraining factor ξ reaches 3.0, that is, the steel plate thickness is 7.5 mm, the steel content ratio of CDSWs is so large that the hysteretic curves of CDSWs are close to those of steel plate shear walls, and the strength still does not degrade significantly until the drift ratio of 1/50.

Except for the constraining factor of boundary elements ξ, the boundary elements should meet the requirements of plate slenderness. Equations (2.1) and (2.2) can only ensure that steel tubes do not buckle before yielding. To ensure that steel tubes do not buckle until the ultimate drift ratio, stricter limits of plate slenderness should be adopted.

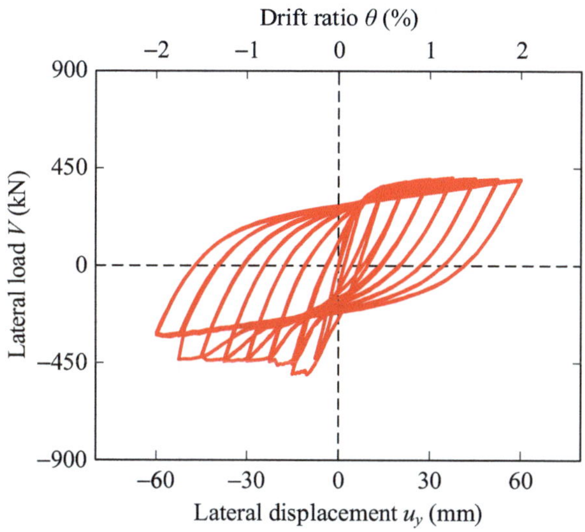

Fig. 6.12 Load–displacement curves of the T-CDSW: $\xi = 0.5$, $n_t = 0.2$

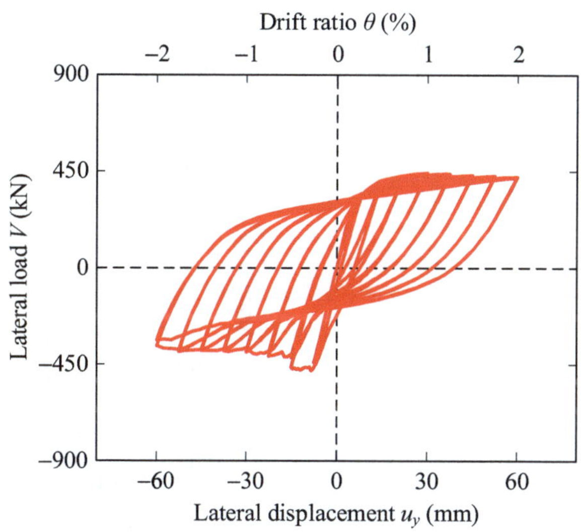

Fig. 6.13 Load–displacement curves of the T-CDSW: $\xi = 0.5$, $n_t = 0.3$

Combining the analyses on Eq. (6.7) and numerical results, the lower limit of the constraining factor of boundary elements ξ can be proposed. To meet the requirements in JGJ/T 380 (MOHURD 2015b) on the elastic-plastic inter-storey drift ratio of CDSWs, considering the upper limit of axial compression raio n is 0.4, it is recommended that the constraining factor of boundary elements should not be less than 1.5, especially for the boundary elements at the web ends of T-CDSWs, the constraining factor must not be less than 1.5. The constraining factor of CDSWs should not be less than 0.5, especially for the web walls of T-CDSWs, the constraining factor must not be less than 0.5.

Fig. 6.14 Load–displacement curves of the T-CDSW: $\xi = 0.5$, $n_t = 0.4$

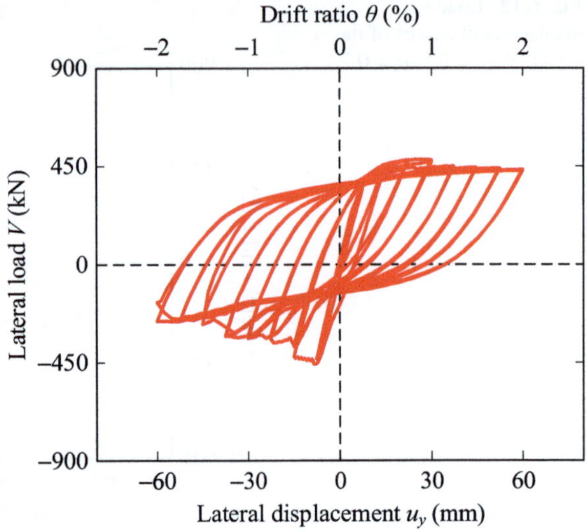

Fig. 6.15 Load–displacement curves of the T-CDSW: $\xi = 1.0$, $n_t = 0$

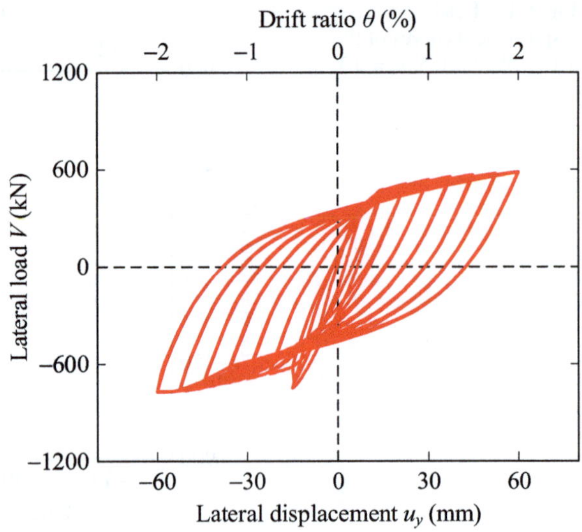

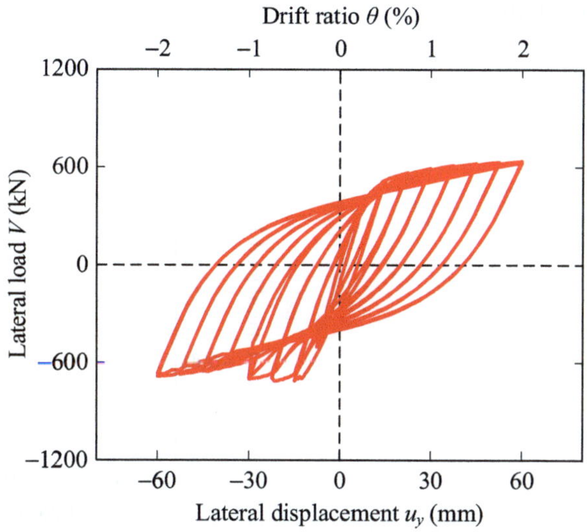

Fig. 6.16 Load–displacement curves of the T-CDSW: $\xi = 1.0$, $n_t = 0.1$

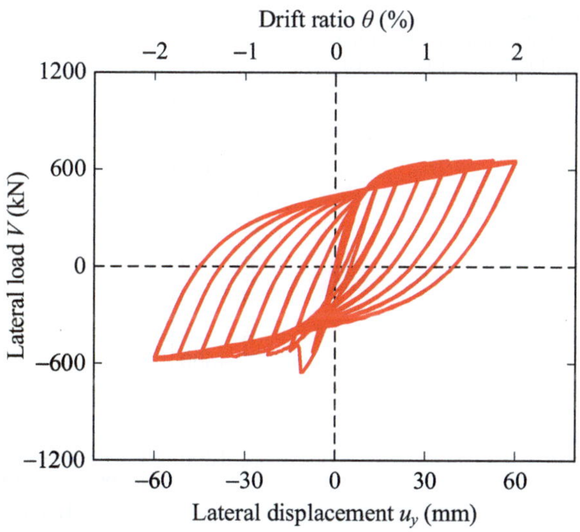

Fig. 6.17 Load–displacement curves of the T-CDSW: $\xi = 1.0$, $n_t = 0.2$

Fig. 6.18 Load–displacement curves of the T-CDSW: $\xi = 1.0$, $n_t = 0.3$

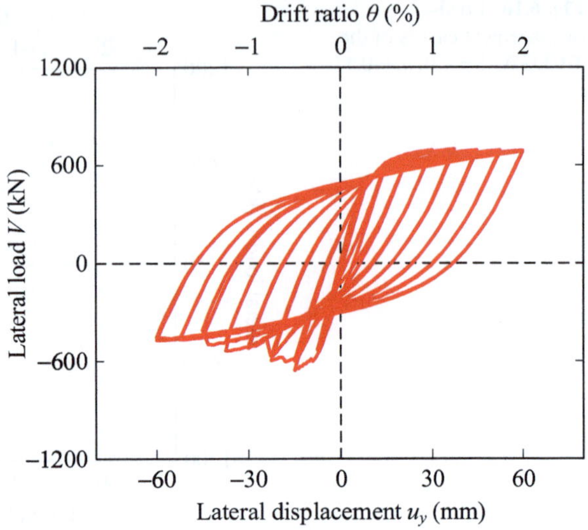

Fig. 6.19 Load–displacement curves of the T-CDSW: $\xi = 1.0$, $n_t = 0.4$

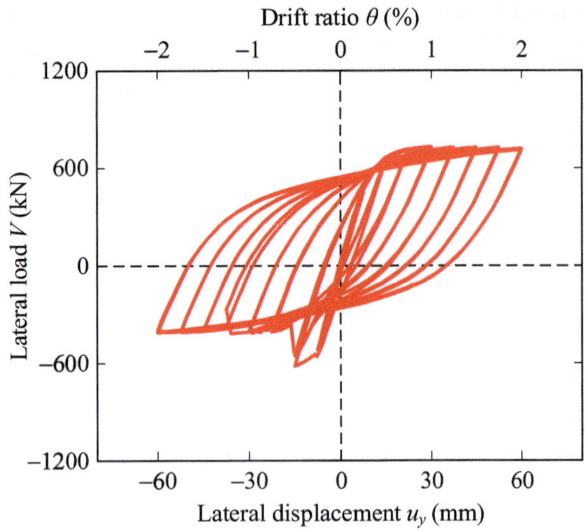

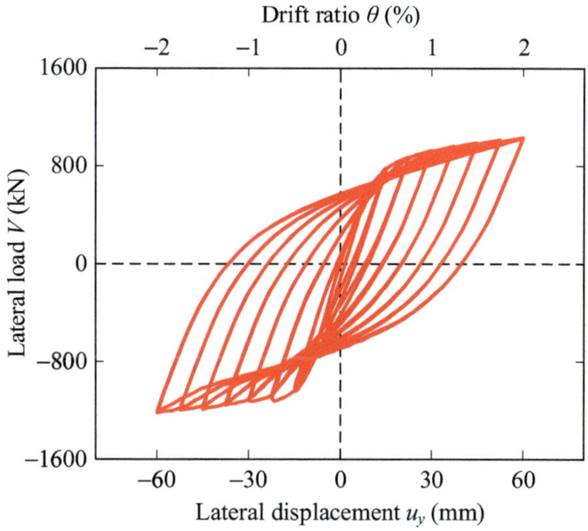

Fig. 6.20 Load–displacement curves of the T-CDSW: $\xi = 1.5$, $n_t = 0$

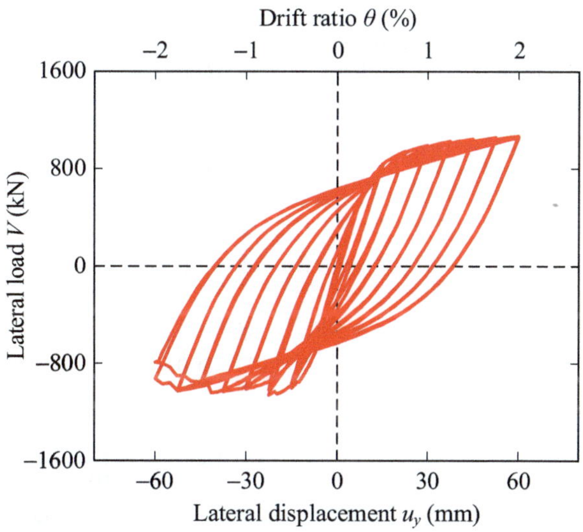

Fig. 6.21 Load–displacement curves of the T-CDSW: $\xi = 1.5$, $n_t = 0.1$

Fig. 6.22 Load–displacement curves of the T-CDSW: $\xi = 1.5$, $n_t = 0.2$

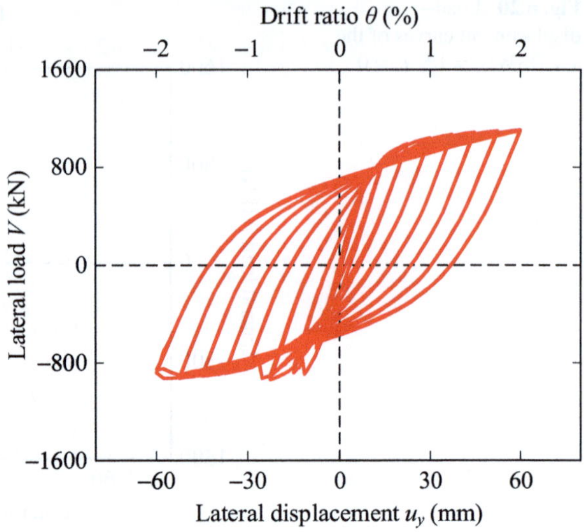

Fig. 6.23 Load–displacement curves of the T-CDSW: $\xi = 1.5$, $n_t = 0.3$

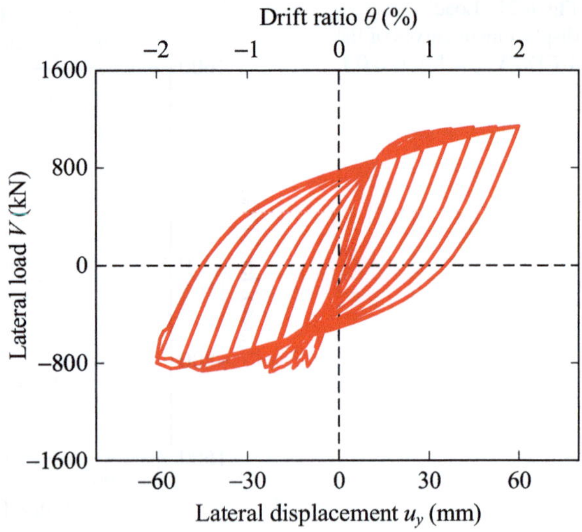

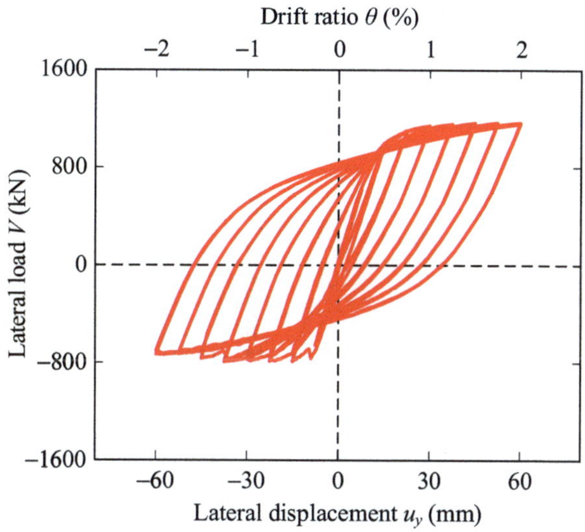

Fig. 6.24 Load–displacement curves of the T-CDSW: $\xi = 1.5$, $n_t = 0.4$

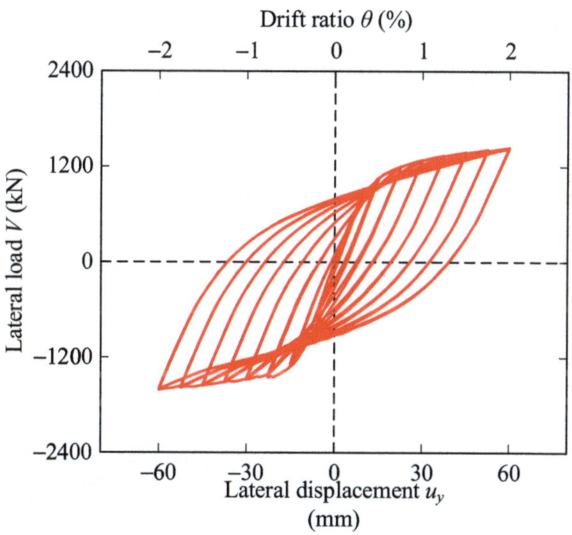

Fig. 6.25 Load–displacement curves of the T-CDSW: $\xi = 2.0$, $n_t = 0$

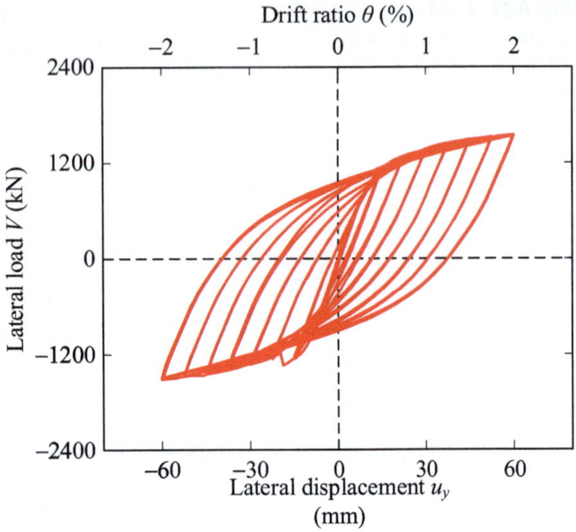

Fig. 6.26 Load–displacement curves of the T-CDSW: $\xi = 2.0$, $n_t = 0.1$

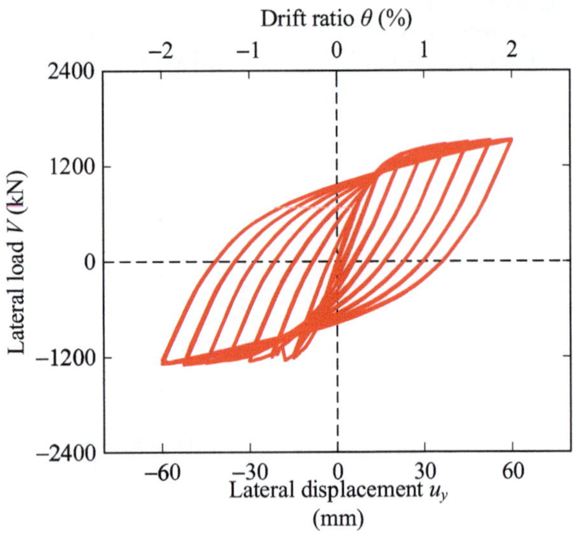

Fig. 6.27 Load–displacement curves of the T-CDSW: $\xi = 2.0$, $n_t = 0.2$

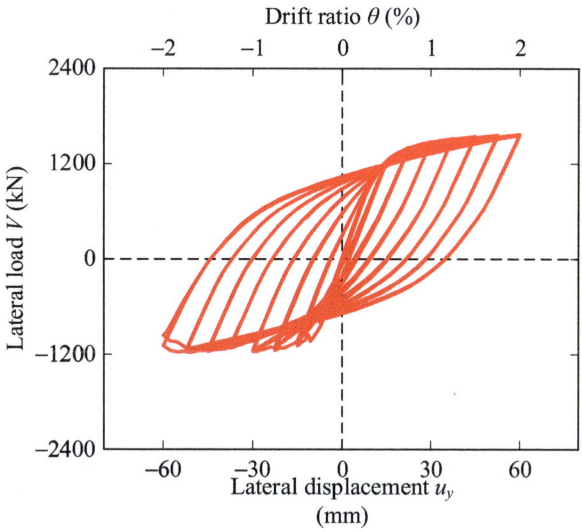

Fig. 6.28 Load–displacement curves of the T-CDSW: $\xi = 2.0$, $n_t = 0.3$

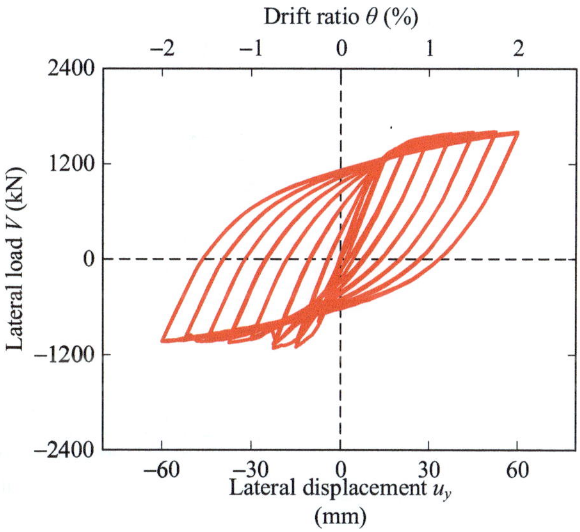

Fig. 6.29 Load–displacement curves of the T-CDSW: $\xi = 2.0$, $n_t = 0.4$

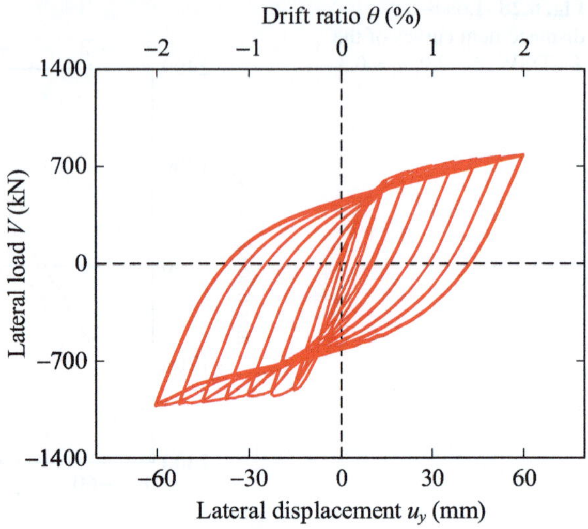

Fig. 6.30 Load–displacement curves of the T-CDSW: $\xi = 3.0$, $n_t = 0$

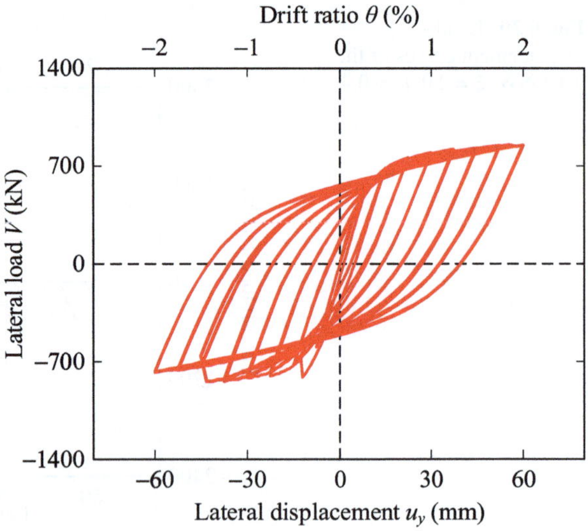

Fig. 6.31 Load–displacement curves of the T-CDSW: $\xi = 3.0$, $n_t = 0.1$

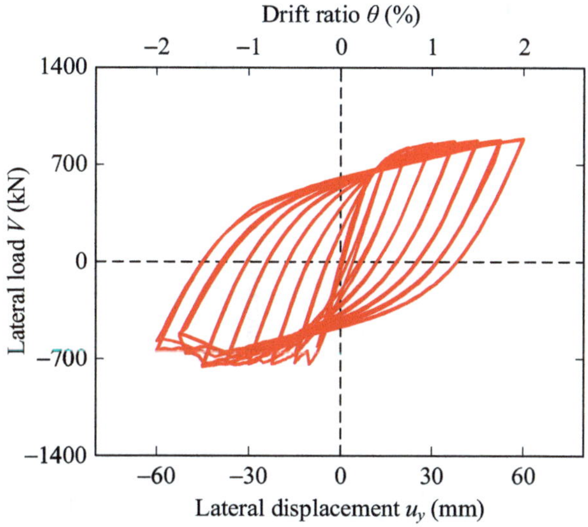

Fig. 6.32 Load–displacement curves of the T-CDSW: $\xi = 3.0$, $n_t = 0.2$

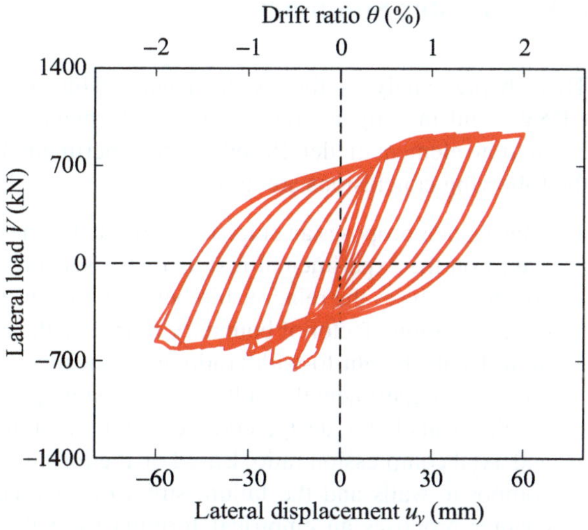

Fig. 6.33 Load–displacement curves of the T-CDSW: $\xi = 3.0$, $n_t = 0.3$

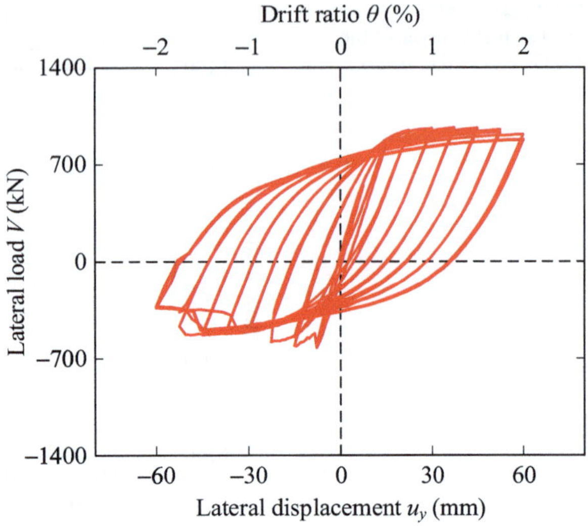

Fig. 6.34 Load–displacement curves of the T-CDSW: $\xi = 3.0$, $n_t = 0.4$

6.3 Chapter Summary

This chapter analyses the experimental results of the seismic performance of CDSWs and investigates the seismic performance of T-CDSWs using the shell–beam finite element model. Based on the experimental results and numerical results, the following conclusions are drawn.

1. This chapter proposes an empirical formula for the ultimate drift ratio of CDSWs. Based on the experimental results of the seismic performance of CDSWs, this chapter analyses the relationship between the ultimate drift ratio of CDSWs and the constraining factor of boundary elements, the constraining factor of composite walls, the height-to-width ratio of composite walls, and the axial compression ratio. The experimental results of the seismic performance validate the seismic requirements in the design code regarding the limits of elastic-plastic drift ratio and axial compression ratio. Based on the formula for the plastic hinge length of composite walls and the failure stress of concrete in boundary elements, this chapter proposes an empirical formula to evaluate the ultimate drift ratio of CDSWs, which agrees well with the experimental results.
2. Combining the empirical formula and numerical results, this chapter proposes the seismic requirements for CDSWs. The numerical results show that increasing the constraining factor of boundary elements and reducing the axial compression ratio can enhance the seismic performance of T-CDSWs. To meet the requirements of the limits of elastic-plastic drift ratio and axial compression ratio in the design code, it is recommended that the constraining factor of the boundary elements in CDSWs should not be less than 1.5, especially for the boundary elements at the web ends of T-CDSWs, the constraining factor must not be less than 1.5. The constraining factor of CDSWs should not be less than 0.5, especially for the web walls of T-CDSWs, the constraining factor must not be less than 0.5.

References

Alzeni Y, Bruneau M (2017) In-plane cyclic testing of concrete-filled Sandwich steel panel walls with and without boundary elements. J Struct Eng 143:4017115. https://doi.org/10.1061/(ASCE)ST.1943-541X.0001791

Chen L, Mahmoud H, Tong S, Zhou Y (2015) Seismic behavior of double steel plate-HSC composite walls. Eng Struct 102:1–12. https://doi.org/10.1016/j.engstruct.2015.08.017

Guo Y, Zhu J, Wang M, Yang X (2017) Research on corrugated-steel-concrete composite wall structure system in high-rise building (in Chinese). Tsinghua University, Beijing

Guo L, Wang Y, Zhang S (2018) Experimental study of rectangular multi-partition steel-concrete composite shear walls. Thin Wall Struct 130:577–592. https://doi.org/10.1016/j.tws.2018.06.011

Guo Q, Zhang P, Fu Y, Ke F (2020) Effects of out-of-plane eccentric axial compression on in-plane hysteretic behavior of SC walls. Struct 27:1570–1581. https://doi.org/10.1016/j.istruc.2020.07.041

Han L, Zhao X, Tao Z (2001) Tests and mechanics model for concrete-filled SHS stub columns, columns and beam-columns. Steel Compos Struct 1:51–74

Hu H, Nie J, Fan J, Tao M, Wang Y, Li S (2016) Seismic behavior of CFST-enhanced steel plate-reinforced concrete shear walls. J Constr Steel Res 119:176–189. https://doi.org/10.1016/j.jcsr.2015.12.010

Huang S, Huang Y, He A, Tang X, Chen Q, Liu X, Cai J (2018) Experimental study on seismic behaviour of an innovative composite shear wall. J Constr Steel Res 148:165–179. https://doi.org/10.1016/j.jcsr.2018.05.003

Ji X, Jiang F, Qian J (2013) Seismic behavior of steel tube-double steel plate-concrete composite walls: experimental tests. J Constr Steel Res 86:17–30. https://doi.org/10.1016/j.jcsr.2013.03.011

Ji X, Cheng X, Jia X, Varma AH (2017) Cyclic in-plane shear behavior of double-skin composite walls in high-rise Buildings. J Struct Eng 143:4017025. https://doi.org/10.1061/(ASCE)ST.1943-541X.0001749

Kang S, Kim J (2014) Evaluation of deformation capacity including yield deformation in displacement-based design of special RC shear wall. Struct Des Tall Spec Build 23:181–209. https://doi.org/10.1002/tal.1030

Kazaz O (2013) Analytical study on plastic hinge length of structural walls. J Struct Eng 139:1938–1950. https://doi.org/10.1061/(ASCE)ST.1943-541X.0000770

Kowalsky MJ (2001) RC STRUCTURAL WALLS DESIGNED ACCORDING TO UBC AND DISPLACEMENT-BASED METHODS. J Struct Eng 127:506–516. https://doi.org/10.1061/(ASCE)0733-9445(2001)127:5(506)

MOHURD (2015a) Specification for seismic test of Buildings (in Chinese). China Architecture and Building Press, Beijing

MOHURD (2015b) Technical specification for steel plate shear walls (in Chinese). China Architecture and Building Press, Beijing

MOHURD (2016) Code for design of composite structures (in Chinese). China Architecture and Building Press, Beijing

MOHURD and AQSIQ (2003) Code for design of steel structures (in Chinese). China Planning Press, Beijing

MOHURD and AQSIQ (2010) Code for design of concrete structures (in Chinese). China Architecture and Building Press, Beijing

MOHURD and AQSIQ (2012) Load code for the design of building structures (in Chinese). China Architecture and Building Press, Beijing

MOHURD and AQSIQ (2016) Code for seismic design of buildings (in Chinese). China Architecture and Building Press, Beijing

MOHURD and AQSIQ (2017) Standard for design of steel structures (in Chinese). China Architecture and Building Press, Beijing

Nie J, Hu H, Fan J, Tao M, Li S, Liu F (2013) Experimental study on seismic behavior of high-strength concrete filled double-steel-plate composite walls. J Constr Steel Res 88:206–219. https://doi.org/10.1016/j.jcsr.2013.05.001

Qiao Q, Cao W, Li X, Dong H, Zhang W, Yin F (2018) Seismic behavior of shear walls with boundary CFST columns and embedded multiple steel plates: experimental investigation. Eng Struct 160:243–256. https://doi.org/10.1016/j.engstruct.2018.01.040

Shen L, Li X (2016) Experimental study of in-plane seismic performance of double steel plate reinforced concrete composite Wall for nuclear power plants (in Chinese). Indust Construct 46:57–61

Susantha KAS, Ge H, Usami T (2001) Uniaxial stress-strain relationship of concrete confined by various shaped steel tubes. Eng Struct 23:1331–1347. https://doi.org/10.1016/S0141-0296(01)00020-7

Tao M, Nie J (2015) Element mesh, section discretization and material hysteretic laws for fiber beam-column elements of composite structural members. Mater Struct 48:2521–2544. https://doi.org/10.1617/s11527-014-0335-2

Thomsen JH, Wallace JW (2004) Displacement-based design of slender reinforced concrete structural walls-experimental verification. J Struct Eng 130:618–630. https://doi.org/10.1061/(ASCE)0733-9445(2004)130:4(618)

Wallace JW, Moehle JP (1992) DUCTILITY AND DETAILING REQUIREMENTS OF BEARING WALL BUILDINGS. J Struct Eng 118:1625–1644. https://doi.org/10.1061/(ASCE)0733-9445(1992)118:6(1625)

Wang B, Shi Q, Cai W, Peng Y, Li H (2020) Research on the calculation method for the deformation capacity of RC shear walls with a flange (in Chinese). Eng Mech 37:167–175

Yan J, Li Z, Wang T (2018) Seismic behaviour of double skin composite shear walls with overlapped headed studs. Constr Build Mater 191:590–607. https://doi.org/10.1016/j.conbuildmat.2018.10.042

Zhang X, Qin Y, Chen Z (2016a) Experimental seismic behavior of innovative composite shear walls. J Constr Steel Res 116:218–232. https://doi.org/10.1016/j.jcsr.2015.09.015

Zhang X, Qin Y, Chen Z, Jie L (2016b) Experimental behavior of innovative T-shaped composite shear walls under in-plane cyclic loading. J Constr Steel Res 120:143–159. https://doi.org/10.1016/j.jcsr.2016.01.008

Zhang W, Wang K, Chen Y, Ding Y (2019) Experimental study on the seismic behaviour of composite shear walls with stiffened steel plates and infilled concrete. Thin Wall Struct 144:106279. https://doi.org/10.1016/j.tws.2019.106279

Zhao Q, Li Y, Tian Y (2020a) Cyclic behavior of double-skin composite walls with flat and corrugated faceplates. Eng Struct 220:111013. https://doi.org/10.1016/j.engstruct.2020.111013

Zhao Q, Li Y, Tian Y, Li Z (2020b) Cyclic behavior of corrugated double-skin composite walls with different aspect ratios. J Struct Eng 146:4020214. https://doi.org/10.1061/(ASCE)ST.1943-541X.0002783

Zhu J, Guo Y, Wang M, Yang X, Pi Y (2020) Seismic performance of concrete-infilled double steel corrugated-plate walls: experimental research. Eng Struct 215:110601. https://doi.org/10.1016/j.engstruct.2020.110601

Chapter 7
Conclusions and Outlook

7.1 Conclusions

This study investigates the load-carrying mechanism and design methods of concrete-infilled double steel corrugated-plate walls with T-section (T-CDSCWs). CDSCWs consist of wall elements and boundary elements. The wall elements consist of outer steel corrugated plates, infilled concrete, and intermediate bolts connecting steel corrugated plates. The boundary elements consist of steel rectangular tubes and infilled concrete. Due to the interaction between steel and concrete, as well as the constraining effect of intermediate bolts, CDSCWs exhibit excellent load-carrying performance.

In high-rise buildings, shear walls are arranged in different cross-sections, such as T-section and L-section, to meet architectural layouts and structural requirements. This study focuses on the in-depth investigation of T-CDSCWs, with an emphasis on the cross-sectional capacity, global stability, and seismic performance. Experimental investigations, finite element simulations, and theoretical analyses are conducted to establish the design methods for the cross-sectional capacity under axial compression and biaxial bending moment, the global stability under axial compression and bending moment in the web plane, and the seismic requirements.

The main conclusions are as follows.

M.-Z. Wang, *Design Method of Concrete-Infilled Double Steel Corrugated-Plate Walls with T-section*, Springer Theses,
https://doi.org/10.1007/978-981-95-7789-7_7

7.1.1 Experimental Investigations on the Cross-Sectional Capacity, Global Stability, and Seismic Performance of T-CDSCWs

Three sets of T-CDSCW tests are conducted, namely the cross-sectional capacity test, the global stability test, and the seismic performance test. This study summarises the design, phenomena, and results of the tests. Based on the experimental results, the following conclusions are drawn.

T-CDSCWs exhibit excellent cross-sectional capacity. To investigate the cross-sectional capacity under axial compression and bending moment in the web plane, specimen T-1 with a wide web is designed. Eccentric compression in the web plane is applied to specimen T-1 to achieve the effect of axial compression and negative bending moment. Under axial compression and bending moment, specimen T-1 fails by flexure and steel corrugated plates and rectangular tubes fail by local buckling.

T-CDSCWs exhibit excellent global stability. To investigate the flexural buckling about the asymmetry axis under axial compression, specimen T-2 with a short web is designed. Under axial compression, specimen T-2 fails by the flexural buckling and the buckling direction is the direction with the flange on the convex side. The local buckling does not occur in steel corrugated plates and rectangular tubes. The strength reduction factor of specimen T-2 is 0.933.

T-CDSCWs exhibit excellent seismic performance. To investigate the seismic performance, specimen T-3 is designed. Under constant axial compression and cyclic horizontal load, specimen T-3 fails by flexure. The failure phenomena include the local buckling of steel corrugated plates and rectangular tubes, the fracture of the steel rectangular tubes at the end of the web wall, and the cracking and crush of infilled concrete. The horizontal load–horizontal displacement hysteretic curve of specimen T-3 exhibits asymmetry: with specimen T-3 under positive loading, the hysteretic curve is plump; with specimen T-3 under negative loading, the curve drops significantly and the load degrades in each level.

7.1.2 Solid–Shell and Shell–Beam Finite Element Models of T-CDSCWs

Two kinds of finite element models of T-CDSCWs are developed, namely the solid–shell finite element model and the shell–beam finite element model. The finite element simulation results are compared with the results of T-CDSCW experiments and other existing experiments to validate the effectiveness of the models.

The solid–shell finite element model can effectively simulate the load-carrying performance of T-CDSCWs and the local buckling performance of steel corrugated plates, especially accurately simulating the cross-sectional capacity of T-CDSCWs and the load-carrying capacity of steel corrugated plates.

The solid–shell finite element model can also effectively simulate the global stability of T-CDSCWs, especially the flexural buckling about the asymmetry axis with T-CDSCWs under axial compression. The load-carrying capacity of T-CDSCWs can be accurately simulated.

The shell–beam finite element model can effectively simulate the seismic performance of T-CDSCWs. For CDSCWs with rectangular and T-sections, the shell–beam finite element model can accurately simulate the maximum horizontal load, failure displacement, unloading, and reloading.

7.1.3 Cross-Sectional Capacity of T-CDSCWs under Axial Compression and Biaxial Bending Moment

Based on the solid–shell finite element model, this study analyses the local buckling performance of steel corrugated plates under axial compression and the cross-sectional capacity of T-CDSCWs under axial compression and bending moment in the web plane, bending moment in the flange plane, and biaxial bending moment. Based on the finite element simulation results and theoretical analyses, the following conclusions are drawn.

This study proposes the stability design method of steel corrugated plates, and thus the local buckling of steel corrugated plates can be considered in the cross-sectional capacity of T-CDSCWs. This study proposes the formulas of bilateral buckling of steel corrugated plates with only intermediate bolt constraints and contact buckling of steel corrugated plates with intermediate bolt and infilled concrete constraints. Based on the normalised slenderness ratio, the stability design curve of steel corrugated plates is developed and agrees well with finite element simulation results. Based on the stability design curve, the upper limit of the normalised slenderness ratio for contact buckling is proposed to be 0.4, below which the effect of the local buckling of steel corrugated plates can be neglected on the cross-sectional capacity of T-CDSCWs.

This study proposes the design method of the cross-sectional capacity of T-CDSCWs under axial compression and bending moment in the web plane. When T-CDSCWs sustain bending moment in the web plane, the direction of bending moment influences the shape of the interaction curves. For positive bending moment, the interaction curve is convex, and the convexity becomes significant as the web breadth increases and the steel plate thickness decreases. For negative bending moment, the interaction curve is basically linear, and the overall shape is independent on the cross-section. The local buckling of steel corrugated plates does not influence the overall shape of the interaction curve. By reducing the full-section plastic bending moment, the pure moment capacity in the web plane of T-CDSCWs can be evaluated, and thus the cross-sectional capacity under axial compression and bending moment in the web plane can be designed according to the bilinear interaction equation for positive bending and the linear interaction equation for negative bending. The design formulas agree well with the finite element simulation results.

This study proposes the design method of the cross-sectional capacity of T-CDSCWs under axial compression and bending moment in the flange plane. When T-CDSCWs sustain bending moment in the flange plane, the overall shape of the interaction curve is basically independent on the cross-section. The local buckling of steel corrugated plates does not influence the overall shape of the interaction curve. By reducing the full-section plastic bending moment, the pure moment capacity in the flange plane of T-CDSCWs can be evaluated, and thus the cross-sectional capacity under axial compression and bending moment in the flange plane can be designed according to the parabolic interaction equation. The design formulas agree well with the finite element simulation results.

This study proposes the design method of the cross-sectional capacity of T-CDSCWs under axial compression and biaxial bending moment. When T-CDSCWs sustain biaxial bending moment, the overall shape of the interaction curve primarily depends on the axial compression ratio. The local buckling of steel corrugated plates does not influence the overall shape of the interaction curve. In the design of the cross-sectional capacity of T-CDSCWs, the corresponding moment capacities are evaluated in the web and flange plane according to the given axial compression ratio, and then the cross-sectional capacity of T-CDSCWs is designed according to the interaction curve of biaxial bending moment. The design formulas agree well with the finite element simulation results.

7.1.4 Global Stability of T-CDSCWs Under Axial Compression and Bending Moment in the Web Plane

Based on the solid–shell finite element model, this study analyses the flexural buckling about the asymmetric axis and the flexural-torsional buckling about the symmetric axis of T-CDSCWs under axial compression and the global instability of T-CDSCWs under axial compression and bending moment in the web plane. Based on the numerical results and theoretical analyses, the following conclusions are drawn.

This study proposes the flexural stability design method of T-CDSCWs under axial compression. This study proposes the formulas of flexural buckling about the asymmetric axis, which can accurately evaluate the elastic flexural buckling load of T-CDSCWs with flexural buckling as the first-order mode. Based on the geometry feature of T-CDSCWs, a buckling mode determination parameter is obtained from the breadth-thickness ratio and the height-breadth ratio to determine the first-order buckling mode. When the buckling mode determination parameter is no greater than 1.1, T-CDSCWs under axial compression can be design according to flexural instability. The strength reduction curve of flexural instability is developed based on the normalised flexural slenderness ratio and agrees well with the numerical results.

This study proposes the flexural-torsional stability design method of T-CDSCWs under axial compression. This study proposes the formulas of flexural-torsional buckling about the symmetric axis, which can accurately evaluate the elastic flexural-torsional buckling load of T-CDSCWs. Based on the geometry feature of T-CDSCWs, a torsional buckling determination parameter is obtained from the breadth-thickness ratio and the height-breadth ratio. When the torsional buckling determination parameter is no greater than 3.0, which applies to most cases, the flexural-torsional buckling is basically governed by the torsional buckling. Taking into consideration the effect of the breadth-thickness ratio of web walls, the strength reduction curve of flexural-torsional instability is developed based on the normalised torsional slenderness ratio and agrees well with the numerical results.

This study proposes the design method of the in-plane global stability of T-CDSCWs under axial compression and bending moment in the web plane. Based on the interaction equations of cross-sectional capacity and the design formulas of flexural instability, this study proposes the interaction equations of global stability of T-CDSCWs under axial compression and bending moment in the web plane. Taking into consideration the secondary effect of axial compression, the interaction curves of global stability are generated by reducing the interaction curves of cross-sectional capacity. For positive bending, the interaction curve is still divided into two parts, and the protruding point is determined by the interaction curve of cross-sectional capacity. The interaction equations of global stability agree well with the numerical results.

7.1.5 Seismic Performance of Concrete Infilled Double Steel Plate Walls with T-Section Under Cyclic Horizontal Load

This study analyses the experimental results of the seismic performance of concrete-infilled double steel plate composite walls (CDSWs) and investigates the seismic performance of CDSWs with T-section (T-CDSWs) using the shell–beam finite element model. Based on the experimental results and numerical results, the following conclusions are drawn.

This study proposes an empirical formula for the ultimate drift ratio of CDSWs. Based on the experimental results of the seismic performance of CDSWs, this study analyses the relationship between the ultimate drift ratio of CDSWs and the constraining factor of boundary elements, the constraining factor of composite walls, the height-to-width ratio of composite walls, and the axial compression ratio. The experimental results of the seismic performance validate the seismic requirements in the design code regarding the limits of elastic-plastic drift ratio and axial compression ratio. Based on the formula for the plastic hinge length of composite walls and the failure stress of concrete in boundary elements, this study proposes an empirical formula to evaluate the ultimate drift ratio of CDSWs, which agrees well with the experimental results.

Combining the empirical formula and numerical results, this study proposes the seismic requirements for CDSWs. The numerical results show that increasing the constraining factor of boundary elements and reducing the axial compression ratio can enhance the seismic performance of T-CDSWs. To meet the requirements of the limits of elastic-plastic drift ratio and axial compression ratio in the design code, it is recommended that the constraining factor of the boundary elements in CDSWs should not be less than 1.5, especially for the boundary elements at the web ends of T-CDSWs, the constraining factor must not be less than 1.5. The constraining factor of CDSWs should not be less than 0.5, especially for the web walls of T-CDSWs, the constraining factor must not be less than 0.5.

7.2 Outlook

Further analyses are required to fully understand the behaviour of T-CDSCWs. To gain a deeper understanding of the load-carrying mechanism and design method of T-CDSCWs and composite walls with different cross-sections, the following research can be conducted in the future.

7.2.1 Cross-Sectional Capacity of T-CDSCWs Under Axial Compression, Bending Moment, and Shear Force

This study investigates the cross-sectional capacity of T-CDSCWs under axial compression and biaxial bending moment, but does not consider the effect of shear force on the cross-sectional capacity. From a qualitative perspective, it can be assumed that the wall elements only sustain shear forces within the plane. Specifically, the flange walls sustain shear force in the flange plane, while the web walls sustain shear force in the web plane.

In the design of CDSCWs with rectangular section, the shear capacity of steel corrugated plates is much higher than that of infilled concrete. Therefore, the contribution of infilled concrete to the shear capacity is not considered. When steel corrugated plates sustain shear forces, the combined effect of bending moment and shear force is considered by reducing the normal cross-sectional capacity of steel corrugated plates. Similarly, the same approach can be applied to T-CDSCWs by reducing the normal cross-sectional capacity of the corresponding wall elements. Nevertheless, specific design methods still need to be validated through experimental investigations and finite element simulations.

7.2.2 *Global Stability of T-CDSCWs Under Axial Compression and Biaxial Bending Moment*

This study investigates the global stability of T-CDSCWs under axial compression and bending moment in the web plane, but does not consider the out-of-plane stability under bending moment in the web plane and the global stability under combined biaxial bending moment.

In the design of out-of-plane stability, the elastic buckling bending moment can be used as the flexural load-carrying capacity, and a linear interaction equation of axial compression and bending moment can be employed as design formulas. However, specific design methods still need to be validated through experimental investigations and finite element simulations.

7.2.3 *Design Method of CDSCWs with Different Cross-Sections*

In high-rise buildings, besides rectangular section and T-section, shear walls can also be arranged in different cross-sections, such as L-section and C-section, according to architectural layout and structural requirements. This study mainly focuses on the load-carrying mechanism and design methods of T-CDSCWs, providing references for the design of CDSCWs with cross-sections different from rectangular section. Further research is needed to investigate the cross-sectional capacity, global stability, and seismic performance of CDSCWs with different cross-sections.

MIX
Papier aus verantwortungsvollen Quellen
Paper from responsible sources
FSC® C105338

If you have any concerns about our products,
you can contact us on
ProductSafety@springernature.com

In case Publisher is established outside the EU,
the EU authorized representative is:
Springer Nature Customer Service Center GmbH
Europaplatz 3, 69115 Heidelberg, Germany

Printed by Libri Plureos GmbH
in Hamburg, Germany